Die Grundlehren der mathematischen Wissenschaften

in Einzeldarstellungen
mit besonderer Berücksichtigung
der Anwendungsgebiete

Band 160

M. M. Agrest · M. S. Maksimov

Theory of Incomplete Cylindrical Functions and their Applications

Translated from the Russian by
H. E. Fettis J. W. Goresh D. A. Lee

With 20 Figures

Springer-Verlag New York Heidelberg Berlin 1971

Professor Matest M. Agrest

Suchumi/USSR

Professor Michail S. Maksimov

Suchumi/USSR

Title of the Russian Original Edition:
Teorija nepolnych zilindritscheskich funkzij i ejo priloshenija
Publisher: Atomizdat, Moscow/USSR, 1965

Translators:

Henry E. Fettis

Mathematician
Applied Mathematics
Research Laboratory

John W. Goresh †

Aerospace Engineer
Hypersonic Research
Facility

David A. Lee

Mathematician
Applied Mathematics
Research Laboratory

Aerospace Research Laboratories
Wright Patterson Air Force Base, Ohio, USA

AMS Subject Classifications (1970)
Primary 3302 — 33A40 — 33A70 — 65A05 — 65D20
Secondary 78A35 — 78A45 — 81A63 — 81A69

ISBN 0-387-05111-2 Springer-Verlag New York-Heidelberg-Berlin
ISBN 3-540-05111-2 Springer-Verlag Berlin-Heidelberg-New York

Library of Congress Catalog Card Number 78-139673

Preface to the English Edition

In preparing the English edition of this unique work, every effort has been made to obtain an easily read and lucid exposition of the material. This has frequently been done at the expense of a literal translation of the original text and it is felt that such liberties as have been taken with the author's language are justified in the interest of ease in reading. None of us pretends to be an authority in the Russian language, and we trust that the original intent of the authors has not been lost.

The equations, which were for the most part taken verbatim from the original work, were checked only cursorily; obvious and previously noted errors have been corrected. Fortunately, the Russian and English mathematical notations are generally in good agreement. An exception is the shortened abbreviations for the hyperbolic functions (e.g. sh for sinh), and the symbol Jm rather that Im to denote the imaginary part. As near as possible, these discrepancies have been corrected.

In preparing the Bibliography, works having an English equivalent have been translated into the English title, but in the text the reference to the Russian work was retained, as it was impractical to attempt to find in each case the corresponding citation in the English edition. Authors' names and titles associated with purely Russian works have been transliterated as nearly as possible to the English equivalent, along with the equivalent English title of the work cited.

The numerical tables appear exactly as they did in the original work, and no attempt was made to check their accuracy. This may be done at a later date.

The translators' interest in this subject goes back many years. One of them first encountered these functions in connection with a problem in unsteady aerodynamics, and later used it as a thesis subject. AGREST and MAKSIMOV have extended the theory much further than could originally have been expected, and have included a wealth of applications in almost all fields of mathematical physics.

Ohio, May 1971

Aerospace Research Laboratories
Wright Patterson Air Force Base

H. E. FETTIS · J. W. GORESH † · D. A. LEE

Contents

List of Symbols

∇_ν	Bessel Differential Operator
$J_\nu(z)$	Bessel Function of First Kind
$N_\nu(z)$	Bessel Function of Second Kind (Neumann Function)
$\mathscr{H}_\nu^{(1)}(z), \mathscr{H}_\nu^{(2)}(z)$	Hankel Function
$I_\nu(z)$	Modified Bessel Function
$K_\nu(z)$	Modified Hankel Function (MacDonald Function)
W	Wronskian
$L_\nu(z)$	Recurrence Operator
$L_\nu'(z)$	Recurrence-Differential Operator
A_ν	$= 2^\nu \Gamma\left(\nu + \frac{1}{2}\right) \Gamma(^1/_2)$
$S_\nu(z)$	Semi-Cylindrical Function
ϑ	$= z \frac{d}{dz}$
Re	Real Part of
Im	Imaginary Part of
$\Gamma(z)$	Gamma Function
$\mathscr{E}_\nu^{(1)}(c, z), \mathscr{E}_\nu^{(2)}(c, z)$	Incomplete Hankel Function
$E_\nu^\pm(w, z)$	Incomplete Cylindrical Function (Poisson Form)
$J_\nu(w, z)$	Incomplete Bessel Function (Poisson Form)
$H_\nu(w, z)$	Incomplete Struve Function
$I_\nu(w, z)$	Modified Incomplete Bessel Function
$L_\nu(w, z)$	Modified Incomplete Struve Function
$F_\nu^+(w, z), F_\nu^-(w, z)$	Alternative Forms of Above
$K_\nu(w, z)$	Incomplete Hankel Function (Alternative Form)
$\mathscr{H}_\nu(w, z)$	Incomplete MacDonald Function
$\varphi_\nu(w, z), \psi_\nu(w, z), q_\nu(w, z)$	Alternative Forms of Incomplete Bessel Functions of Poisson Form
$C_{m,\nu}(w)$	Trigonometric Integral [See Chap. II, Eq. (8.1)]
$P_\nu^\mu(x), Q_\nu^\mu(x)$	Associated Legendre Functions
${}_2F_1(\alpha, \beta; \gamma; z) \equiv F(\alpha, \beta; \gamma; z)$	Hypergeometric Function
${}_1F_1(\alpha; \gamma; z) = F(\alpha; \gamma; z)$	Confluent Hypergeometric Function
$Je_\nu(a, z), Ne_\nu(a, z)$	Incomplete Lipschitz-Hankel Integrals (Exponential Form)
$Jc_\nu(a, z), Js_\nu(a, z)$	Incomplete Lipschitz-Hankel Integrals (Trigonometric Form)
$Ie_\nu(a, z), Ke_\nu(a, z)$	Incomplete Lipschitz-Hankel Integrals (Modified Form)
$R_{m,\mu}(z)$	Lommel Polynomial

$\gamma(p, y)$	Incomplete Gamma Function
$\Psi_\nu(w, z)$	See Chap. II, Eq. (10.5)
$\Phi_\nu(b, z)$	See Chap. III, Eq. (6.5)
$T_m^{-\nu}(x)$	Gegenbauer Polynomial
$\varepsilon_\nu(w, z)$, $j_\nu(w, z)$, $h_\nu(w, z)$, $i_\nu(w, z)$, $k_\nu(w, z)$	Incomplete Cylindrical Functions (Bessel Form)
$P_k^{(\alpha,\beta)}(x)$	Jacobi Polynomials
$A_\nu(\beta, z)$, $B_\nu(\beta, z)$	Incomplete Anger-Weber Functions
$Ei(\alpha, z)$, $Ec(\alpha, z)$, $Es(\alpha, z)$	Incomplete Airy Integrals
$S_\nu(p, q; z)$	Incomplete Cylindrical Function (Sonine-Schlaefli Form)
$P_\nu(x, z)$, $Q_\nu(x, z)$	Incomplete Weber Integrals
$\widetilde{P}_\nu(x, z)$, $\widetilde{Q}_\nu(x, z)$	Incomplete Weber Integrals (Modified Form)
$U_\nu(w, z)$, $V_\nu(w, z)$	Lommel's Functions of Two Variables
$\Omega(p, q)$	Unit Step Function
$\delta(x)$	Dirac Delta Function
$Z_{\mu,\nu}(a, z)$	Sonine's Integral
$u_{k,\nu}(w, z)$, $q_\nu(w, y, z)$	Integrals Involving the Incomplete Cylindrical Functions of Poisson Form (See Chap. VI, § 6)
$\Pi(r, \varphi, z)$	Hertz Vector (See Chap. VII, § 3)
$G_{\mu,\nu}(a, z)$	Gallop Integral
C	Euler's constant
$H_\nu(z)$	Struve Function
$L_\nu(z)$	Modified Struve Function
$S(a, x)$, $C(a, x)$, $E(a, x)$	Generalized Sine, Cosine and Exponential Integrals
$W(z)$	Dawson's Integral $= e^{-z^2}\int_0^z e^{x^2}\,dx$
$(\alpha)_m$	Pochhammer Symbol: $(\alpha)_m = \dfrac{\Gamma(\alpha+m)}{\Gamma(\alpha)}$

$$ {}_pF_q(\alpha_1, \alpha_2, \ldots, \alpha_p;\ \gamma_1, \gamma_2, \ldots, \gamma_q; z) = \sum_{m=0}^{\infty} \frac{(\alpha_1)_m (\alpha_2)_m \cdots (\alpha_p)_m}{(\gamma_1)_m (\gamma_2)_m \cdots (\gamma_q)_m} \cdot \frac{z^m}{m!} $$

Introduction

Cylindrical functions have found wide application in various branches of science and technology. At the present time it would be difficult to find any area of Applied Mathematics in which one would not encounter Bessel, Hankel, Neuman and other similar functions.

In both theoretical and applied problems one frequently encounters integrals having variable limits which, when the limits are given certain specific values, can be expressed by cylindrical functions. By analogy with incomplete elliptic integrals, or incomplete gamma or beta functions, such integrals may well be called "incomplete cylindrical functions".

To develop a general theory of such integrals, the authors have introduced a generalized concept of incomplete cylindrical functions. The functions are defined in such a way that, when the limits of integration assume certain specific values, the integrals become the corresponding cylindrical functions. In order to arrive at a comprehensive distinction between the various classes of incomplete cylindrical functions, consideration is given to incomplete cylindrical functions of the forms of the integrals of Poisson, Bessel and Sonine-Schlaefli. Similarly, incomplete Lipschitz-Hankel and Weber integrals are introduced. These are some of the various forms which represent different elements of the incomplete cylindrical functions.

In the general case all of the functions depend on three variables. Two of these play the same role as in the conventional cylindrical functions, while the third appears as one of limits of integration and characterizes the incomplete nature of the functions. By giving this variable certain specific values, we obtain all of the known cylindrical functions as well as the widely known and frequently applied functions of Struve and Anger-Weber, probability integrals, Lommel's function of one and two variables, etc. The theory of incomplete cylindrical functions therefore permits the investigation of all of the aforementioned functions. The incomplete cylindrical functions will be considered in order of their importance in many problems in physics and technology. It is the authors' objective to develop in detail and to arrange systematically the general theory of these functions and especially to illustrate their relation with known functions and special integrals and, further, to consider examples of their application to various problems of physics and technology.

The study of the various properties of integrals similar to these had its origin in the somewhat distant past. Some of such integrals can be found, for example, in works of Binet [1] and Nagaoka [2] dealing with the diffraction of light, which appeared in the previous century. One can also find them in more recent work in connection with the same problem in diffraction [3]. Schwartz [4] studied in detail different elements of the incomplete cylindrical functions which may be called incomplete Lipschitz-Hankel integrals. In the last two decades integrals related to the incomplete cylindrical functions have occurred repeatedly in the solution of various problems in mathematical physics. (See for example references [5] through [7] in the bibliography.) Finally, in references [8] and [9] the concept of incomplete functions of Bessel, Struve and Anger-Weber is introduced, together with a rather complete investigation of their properties. Illustrated there also are various problems leading to these functions, as well as tables of numerical values. Additional properties of these incomplete integrals can be found in references mentioned throughout this book.

The authors first encountered incomplete cylindrical functions of the Poisson form in 1956 in connection with the solution of a certain physical problem (see Chapter 8, Paragraph 3) proposed by N. I. Leont'ev. This motivated a more comprehensive study of their properties which forms the basis for this book.

The work is comprised of 11 Chapters. Chapter one presents the general theory of cylindrical functions. The next three chapters deal with the theory of incomplete cylindrical functions of the form of Poisson, Bessel, and Sonine-Schlaefli as well as with various elements of them-incomplete Lipschitz-Hankel and Weber integrals. In Chapter V, wide use is made of the so-called unit step function, which permits one to readily obtain a series of additional properties of other integral representations for incomplete cylindrical functions and to relate them to some classes of indefinite integrals of Bessel functions.

In Chapter VI we develop methods of evaluating definite integrals with finite limits as well as improper integrals involving incomplete cylindrical functions.

In Chapters VII through X applications of incomplete cylindrical functions to several problems of physics and technology are discussed. For this purpose the areas of atomic, nuclear and solid state physics are representative although the functions also find application in other areas. We note that because of the nature of the contents of this book it is difficult to draw a sharp line between the theory of incomplete cylindrical and their applications. Every relation between the functions and integrals established in one chapter and these of another widens the area of their application; on the other hand, the same relation permits the development of further and more precise knowledge of the theory.

For convenience in practical applications a list of computational formulae relating the most often used incomplete cylindrical functions is collected in the last chapter, together with references to tables of special integrals available in the literature by means of which they can be evaluated. However, the tables do not permit one to evaluate the functions for all values of their arguments. In order to fill this gap and to enlarge the area of applications, the authors, together with M. M. Rikenglas, N. V. Khaykhyan, I. N. Bekaury, L. A. Orlova, and Ts. Sh. Chachibaya have constructed tables of the incomplete cylindrical functions of order zero and one, which are given in condensed form at the end of this last chapter[1].

Inasmuch as the book is a first edition, it can be neither exhaustive nor devoid of various other insufficiencies. This comment applies first of all to the historical information on incomplete cylindrical functions and to the bibliography. Furthermore, the choice of applied problems may have suffered from the lack of strict systematization, and it is entirely possible that some important problems have been neglected. In addition there may well be some errors in the physical formulations of the problems, since the details of the analysis of some of them lie outside the limits of our competence. In each specific case our aim has been to show how incomplete cylindrical functions arise and to exploit the general theory of these functions for mathematical analysis of the final results.

In writing this book the authors received much assistance from many colleagues. M. M. Rikenglas and N. V. Khaykhyan gave valuable aid in the first stages of the work by carrying out a series of analytic transformations. M. M. Rikenglas' help in establishing the relations between incomplete cylindrical functions and various Sonine integrals was of the greatest importance. L. A. Orlova, Ts. Sh. Chachibaya, B. B. Tavdidishvili, S. I. Labakhua, T. S. Tsulaya, and T. I. Lapauri rendered much help in preparing the manuscript for publication. Valuable technical assistance was also given by O. A. Venediktova and N. G. Fursa. To all these colleagues the authors extend their warmest thanks. The authors are grateful to N. I. Leont'ev for suggesting the physical problem which served to stimulate their work on incomplete cylindrical functions, and for his interest in the work as a whole.

---

[1] See the foot-note p. 288.

The authors hope that this book will be useful to mathematicians, physicists and engineers who are interested in methods of computation, as well as to advanced students concerned with special functions, and they will be grateful to any readers who will forward their critical comments and suggestions concerning its contents.

Chapter I

Some Information from the Theory of Cylindrical Functions

1. The Differential Equation and Recursion Relationships for Cylindrical Functions

We shall consider the class of incomplete cylindrical functions as including the incomplete functions of Bessel, Neumann, Hankel and MacDonald; the incomplete functions of Struve, Anger and Weber, and also some other special integrals. They posess, as will be shown, an entire series of properties which are similar to those of the cylindrical functions and during the process of describing this material, we shall frequently take advantage of the basic properties of cylindrical functions and also of their different analytical representations.

To avoid frequent references to special monographs, we shall simplify matters by presenting basic and necessary derivations from the theory of these functions following mainly the monograph of Watson [10].

As is well known, the following three function classes comprise the cylindrical ones: the Bessel Functions, $J_\nu(z)$, Neumann Functions, $N_\nu(z)$, and Hankel Functions, $\mathscr{H}_\nu^{(1)}(z)$ and $\mathscr{H}_\nu^{(2)}(z)$; or as they are also called, Bessel Functions of the first, second and third kind respectively. All of these functions represent solutions of Bessel's differential equation:

$$z^2 \frac{d^2E}{dz^2} + z \frac{dE}{dz} + (z^2 - \nu^2) E = 0, \tag{1.1}$$

which for the sake of brevity is customarily written in the form

$$\nabla_\nu E = 0,$$

where ∇_ν is the differential operator of the second order:

$$\nabla_\nu = z^2 \frac{d^2}{dz^2} + z \frac{d}{dz} + z^2 - \nu^2. \tag{1.2}$$

Equation (1.1) is a linear differential equation of second order. To obtain a general solution it is necessary to know any two linearly independent solutions. As such a pair of fundamental solutions, it is customary to choose either Bessel functions with positive and negative indicies $[J_\nu(z), J_{-\nu}(z)]$ for non-integer ν or Bessel and Neumann functions $[J_\nu(z), N_\nu(z)]$ or Hankel functions of the first and second kind $[\mathscr{H}_\nu^{(1)}(z), \mathscr{H}_\nu^{(2)}(z)]$. Each pair of these solutions is, of course, a linearly independent set, since the Wronskian, W, of these solutions is different from zero and is equal respectively to

$$W[J_\nu(z), J_{-\nu}(z)] = \frac{2 \sin \nu\pi}{\pi z}; \tag{1.3}$$

$$W[J_\nu(z), N_\nu(z)] = \frac{2}{\pi z}; \tag{1.4}$$

$$W[\mathscr{H}_\nu^{(1)}(z), \mathscr{H}_\nu^{(2)}(z)] = -\frac{4i}{\pi z}. \tag{1.5}$$

Any general solution of the Bessel Eq. (1.1) can be given in the form of a linear combination of each pair of fundamental solutions. Among these pairs there exists a connection which can be obtained, for example, from the following relationships:

$$N_\nu(z) = \frac{J_\nu(z) \cos \nu\pi - J_{-\nu}(z)}{\sin \nu\pi}; \tag{1.6}$$

$$\mathscr{H}_\nu^{(1)}(z) = J_\nu(z) + iN_\nu(z); \tag{1.7}$$

$$\mathscr{H}_\nu^{(2)}(z) = J_\nu(z) - iN_\nu(z), \tag{1.8}$$

which retain their meaning in the limit as when the index ν tends to an integer.

If, in the relationships (1.7), (1.8), the index ν is changed to $-\nu$, then by taking advantage of (1.6) one can easily obtain the following relations between the functions $\mathscr{H}_\nu(z)$ and $\mathscr{H}_{-\nu}(z)$:

$$\left.\begin{aligned} \mathscr{H}_{-\nu}^{(1)}(z) &= e^{\pi\nu i} \mathscr{H}_\nu^{(1)}(z); \\ \mathscr{H}_{-\nu}^{(2)}(z) &= e^{-\pi\nu i} \mathscr{H}_\nu^{(2)}(z). \end{aligned}\right\} \tag{1.9}$$

The Bessel function $J_\nu(z)$ has the the following representation as a power series in z:

$$J_\nu(z) = \sum_{m=0}^{\infty} \frac{(-1)^m (z/2)^{\nu+2m}}{m!\Gamma(\nu + m + 1)}. \tag{1.10}$$

This series converges absolutely and uniformly for all values of the variable z and the index ν. By making use of the relationships (1.6) through (1.9) one can easily obtain the corresponding expansions of other cylindrical functions. In particular, for the Neumann function

$N_\nu(z)$, the expansion has the form

$$N_\nu(z) = \frac{1}{\sin \nu\pi}\left[(z/2)^\nu \cos \nu\pi \sum_{k=0}^{\infty}(-1)^k (z/2)^{2k} \frac{1}{k!\Gamma(\nu+k+1)} - (z/2)^{-\nu}\sum_{k=0}^{\infty}(-1)^k (z/2)^{2k}\frac{1}{k!\Gamma(\nu+k+1)}\right], \quad (1.11)$$

and in the case where $\nu = n$, an integer, this expansion goes into

$$\pi N_n(z) = 2J_n(z)\left(\ln\frac{z}{2} + C\right) - \sum_{k=0}^{n-1}\frac{(n-k-1)!}{k!}(z/2)^{2k-n} - \left(\frac{z}{2}\right)^n \frac{1}{n!}\sum_{k=1}^{n}\frac{1}{k} - \sum_{k=1}^{\infty}\frac{(-1)^k (z/2)^{n+2k}}{k!(n+k)!}\left[\sum_{m=1}^{n+k}\frac{1}{m} + \sum_{m=1}^{k}\frac{1}{m}\right], \quad (1.12)$$

where $C = .57721\ldots$ is Euler's constant.

In the case of purely imaginary values of the variable z, a fundamental system of solutions of Eq. (1.1) can be obtained from the functions

$$\left.\begin{aligned} I_\nu(z) &= e^{-\pi\nu i/2} J_\nu(iz); \\ K_\nu(z) &= \frac{\pi i}{2} e^{\pi\nu i/2}\mathscr{H}_\nu^{(1)}(iz). \end{aligned}\right\} \quad (1.13)$$

These functions are conventionally called modified Bessel functions. In addition $K_\nu(z)$ is sometimes referred to as the MacDonald function, and is defined by the relation:

$$K_\nu(z) = \frac{\pi}{2}\frac{I_{-\nu}(z) - I_\nu(z)}{\sin \nu\pi}. \quad (1.14)$$

The Bessel function $J_\nu(z)$ satisfies the following recurrence relations

$$J_{\nu-1}(z) + J_{\nu+1}(z) = \frac{2\nu}{z}J_\nu(z); \quad (1.15)$$

$$J_{\nu-1}(z) - J_{\nu+1}(z) = 2\frac{dJ_\nu(z)}{dz}, \quad (1.16)$$

from which it is easy to obtain the formula

$$z\frac{d}{dz}J_\nu(z) + \nu J_\nu(z) = zJ_{\nu-1}(z); \quad (1.17)$$

$$z\frac{d}{dz}J_\nu(z) - \nu J_\nu(z) = -zJ_{\nu+1}(z), \quad (1.18)$$

or, more generally

$$\left(\frac{d}{z\,dz}\right)^m [z^\nu J_\nu(z)] = z^{\nu-m}J_{\nu-m}(z); \quad (1.19)$$

$$\left(\frac{d}{z\,dz}\right)^m [z^{-\nu} J_\nu(z)] = (-1)^m z^{-\nu-m}J_{\nu+m}(z). \quad (1.20)$$

The Neumann function $N_\nu(z)$ satisfies this same recurrence relations. The modified functions $I_\nu(z)$ and $K_\nu(z)$, on the other hand, satisfy the following relations:

$$I_{\nu-1}(z) - I_{\nu+1}(z) = \frac{2\nu}{z} I_\nu(z); \tag{1.21}$$

$$I_{\nu-1}(z) + I_{\nu+1}(z) = 2\frac{d}{dz} I_\nu(z); \tag{1.22}$$

$$\left(\frac{d}{z\,dz}\right)^m [z^\nu I_\nu(z)] = z^{\nu-m} I_{\nu-m}(z); \tag{1.23}$$

$$\left(\frac{d}{z\,dz}\right)^m [z^{-\nu} I_\nu(z)] = z^{-\nu-m} I_{\nu+m}(z); \tag{1.24}$$

$$K_{\nu-1}(z) - K_{\nu+1}(z) = -\frac{2\nu}{z} K_\nu(z); \tag{1.25}$$

$$K_{\nu-1}(z) + K_{\nu+1}(z) = -2\frac{d}{dz} K_\nu(z); \tag{1.26}$$

$$\left(\frac{d}{z\,dz}\right)^m [z^\nu K_\nu(z)] = (-1)^m z^{\nu-m} K_{\nu-m}(z); \tag{1.27}$$

$$\left(\frac{d}{z\,dz}\right)^m [z^{-\nu} K_\nu(z)] = (-1)^m z^{-\nu-m} K_{\nu+m}(z). \tag{1.28}$$

2. Generalized System of Functional Equations for Cylindrical Functions and the Inhomogenious Bessel Differential Equation

For the sake of brevity we denote by L_ν and L'_ν the following operators:

$$L_\nu E = L_\nu E_\nu = E_{\nu-1}(z) + E_{\nu+1}(z) - \frac{2\nu}{z} E_\nu(z); \tag{2.1}$$

$$L'_\nu E = L'_\nu E_\nu = E_{\nu-1}(z) - E_{\nu+1}(z) - 2\frac{d}{dz} E_\nu(z). \tag{2.2}$$

Using this notation, the recurrence relations (1.15) and (1.16) can be written in the form:

$$L_\nu J_\nu(z) = 0; \; L'_\nu J_\nu(z) = 0.$$

One can easily prove that any function $E_\nu(z)$ which satisfies the functional equations

$$L_\nu E_\nu(z) = 0; \tag{2.3}$$

$$L'_\nu E_\nu(z) = 0, \tag{2.4}$$

also satisfies Bessel's differential equation

$$\nabla_\nu E_\nu = z^2 \frac{d^2 E_\nu}{dz^2} + z\frac{dE_\nu}{dz} + (z^2 - \nu^2) E_\nu = 0.$$

Therefore, the system of functional equations (2.3) and (2.4) can be used as a basis for defining the cylindrical functions. In view of this definition for the cylindrical functions, functions $S_\nu(z)$ which satisfy only one recurrence relationship (2.4), and the condition

$$S_1(z) = -\frac{dS_0(z)}{dz},$$

are sometimes called semi-cylindrical functions.

In the following, we shall frequently encounter the non-homogenious Bessel differential equation

$$\nabla_\nu E_\nu = z\omega_\nu(z), \tag{2.5}$$

as well as a system of functional equations of the form (2.3) and (2.4) in which the right sides are different from zero:

$$L_\nu E_\nu(z) = \frac{2g_\nu(z)}{z}; \tag{2.6}$$

$$L'_\nu E_\nu(z) = \frac{2f_\nu(z)}{z}, \tag{2.7}$$

where $\omega_\nu(z)$, $f_\nu(z)$ and $g_\nu(z)$ are arbitrary functions of the variables ν and z. We shall show the connection between the general solution of Eq. (2.5) and the system of non-homogenious functional equations (2.6), and (2.7). It is evident that the system (2.6), (2.7) constitutes a generalization of the system (2.3), (2.4) which defined the cylindrical functions. One can further show (see, e.g. [10, p. 387]) that if the functions $f_\nu(z)$ and $g_\nu(z)$ satisfy Nielson's condition:

$$f_{\nu-1}(z) + f_{\nu+1}(z) - \frac{2\nu}{z} f_\nu(z) = g_{\nu-1}(z) - g_{\nu+1}(z) - 2\frac{d}{dz} g_\nu(z), \tag{2.8}$$

then the system (2.6), (2.7) is equivalent to a non-homogeneous Bessel differential equation in which the right side can be expressed in terms of the functions $f_\nu(z)$ and $g_\nu(z)$.

For brevity let us denote the sum and difference of these functions by

$$\left.\begin{aligned} \alpha_\nu(z) &= f_\nu(z) + g_\nu(z); \\ \beta_\nu(z) &= f_\nu(z) - g_\nu(z) \end{aligned}\right\} \tag{2.9}$$

and by ϑ the differential operator $\vartheta = z\frac{\mathrm{d}}{\mathrm{d}z}$. It is then easy to transform the system of functional equations (2.6), (2.7) into the form

$$\left.\begin{aligned} (\vartheta + \nu)\, E_\nu(z) &= zE_{\nu-1}(z) - \alpha_\nu(z); \\ (\vartheta - \nu)\, E_\nu(z) &= -zE_{\nu+1}(z) - \beta_\nu(z). \end{aligned}\right\} \tag{2.10}$$

By applying the operator $\vartheta - \nu$ to the first of these equations we obtain

$$[(\vartheta)^2 - \nu^2]\, E_\nu(z) = (\vartheta - \nu)\, [zE_{\nu-1}(z) - \alpha_\nu(z)],$$

and by making use of the second Eq. (2.10) we find

$$[(\vartheta)^2 - \nu^2 + z^2]\, E_\nu(z) = -z\beta_{\nu-1}(z) - (\vartheta - \nu)\, \alpha_\nu(z)$$

or

$$\nabla_\nu E_\nu(z) = -z\beta_{\nu-1}(z) - (\vartheta - \nu)\, \alpha_\nu(z)\,. \tag{2.11}$$

On the other hand, by applying the operator $\vartheta + \nu$ to the second equation (2.10) and proceeding in an analogous manner we find for the function $E_\nu(z)$ a non-homogenious Bessel Equation in the form

$$\nabla_\nu E_\nu(z) = z\alpha_{\nu+1}(z) - (\vartheta + \nu)\, \beta_\nu(z)\,.$$

In order that this equation and Eq. (2.11) define the same function $E_\nu(z)$ it is necessary to have equality of the right sides, i.e.

$$z\alpha_{\nu+1}(z) - (\vartheta + \nu)\, \beta_\nu(z) = -z\beta_{\nu-1}(z) - (\vartheta - \nu)\, \alpha_\nu(z)$$

or, if we express α_ν and β_ν in terms of the original functions $f_\nu(z)$ and $g_\nu(z)$, we obtain immediately the Nielson condition (2.8).

In this way a system of generalized functional equations (2.6), (2.7) has been obtained which is equivalent to the non-homogeneous Bessel differential equation

$$\nabla_\nu E_\nu(z) = z\,\omega_\nu(z)\,,$$

where the function $\omega_\nu(z)$ is connected with the right sides of Eqs. (2.6) and (2.7) by the following relation

$$z\,\omega_\nu(z) = -z\,[f_{\nu-1}(z) - g_{\nu-1}(z)] - \left(z\frac{d}{dz} - \nu\right)[f_\nu(z) + g_\nu(z)]\,. \tag{2.12}$$

According to the theory of differential equations a general solution of this equation can be written as

$$\begin{aligned} E_\nu(z) = J_\nu(z)\left[c_\nu - \frac{\pi}{2}\int_a^z N_\nu(t)\,\omega_\nu(t)\,dt\right] \\ + N_\nu(z)\left[d_\nu - \frac{\pi}{2}\int_b^z J_\nu(t)\,\omega_\nu(t)\,dt\right], \end{aligned} \tag{2.13}$$

where $J_\nu(z)$ and $N_\nu(z)$ are Bessel and Neumann functions, a and b are arbitiary constants, and the quantities c_ν and d_ν depend in general on the index ν but are independent of the variable z. On the other hand, if one has already found the solution for $E_\nu(z)$, he may substitute it in the generalized system of functional equations (2.6), (2.7) or, equivalently, in the system (2.10) and can obtain for c_ν and d_ν the recurrence relations:

$$c_\nu - c_{\nu-1} = -\frac{\pi}{2}\,[N_{\nu-1}(a)\,\beta_{\nu-1}(a) - N_\nu(a)\,\alpha_\nu(a)]\,; \tag{2.14}$$

$$d_\nu - d_{\nu-1} = \frac{\pi}{2}\,[J_{\nu-1}(b)\,\beta_{\nu-1}(b) - J_\nu(b)\,\alpha_\nu(b)] \tag{2.15}$$

where the quantities $\alpha_\nu(z)$ and $\beta_\nu(z)$ are defined as above.

3. Integral Representations of Cylindrical Functions in the Poisson Form

The cylindrical functions have various integral representations. The most useful and freqently encountered are those of Poisson, Bessel, and Sonine-Schlaefli. As an example, we shall consider an integral of the Poisson type

$$z^{\nu} \int_a^b e^{izt} T(t)\, dt, \tag{3.1}$$

where the limits of integration a and b are arbitrary complex numbers and the function $T(t)$ depends on t but not on z. We shall find conditions for which this integral satisfies Bessel's equation. Applying the operator (1.2) to the integral, we obtain

$$\nabla_{\nu}\left[z^{\nu} \int_a^b e^{izt} T(t)\, dt\right] = z^{\nu+2} \int_a^b e^{izt} (1 - t^2)\, T(t)\, dt$$
$$+ i(2\nu + 1)\, z^{\nu+1} \int_a^b e^{izt} T(t)\, dt.$$

Integrating the first term by parts we get

$$\begin{aligned}\nabla_{\nu}\left[z^{\nu} \int_a^b e^{izt} T(t)\, dt\right] &= i z^{\nu+1}\, [(t^2 - 1)\, e^{izt} T(t)]_{t=a}^{t=b} \\ &+ i z^{\nu+1} \int_a^b e^{izt} \left\{(2\nu + 1)\, T(t) - \frac{d}{dt}[(t^2 - 1)\, T(t)]\right\} dt,\end{aligned} \tag{3.2}$$

from which it is evident that (3.1) is a solution of Bessel's equation provided $T(t)$ fulfills the following conditions:

$$\frac{d}{dt}[(t^2 - 1)\, T(t)] = (2\nu + 1)\, t T(t); \tag{3.3}$$

$$e^{izt} (t^2 - 1)\, T(t)\, \big|_{t=a}^{t=b} = 0. \tag{3.4}$$

From (3.3) it is clear that, except for a multiplicative factor, $T(t)$ must be of the form

$$T(t) = \frac{1}{A_{\nu}} (t^2 - 1)^{\nu - 1/2}. \tag{3.5}$$

Equation (3.4) shows that the contour of integration in Eq. (3.1) may be either: (1) a closed one, so that the function $e^{izt} (t^2 - 1)^{\nu-1/2}$ returns to its original value as the variable of integration describes its complete path, or (2) such that this expression vanishes at both ends of the path. One contour of the first type, denoted L_1, has the form of a figure eight encircling the point $(t = 1)$ in the counter-clockwise direction and the point $t = -1$ in the clockwise direction [Fig. 1]. A contour of the second type denoted L_2, can be described, in the case $\mathrm{Re}\,(z) > 0$, by a loop in the counter-clockwise direction which contains the segment $(-1, +1)$, and which extends to $i\infty$ (Fig. 2). These contours define two integrals of the type (3.1)

$$\frac{z^\nu}{A_\nu} \int_{L_1} e^{izt} (t^2 - 1)^{\nu-1/2} \, dt; \tag{3.6}$$

$$\frac{z^\nu}{A_\nu} \int_{L_2} e^{izt} (t^2 - 1)^{\nu-1/2} \, dt, \tag{3.7}$$

which yield two linearly independent solutions of Bessel's equation. The first defines Bessel functions of the first kind, while the second defines either Bessel functions with negative indices or Neumann functions.

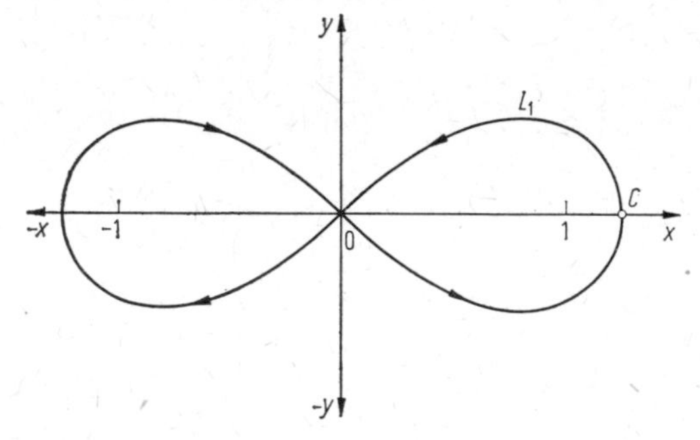

Fig. 1. Integration contour for the Bessel function.

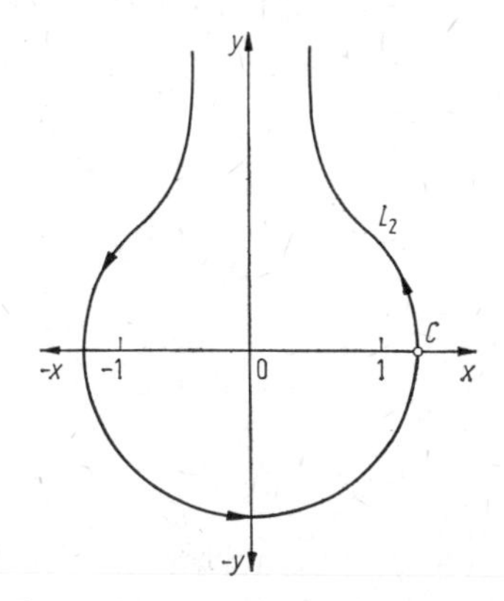

Fig. 2. Integration contour for the Neumann function.

By choosing special forms for the contours L_1 and L_2, one can obtain different integral representations for cylindrical functions. We shall direct our attention to the integral representations which are most frequently encountered, and which will appear in the theory of incomplete cylindrical functions.

From formula (3.6) for Re $(\nu + 1/2) > 0$, we obtain the following representation for Bessel functions of the Poisson form:

$$J_\nu(z) = \frac{z^\nu}{A_\nu} \int_0^\pi e^{\pm iz\cos\theta} \sin^{2\nu}\theta \, d\theta; \tag{3.8}$$

$$J_\nu(z) = \frac{z^\nu}{A_\nu} \int_{-1}^{1} e^{\pm izt} (1 - t^2)^{\nu - 1/2} \, dt, \tag{3.9}$$

where for brevity wc have introduced the following notation to be used in the future:

$$A_\nu = 2^\nu \Gamma\left(\nu + \frac{1}{2}\right) \Gamma\left(\frac{1}{2}\right). \tag{3.10}$$

The expressions (3.8) and (3.9) are valid for any z provided Re $(\nu + 1/2) > 0$. Replacing z by iz and introducing the relation

$$I_\nu(z) = e^{-i\nu\pi/2} J_\nu(iz),$$

we immediately find integral representations for the modified Bessel functions $I_\nu(z)$ in Poisson form:

$$I_\nu(z) = \frac{z^\nu}{A_\nu} \int_0^\pi e^{\pm z\cos\theta} \sin^{2\nu}\theta \, d\theta; \tag{3.11}$$

$$I_\nu(z) = \frac{z^\nu}{A_\nu} \int_{-1}^{1} e^{\pm zt} (1 - t^2)^{\nu - 1/2} \, dt, \tag{3.12}$$

which are valid for all z, provided Re $(\nu + 1/2) > 0$. Integral representations of Bessel function of the second kind, i.e. the Neumann functions, may be obtained from integrals of the Poisson type by deforming the contours so as to obtain the following:

$$N_\nu(z) = \frac{2z^\nu}{A_\nu} \left[\int_0^{\pi/2} \sin(z\cos\theta) \sin^{2\nu}\theta \, d\theta - \int_0^\infty e^{-z\sinh\theta} \cosh^{2\nu}\theta \, d\theta \right], \tag{3.13}$$

$$N_\nu(z) = \frac{2z^\nu}{A_\nu} \left[\int_0^1 (1 - t^2)^{\nu - 1/2} \sin zt \, dt - \int_0^\infty e^{-zt} (1 + t^2)^{\nu - 1/2} \, dt \right]. \tag{3.14}$$

These formulae are valid for Re $(z) > 0$ and Re $(\nu + 1/2) > 0$. If we make use of the relations (1.7) and (1.8), we obtain the corresponding expression for the Hankel functions when $-\pi < \arg z < 2\pi$:

$$\mathscr{H}_\nu^{(1)}(z) = \frac{\Gamma\left(\frac{1}{2} - \nu\right)\left(\frac{z}{2}\right)^\nu}{\pi i \Gamma\left(\frac{1}{2}\right)} \int_{1+i\infty}^{(1+)} e^{izt} (t^2 - 1)^{\nu - 1/2} \, dt \tag{3.15}$$

and

$$\mathscr{H}_\nu^{(2)}(z) = \frac{\Gamma\left(\frac{1}{2}-\nu\right)\left(\frac{z}{2}\right)^\nu}{\pi i \Gamma\left(\frac{1}{2}\right)} \int\limits_{-1+i\infty}^{(-1-)} e^{izt}(t^2-1)^{\nu-1/2}\,dt. \tag{3.16}$$

when $-2\pi < \arg z < \pi$. In this case the path of integration is as shown in Fig. 3.

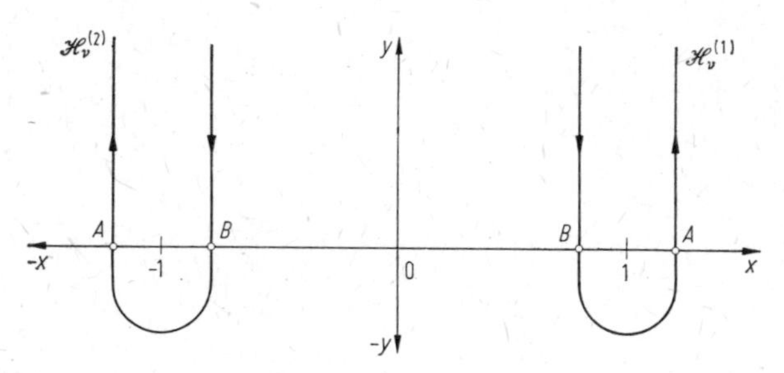

Fig. 3. Integration contours for the Hankel functions of first and second kinds.

By the principle of analytic continuation, the expressions (3.15) and (3.16) can be written in a more general form, provided the path of integration is changed appropriately:

$$\mathscr{H}_\nu^{(1)}(z) = \frac{\Gamma\left(\frac{1}{2}-\nu\right)\left(\frac{z}{2}\right)^\nu}{\pi i \Gamma\left(\frac{1}{2}\right)} \int\limits_{\infty i \exp(-i\omega)}^{(1+)} e^{izt}(t^2-1)^{\nu-1/2}\,dt, \tag{3.17}$$

where the angle of rotation ω lies in the interval $-\pi/2 < \omega < 3\pi/2$, and

$$\mathscr{H}_\nu^{(2)}(z) = \frac{\Gamma\left(\frac{1}{2}-\nu\right)\left(\frac{z}{2}\right)^\nu}{\pi i \Gamma\left(\frac{1}{2}\right)} \int\limits_{\infty i \exp(-i\omega)}^{(-1-)} e^{izt}(t^2-1)^{\nu-1/2}\,dt \tag{3.18}$$

where $-3\pi/2 < \omega < \pi/2$.

These integrals converge for a given ω only if the argument of the variable z remains within the interval $-\pi/2 + \omega < \arg z < \pi/2 + \omega$.

The contour integrals in these formulae can be converted to real integrals provided $\operatorname{Re}(iz) < 0$; in such a case it is sufficient to take $\omega = \pi/2$ and $\omega = -\pi/2$ and to take into account the variation of the phase of the quantity $(t^2 - 1)$, taking its argument as -2π at the branch point (1) and 2π at the point (-1), and lastly by assuming the argument to be zero at the point A. For this transformation it is necessary that $0 < \arg z < \pi$ and $\operatorname{Re}(\nu + 1/2) > 0$. We find

$$\mathscr{H}_\nu^{(1)}(z) = \frac{-2ie^{-i\nu\pi}z^\nu}{A_\nu} \int\limits_1^\infty e^{izt}(t^2-1)^{\nu-1/2}\,dt; \tag{3.19}$$

For $\pi < \arg z < 2\pi$ and Re $(\nu + 1/2) > 0$, we have

$$\mathscr{H}_\nu^{(2)}(z) = \frac{2ie^{i\nu\pi} z^\nu}{A_\nu} \int_1^\infty e^{-izt} (t^2 - 1)^{\nu - 1/2}\, dt. \tag{3.20}$$

In deriving these formulae use was made of the following relation involving Gamma functions (see [11, p. 951]:

$$\Gamma\left(\frac{1}{2} + \nu\right) \Gamma\left(\frac{1}{2} - \nu\right) = \frac{\pi}{\cos \nu\pi}. \tag{3.21}$$

By making the change of variable $t = \cosh u$ in formula (3.19) and (3.20), other representations can be obtained. In particular, for $H_\nu^{(1)}(z)$, with $0 < \arg z < \pi$, and Re $(\nu + 1/2) > 0$:

$$\mathscr{H}_\nu^{(1)}(z) = \frac{-2ie^{-i\nu\pi} z^\nu}{A_\nu} \int_0^\infty e^{iz\cosh u} \sinh^{2\nu} u\, du. \tag{3.22}$$

Relations (3.19) and (3.20) remain valid for $-1/2 < \mathrm{Re}(\nu) < 1/2$, provided the variable $z = x$ is real and positive. Replacing ν by $-\nu$ and using the relation (1.9) we obtain the integral representations of Mehler-Sonine:

$$\mathscr{H}_\nu^{(1)}(x) = \frac{2}{i\Gamma\left(\frac{1}{2} - \nu\right) \Gamma\left(\frac{1}{2}\right)} \left(\frac{2}{x}\right)^\nu \int_1^\infty \frac{e^{ixt}}{(t^2 - 1)^{\nu + 1/2}}\, dt; \tag{3.23}$$

$$\mathscr{H}_\nu^{(2)}(x) = \frac{2i}{\Gamma\left(\frac{1}{2} - \nu\right) \Gamma\left(\frac{1}{2}\right)} \left(\frac{2}{x}\right)^\nu \int_1^\infty \frac{e^{-ixt}}{(t^2 - 1)^{\nu + 1/2}}\, dt \tag{3.24}$$

valid for $x > 0$ and $-1/2 < \mathrm{Re}(\nu) < 1/2$.

In addition, if the relations (1.7) and 1.8) among the Hankel, Neumann and Bessel functions are used, one obtains, for the same range of the variable z and index ν, the following formulae:

$$J_\nu(x) = \frac{2}{\Gamma\left(\frac{1}{2} - \nu\right) \Gamma\left(\frac{1}{2}\right)} \left(\frac{2}{x}\right)^\nu \int_1^\infty \frac{\sin xt}{(t^2 - 1)^{\nu + 1/2}}\, dt; \tag{3.25}$$

$$N_\nu(x) = \frac{2}{\Gamma\left(\frac{1}{2} - \nu\right) \Gamma\left(\frac{1}{2}\right)} \left(\frac{2}{x}\right)^\nu \int_1^\infty \frac{\cos xt}{(t^2 - 1)^{\nu + 1/2}}\, dt. \tag{3.26}$$

Further, making use of the expression (1.13) between the MacDonald function and Hankel function of the first kind, it is easy to obtain a different integral representation of the Poisson type for the MacDonald function. From Eqs. (3.22) and (3.19) one finds immediately

$$K_\nu(z) = \frac{\pi z^\nu}{A_\nu} \int_0^\infty e^{-z\cosh u} \sinh^{2\nu} u \, du; \tag{3.27}$$

$$K_\nu(z) = \frac{\pi z^\nu}{A_\nu} \int_1^\infty e^{-zt} (t^2 - 1)^{\nu - 1/2} \, dt \tag{3.28}$$

valid for $\operatorname{Re}(z) > 0$ and $\operatorname{Re}(\nu + 1/2) > 0$.

Further considerations require the introduction of the following representation of this function, valid for $\operatorname{Re}(\nu + 1/2) \geq 0$ and $\operatorname{Re}(z) > 0$

$$K_\nu(z) = \frac{A_\nu z^\nu}{\pi} \int_0^\infty \frac{\cos u \, du}{(u^2 + z^2)^{\nu + 1/2}} \tag{3.29}$$

which can be obtained, for example, in the following way.
In the expression (3.17) for $\mathscr{H}_\nu^{(1)}(z)$ we set $\omega = \pi/2$ and replace ν by $-\nu$. The resulting contour (solid line in Fig. 4) is closed on the imaginary axis (dotted line). The integral along the dotted line goes to zero as $\varrho \to \infty$ provided $\operatorname{Re}(z) > 0$, and the integration along the chosen contour reduces to integration along the imaginary axis. The expression (3.29) now follows directly from the relation

$$K_\nu(z) = \frac{\pi i}{2} e^{-i\nu\pi/2} \mathscr{H}_\nu^{(1)}(z e^{i\pi/2}). \tag{3.30}$$

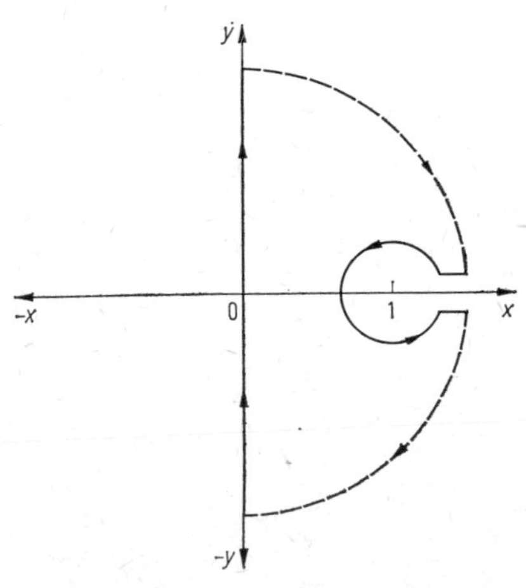

Fig. 4. Deformation of the contour for the Hankel function of the first kind.

There also exist other methods which demonstrate the validity of formula (3.29) (for details, see Ref. [10, pp. 191, 206, 209]). For future reference the following integral representations for Hankel functions of the first and second kind are included:

For $-\pi/2 + \lambda < \arg(z) < 3\pi/2 + \lambda$:

$$\mathscr{H}_\nu^{(1)}(z) = \sqrt{\frac{2}{\pi z}} \frac{e^{i\left(z - \frac{\nu\pi}{2} - \frac{\pi}{4}\right)}}{\Gamma\left(\nu + \frac{1}{2}\right)} \int_0^{\infty \exp i\lambda} e^{-u} u^{\nu - 1/2} \left(1 - \frac{u}{2iz}\right)^{\nu - 1/2} du \tag{3.31}$$

and for $-3\pi/2 + \lambda < \arg(z) < \pi/2 + \lambda$

$$\mathscr{H}_\nu^{(2)}(z) = \sqrt{\frac{2}{\pi z}} \frac{e^{-i\left(z - \frac{\nu\pi}{2} - \frac{\pi}{4}\right)}}{\Gamma\left(\nu + \frac{1}{2}\right)} \int_0^{\infty \exp i\lambda} e^{-u} u^{\nu - 1/2} \left(1 + \frac{u}{2iz}\right)^{\nu - 1/2} du, \tag{3.32}$$

which, for $\mathrm{Re}(\nu + 1/2) > 0$ result from formulas (3.17), (3.20), by an obvious change of the variable.

4. Integral Representations of Cylindrical Functions of the Bessel-Schlaefli and Sonine Form

Together with the representation of cylindrical functions in Poisson form, there is wide use made of integral representations in the Bessel-Schlaefli form. For integral indices $\nu = n$ this representation is

$$J_n(z) = \frac{1}{2\pi} \int_0^{2\pi} \cos(n\theta - z \sin\theta)\, d\theta, \tag{4.1}$$

or, what is equivalent,

$$J_n(z) = \frac{1}{2\pi} \int_0^{2\pi} e^{i(n\theta - z\sin\theta)}\, d\theta. \tag{4.2}$$

The validity of these formulae is easily proved by using the generating function in the form

$$\exp\frac{z}{2}\left(t - \frac{1}{t}\right) = \sum_{m=-\infty}^{\infty} J_m(z)\, t^m \tag{4.3}$$

multiplying both sides by t^{-n-1} and integrating along a contour which travels about the origin in the positive direction:

$$\frac{1}{2\pi i} \int^{(0+)} t^{-n-1} \exp\frac{z}{2}\left(t - \frac{1}{t}\right) dt = \sum_{-\infty}^{\infty} J_m(z) \frac{1}{2\pi i} \int^{(0+)} t^{m-n-1}\, dt.$$

All of the integrals on the right side are zero except that for which $m = n$. Consequently

$$J_n(z) = \frac{1}{2\pi i} \int^{(0+)} t^{-n-1} \exp\frac{z}{2}\left(t - \frac{1}{t}\right) dt \tag{4.4}$$

and, on setting the variable of integration $t = e^{i\theta}$, this reduces directly to (4.2).

The expression (4.4) can be generalized to non-integral values of ν. The complete picture is given by formula (1.10):

$$J_\nu(z) = \sum_{m=0}^{\infty} \frac{(-1)^m (z/2)^{\nu+2m}}{m!\,\Gamma(\nu+m+1)},$$

which is valid for arbitrary ν. Making use of Euler's second integral

$$\frac{1}{\Gamma(\nu+m+1)} = \frac{1}{2\pi i} \int_{-\infty}^{(0+)} t^{-\nu-m-1}\, e^t\, dt, \tag{4.5}$$

one obtains

$$J_\nu(z) = \frac{(z/2)^\nu}{2\pi i} \sum_{m=0}^{\infty} \int_{-\infty}^{(0+)} \frac{(-1)^m}{m!} \left(\frac{z^2}{4t}\right)^m \frac{e^t}{t^{\nu+1}}\, dt.$$

Summing under the integral sign gives $\exp(-z^2/4t)$ and as a result

$$J_\nu(z) = \frac{(z/2)^\nu}{2\pi i} \int_{-\infty}^{(0+)} t^{-\nu-1} \exp\left(t - \frac{z^2}{4t}\right) dt. \tag{4.6}$$

Now, making the change of the variable of integration $t = (z/2)\exp w$, and assuming that $|\arg(z)| < \pi/2$, we find

$$J_\nu(z) = \frac{1}{2\pi i} \int_{\infty - i\pi}^{\infty + i\pi} e^{z \sinh w - \nu w}\, dw. \tag{4.7}$$

The path of integration is shown in Fig. 5.

If we carry out the integration separately on the branches of this contour, it is possible to obtain the following Schlaefli generalization of the Bessel integral:

$$J_\nu(z) = \frac{1}{2\pi} \int_0^{2\pi} \cos(\nu\theta - z\sin\theta)\, d\theta - \frac{\sin\nu\pi}{\pi} \int_0^\infty e^{-\nu t - z\sinh t}\, dt, \tag{4.8}$$

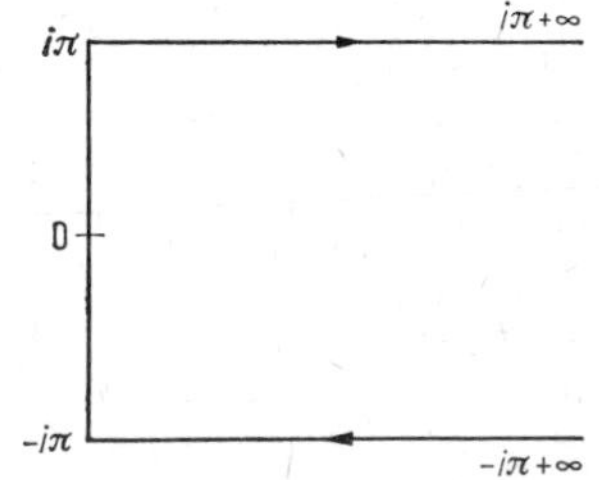

Fig. 5. Schlaefli contour for the Bessel function.

valid for $|\arg(z)| < \pi/2$. This formula reduces directly to (4.1) for integral $\nu = n$.

The corresponding representations of other cylindrical functions $N_\nu(z)$, $H_\nu^{(1)}(z)$ and $H_\nu^{(2)}(z)$ are easily obtained if we consider their connection with the Bessel functions $J_\nu(z)$, according to Eqs. (1.6)—(1.8). For these we have

$$N_\nu(z) = \frac{1}{\pi}\int_0^\pi \sin(z\sin\theta - \nu\theta) - \frac{1}{\pi}\int_0^\infty (e^{\nu t} + e^{-\nu t}\cos\nu\pi)\, e^{-z\sinh t}\, dt; \tag{4.9}$$

$$\mathscr{H}_\nu^{(1)}(z) = \frac{1}{\pi i}\int_{-\infty+i\lambda}^{\infty+i(\pi-\lambda)} e^{z\sinh w - \nu w}\, dw; \tag{4.10}$$

$$\mathscr{H}_\nu^{(2)}(z) = -\frac{1}{\pi i}\int_{-\infty+i\lambda}^{\infty-i(\pi+\lambda)} e^{z\sinh w - \nu w}\, dw. \tag{4.11}$$

For values of the parameter $|\lambda| < \pi$, the convergence of each integral is guaranteed for any z for which $|\arg(z) - \lambda| < \pi/2$. This results in the following expressions for the Hankel functions for real $z = x$ and $-1 < \mathrm{Re}(\nu) < 1$: [10, p. 199]:

$$\mathscr{H}_\nu^{(1)}(x) = \frac{2e^{-i\nu\pi/2}}{\pi i}\int_0^\infty e^{ix\cosh t}\cosh\nu t\, dt; \tag{4.12}$$

$$\mathscr{H}_\nu^{(2)}(x) = -\frac{2e^{i\nu\pi/2}}{\pi i}\int_0^\infty e^{-ix\cosh t}\cosh\nu t\, dt. \tag{4.13}$$

In order to obtain corresponding expressions for the function $I_\nu(z)$, we make use of the definition $I_\nu(z) = J_\nu(iz)\exp(-i\nu\pi/2)$, together with Eq. (4.6). The result is

$$I_\nu(z) = \frac{\left(\frac{z}{2}\right)^\nu}{2\pi i}\int_{-\infty}^{(0+)} t^{-\nu-t}\exp\left(t + \frac{z^2}{4t}\right) dt. \tag{4.14}$$

When $\mathrm{Re}(z) > 0$, this expression, with the aid of the substitution $t = (z/2)\exp(w)$ may be transformed into

$$I_\nu(z) = \frac{1}{2\pi i}\int_{\infty-i\pi}^{\infty+i\pi} e^{z\cosh w - \nu w}\, dw, \tag{4.15}$$

where the contour of integration is shown in Fig. 5. Carrying out the integrations along the separate branches of this contour, we find directly that

$$I_\nu(z) = \frac{1}{\pi}\int_0^\pi e^{z\cos\theta}\cos\nu\theta\, d\theta - \frac{\sin\nu\pi}{\pi}\int_0^\infty e^{-z\cosh t - \nu t}\, dt. \tag{4.16}$$

From the expression thus obtained, and with the aid of (1.14), we find the following expression

$$K_\nu(z) = \frac{1}{2}\int_{-\infty}^\infty e^{-z\cosh t - \nu t}\, dt, \tag{4.17}$$

valid for $|\arg(z)| < \pi/2$. If, in the latter integral, we make a change of the variable according to $\tau = (z/2)\exp(t)$, we obtain[1]

$$K_\nu(z) = \frac{1}{2}(z/2)^\nu \int_0^\infty \tau^{-\nu-1} \exp\left(-\tau - \frac{z^2}{4\tau}\right) d\tau \tag{4.18}$$

valid for $\mathrm{Re}(z^2) > 0$.

To conclude this section, we present one more expression for $J_\nu(z)$, due to Sonine:

$$J_\nu(z) = \frac{\left(\frac{z}{2}\right)^\nu}{2\pi i} \int_{c-i\infty}^{c+i\infty} t^{-\nu-1} \exp\left(t - \frac{z^2}{4t}\right) dt, \tag{4.19}$$

valid for $\mathrm{Re}(\nu) > -1$. The constant c may be any positive number. This expression is quite convenient for many applications. It may be obtained from Eq. (4.6) provided it is possible to deform the path of integration in such a way that it becomes a straight line parallel to the imaginary axis. For $\mathrm{Re}(\nu) > -1$ such a deformation is legitimate since the convergence of the integral at the upper and lower limits is guaranteed.

In all of the integral representations discussed up to this point certain limitations on the variable z were imposed. In the majority of cases, such limitations were that $|\arg(z)| < \pi/2$, so that the corresponding integration could be performed along the real axis. However, by variations of the contours, all of these expressions may be generalized to the case of arbitrary z. See, for example, formulae (4.10), (4.11) and (3.17), (3.18).

[1] In [10, p. 203] as well as in [11, p. 973], the lower limit, due to an oversight, is given as $-\infty$.

Chapter II

General Theory of Incomplete Cylindrical Functions Expressed in Poisson Form

Basic information about the cylindrical functions of Bessel, Neumann, and Hankel was given in the preceeding chapter. These functions are all solutions of Bessel's differential equation, and have integral representations in the Poisson and Bessel forms for which the contours of integration are completely determined. In practice, however, it frequently becomes necessary to study analogous integrals in which the contours are indeterminate. Such functions, by analogy to the incomplete elliptic integrals of Legendre or to the incomplete gamma function [12], may be called incomplete cylindrical functions.

As has already been noted, a motivation for the study of incomplete Bessel functions was first introduced in [8]. In this paper numerous physical problems were presented which gave rise to the investigation of these functions, and which have been studied by a number of authors.

For example, in [5, p. 699] many properties of such functions were presented in quite general form. Prior to this, similar functions appeared in [3] in connection with the expansion of the integral $\int_y^\infty t^\nu \exp\,[x(t \pm t^{-\mu})]\,dt$.

There are still earlier papers in which properties of these functions were studied. One should note in particular the paper of Binet [1] which was devoted to a study of an integral of the form $\int_a^b y^{2\nu} \exp\,(-qy^2 - p/y^2)\,dy$ with arbitrary limits, and also the investigations of Nagaoka in connection with a problem in the diffraction of light [2], (see also [10, p. 624]).

In this chapter the general theory underlying incomplete cylindrical functions of the Poisson form is introduced and their properties are systematically studied.

1. Definitions of Incomplete Cylindrical Functions as Poisson Integrals

Incomplete cylindrical functions of the Poisson form may be introduced in numerous ways. It is sufficient, in any of the integral representations for cylindrical functions (as given in Chapter I), to perform the integration over only a portion of the contour. In other words, any function of the form

$$z^\nu \int e^{\pm izt} (t^2 - 1)^{\nu - 1/2}\, dt,$$

in which the integration is performed along an arbitrary contour represents, up to a constant factor, an incomplete cylindrical function. We will construct these functions in such a manner that, for limiting cases, they reduce to cylindrical functions.

We first consider a function of the form

$$\mathscr{E}_\nu^{(1)}(c, z) = \frac{\Gamma\left(\frac{1}{2} - \nu\right)}{\pi i \Gamma\left(\frac{1}{2}\right)} \left(\frac{z}{2}\right)^\nu \int_c^{(1+)} e^{izt} (t^2 - 1)^{\nu - 1/2}\, dt, \qquad (1.1)$$

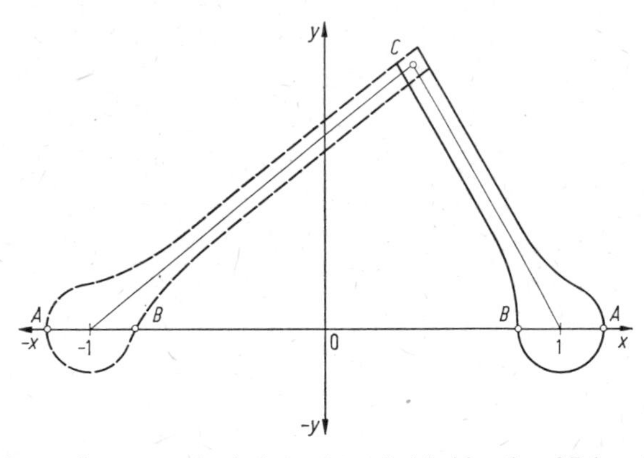

Fig. 6. Integration contours for the incomplete cylindrical function of Poisson form.

where ν, c, and z are arbitrary complex numbers, and the contour of integration is selected as shown in Fig. 6 by the solid line. It begins at the point c in the complex t plane, moves around the branch point $t = 1$ and returns to the original point c. For definiteness we will assume that, as the contour is traversed, the argument of $(t^2 - 1)$ is zero at the point A [Fig. 6] and is equal to -2π at the point B. The resulting function $\mathscr{E}_\nu^{(1)}(c, z)$ is analytic in all of its variables. In addition, for $|\arg z| < \pi$ the integral of Eq. (1.1) retains its meaning for $c \to 1 + i\infty$. However, for this case, the integral represents the Hankel function of the first kind, (see Eq. (3.15), Chapter I) that is,

$$\mathscr{E}_{\nu\ c \to 1 + i\infty}^{(1)}(c, z) = \mathscr{H}_\nu^{(1)}(z).$$

In an analogous manner one can introduce the function

$$\mathscr{E}_\nu^{(2)}(c, z) = \frac{\Gamma\left(\frac{1}{2} - \nu\right)}{\pi i \Gamma\left(\frac{1}{2}\right)} \left(\frac{z}{2}\right)^\nu \int_c^{(-1-)} e^{izt} (t^2 - 1)^{\nu - 1/2} \, dt. \tag{1.2}$$

in which the contour is shown by the dotted line, Fig. 6. This function, for $|\arg z| < \pi$ and $c \to (-1 + i\infty)$ reduces to the Hankel function of the second kind. That is,

$$\mathscr{E}_\nu^{(2)}(c, z)_{c \to -1 + i\infty} = \mathscr{H}_\nu^{(2)}(z).$$

It is not difficult to verify that the sum of these functions is the Bessel function $J_\nu(z)$. Thus, it is sufficient to consider only the function $\mathscr{E}_\nu^{(1)}(c, z)$. Let us obtain a somewhat different representation for this function. If in Eq. (1.1) we complete the integration along the contour and take into account the indicated change in the argument of $(t^2 - 1)$ at the branch points A and B, we have

$$\mathscr{E}_\nu^{(1)}(c, z) = \frac{\Gamma\left(\frac{1}{2} - \nu\right)}{\pi i \Gamma\left(\frac{1}{2}\right)} \left(\frac{z}{2}\right)^\nu$$

$$\times \left[\int_c^1 e^{izt - 2\pi i\nu + \pi i} (t^2 - 1)^{\nu - 1/2} \, dt + \int_1^c e^{izt} (t^2 - 1)^{\nu - 1/2} \, dt\right]$$

$$= 2i \frac{e^{-i\nu\pi} \cos \nu\pi \Gamma\left(\frac{1}{2} - \nu\right)}{\pi \Gamma\left(\frac{1}{2}\right)} \left(\frac{z}{2}\right)^\nu \int_c^1 e^{izt} (t^2 - 1)^{\nu - 1/2} \, dt.$$

Or, making use of the relationship for the gamma function

$$\Gamma\left(\nu - \frac{1}{2}\right) \Gamma\left(\nu + \frac{1}{2}\right) \cos \nu\pi = \pi$$

and introducing the notation $A_\nu = 2^\nu \Gamma(\nu + 1/2) \Gamma(1/2)$, we obtain

$$\mathscr{E}_\nu^{(1)}(c, z) = 2i \frac{e^{-i\nu\pi} z^\nu}{A_\nu} \int_c^1 e^{izt} (t^2 - 1)^{\nu - 1/2} \, dt. \tag{1.3}$$

In order to guarantee convergence of this integral, it is necessary that Re $(\nu + 1/2) > 0$. This function could have been called an incomplete Hankel function, and used as a basis for constructing other incomplete cylindrical functions. However, for convenience of presentation we will introduce a somewhat different function. In the right hand side of Eq. (1.3) we make the change of variable $t = \cos u$ and consider in the w plane a function of the following form

$$E_\nu^\pm(w, z) = \frac{2z^\nu}{A_\nu} \int_0^w e^{\pm iz\cos u} \sin^{2\nu} u \, du. \tag{1.4}$$

For $w = \pi$, the right hand side is none other than the Bessel function in Poisson form (see Chapter I, 3.8). Therefore the analytic function $E_\nu^\pm(w, z)$ will, for values of $w \neq \pi$ be called an incomplete cylindrical function of the Poisson form. Between these functions and the function $\mathscr{E}_\nu^{(1)}(c, z)$ there exists the following easily verified relation

$$\mathscr{E}_\nu^{(1)}(c, z) = E_\nu^+(-i \cosh^{-1} c, z). \tag{1.5}$$

It is also convenient to represent (1.4) in the following form

$$E_\nu^\pm(w, z) = J_\nu(w, z) \pm i H_\nu(w, z), \tag{1.6}$$

where

$$J_\nu(w, z) = \frac{E_\nu^+(w, z) + E_\nu^-(w, z)}{2} = 2\frac{z^\nu}{A_\nu}\int_0^w \cos(z\cos\theta)\sin^{2\nu}\theta\, d\theta; \tag{1.7}$$

$$H_\nu(w, z) = \frac{E_\nu^+(w, z) - E_\nu^-(w, z)}{2i} = 2\frac{z^\nu}{A_\nu}\int_0^w \sin(z\cos\theta)\sin^{2\nu}\theta\, d\theta. \tag{1.8}$$

For $w = \pi/2$, the right hand side represents the familiar expressions for Bessel and Struve functions (see [10, p. 357]).

$$J_\nu(z) = 2\frac{z^\nu}{A_\nu}\int_0^{\pi/2} \cos(z\cos\theta)\sin^{2\nu}\theta\, d\theta; \tag{1.9}$$

$$H_\nu(z) = 2\frac{z^\nu}{A_\nu}\int_0^{\pi/2} \sin(z\cos\theta)\sin^{2\nu}\theta\, d\theta. \tag{1.10}$$

In other words

$$J_\nu\left(\frac{\pi}{2}, z\right) \equiv J_\nu(z); \tag{1.11}$$

$$H_\nu\left(\frac{\pi}{2}, z\right) \equiv H_\nu(z). \tag{1.12}$$

Therefore, as in [8] we will call the function $J_\nu(w, z)$ an incomplete Bessel function and $H_\nu(w, z)$ an incomplete Struve function.

Now, after substituting iz for z in Eqs. (1.6) and (1.7), we have

$$J_\nu(w, iz) = e^{i\nu\pi/2} I_\nu(w, z); \tag{1.13}$$

$$H_\nu(w, iz) = i e^{i\nu\pi/2} L_\nu(w, z), \tag{1.14}$$

where $\operatorname{Re}\left(\nu + \frac{1}{2}\right) > 0$ and

$$I_\nu(w, z) = \frac{2z^\nu}{A_\nu}\int_0^w \cosh(z\cos\theta)\sin^{2\nu}\theta\, d\theta; \tag{1.15}$$

$$L_\nu(w, z) = \frac{2z^\nu}{A_\nu}\int_0^w \sinh(z\cos\theta)\sin^{2\nu}\theta\, d\theta. \tag{1.16}$$

In an analogous manner, we will call the function $I_\nu(w, z)$ an incomplete Bessel function of purely imaginary argument and $L_\nu(w, z)$ an incomplete Struve function of purely imaginary argument. For $w = \pi/2$ we have

$$I_\nu(\pi/2, z) \equiv I_\nu(z); \tag{1.17}$$

$$L_\nu(\pi/2, z) \equiv L_\nu(z). \tag{1.18}$$

The latter function was investigated by Nicholson (see [10, p. 360]). The above definitions of the incomplete Bessel and Struve functions are convenient in that we can recover from them the relationships (1.13), (1.14) just as we did for the Bessel functions.

The functions $I_\nu(w, z)$ and $L_\nu(w, z)$ may also be represented directly in terms of the incomplete cylindrical function $E_\nu^+(w, z)$.

If relationship (1.14) is employed, then

$$I_\nu(w, z) = \frac{1}{2} e^{i\nu\pi/2} E_\nu^+(w, -iz) + \frac{1}{2} e^{-i\nu\pi/2} E_\nu^+(w, iz); \tag{1.19}$$

$$L_\nu(w, z) = \frac{1}{2} e^{i\nu\pi/2} E_\nu^+(w, -iz) - \frac{1}{2} e^{-i\nu\pi/2} E_\nu^+(w, iz). \tag{1.20}$$

In place of these functions, it is sometimes convenient to consider also functions of the form

$$F_\nu^+(w, z) = \frac{1}{2} e^{i\nu\pi/2} E_\nu^+(w, -iz) = \frac{z^\nu}{A_\nu} \int_0^w e^{z\cos\theta} \sin^{2\nu}\theta \, d\theta; \tag{1.21}$$

$$F_\nu^-(w, z) = \frac{1}{2} e^{-i\nu\pi/2} E_\nu^+(w, iz) = \frac{z^\nu}{A_\nu} \int_0^w e^{-z\cos\theta} \sin^{2\nu}\theta \, d\theta, \tag{1.22}$$

so that Eqs. (1.19) and (1.20) take the form

$$I_\nu(w, z) = F_\nu^+(w, z) + F_\nu^-(w, z); \tag{1.23}$$

$$L_\nu(w, z) = F_\nu^+(w, z) - F_\nu^-(w, z). \tag{1.24}$$

We will now turn to the construction of incomplete Hankel and MacDonald functions. From the definition of the incomplete cylindrical functions as given by Eqs. (1.3) and (1.4), it is seen that if $c \to +i\infty$, in Eq. (1.3) and $w \to -i\infty$ in Eq. (1.4), we obtain Hankel functions of the first kind. It is natural, therefore, to regard the former as forms of incomplete cylindrical functions and write

$$\mathscr{H}_\nu(w, z) = \mathscr{E}_\nu^{(1)}(\cosh w, z) = E_\nu^+(-iw, z). \tag{1.25}$$

From the expression (1.25) we can obtain a representation for $\mathscr{H}_\nu(w, z)$. For real $c = \cosh\beta > 1$ and $\operatorname{Re}(\nu + 1/2) > 0$ we have

$$\mathscr{E}_\nu^{(1)}(\cosh\beta, z) = \mathscr{H}_\nu(\beta, z) = -\frac{2ie^{-i\nu\pi} z^\nu}{A_\nu} \int_1^{\cosh\beta} e^{izt} (t^2 - 1)^{\nu - 1/2} \, dt \tag{1.26}$$

or, equivalently

$$E_\nu^+(-i\beta, z) = \mathscr{H}_\nu(\beta, z) = -\frac{2ie^{-i\nu\pi} z^\nu}{A_\nu} \int_0^\beta e^{iz\cosh u} \sinh^{2\nu} u \, du. \tag{1.27}$$

Consider now the function

$$K_\nu(w, z) = i\frac{\pi}{2} e^{i\nu\pi/2} \mathscr{E}_\nu^{(1)}(\cosh w, iz) = \frac{i\pi}{2} e^{i\nu\pi/2} E_\nu^+(-iw, iz). \qquad (1.28)$$

or, as is clear from Eqs. (1.3) and (1.4) for Re $(\nu + 1/2) > 0$

$$K_\nu(w, z) = \pi \frac{z^\nu}{A_\nu} \int_1^{\cosh w} e^{-zt} (t^2 - 1)^{\nu - 1/2} \, dt; \qquad (1.29)$$

$$K_\nu(w, z) = \pi \frac{z^\nu}{A_\nu} \int_0^w e^{-z\cosh t} \sinh^{2\nu} t \, dt. \qquad (1.30)$$

For $w \to \infty$ and Re $z > 0$ this function reduces to the MacDonald function, and therefore for arbitrary w we will call it the incomplete MacDonald function.

From the discussion above it follows that the representations (1.3) and (1.4) enable us, with the aid of a simple relation, to also introduce the incomplete functions of Bessel, Struve, Hankel and MacDonald. By taking into account the relation (1.5) between the functions $\mathscr{E}_\nu^{(1)}(c, z)$ and $E_\nu^+(w, z)$, we will concentrate on the characteristics of incomplete cylindrical functions of the Poisson form using the representation (1.4).

The following additional remarks should be made. In order to obtain the differential equations, recurrence relations etc. for the incomplete Bessel or Struve functions from those already derived for the function $E_\nu^+(w, z)$, we need only formally replace the function $E_\nu^+(w, z)$ according to Eq. (1.6) by the expression $J_\nu(w, z) + iH_\nu(w, z)$ separate real and imaginary parts, formally considering ν, w and z as real quantities. The real part of the resulting expression represents the desired relation for the incomplete Bessel function and the imaginary part the corresponding relation for the incomplete Struve function. By setting $w = \pi/2$, these become the corresponding relations for the conventional Bessel and Struve functions.

In order to obtain corresponding relations for the incomplete Hankel functions, it is sufficient to replace w everywhere by $(-iw)$ according to Eq. (1.25). Here the known relations for Hankel functions are formally obtained by allowing w to approach $(\infty + i\lambda)$. Finally, for the incomplete MacDonald function $K_\nu(w, z)$ it is necessary, according to Eq. (1.28), to replace w by $-iw$; z by iz in the corresponding relationship for $E_\nu^+(w, z)$, and to note that

$$E_\nu^+(-iw, iz) = \frac{2}{\pi i} e^{-i\nu\pi/2} K_\nu(w, z). \qquad (1.31)$$

To conclude this section, we point out that, in the class of functions under consideration, we could also include functions of the following form

$$\varphi_\nu(w, z) = \frac{2z^\nu}{A_\nu} \int_0^w e^{iz\sin t} \cos^{2\nu} t \, dt; \tag{1.32}$$

$$\psi_\nu(w, z) = \frac{2z^\nu}{A_\nu} \int_0^w e^{iz\cos t} \cos^{2\nu} t \, dt; \tag{1.33}$$

$$q_\nu(w, z) = \frac{2z^\nu}{A_\nu} \int_0^w e^{iz\sin t} \sin^{2\nu} t \, dt. \tag{1.34}$$

In fact, by introducing a new variable of integration $\theta = \pi/2 - t$ into Eq. (1.32) we find

$$\varphi_\nu(w, z) = E_\nu^+\left(\frac{\pi}{2}, z\right) - E_\nu^+\left(\frac{\pi}{2} - w, z\right). \tag{1.35}$$

Further, it is not difficult to see that there is a connection between ψ_ν and q_ν which is analogous to (1.35)

$$q_\nu(w, z) = \psi_\nu\left(\frac{\pi}{2}, z\right) - \psi_\nu\left(\frac{\pi}{2} - w, z\right). \tag{1.36}$$

Therefore, it is sufficient to investigate only one of these functions, for example, $\psi_\nu(w, z)$. Also it is not difficult to see that if $2\nu = n$, an integer, (1.36) may be expressed in the form of a finite sum of incomplete cylindrical functions. Actually, in this case according to Eq. (1.33), the function $\frac{A_\nu}{z^\nu}\psi_\nu(w, z)$ for $n = 2\nu$ is the n-th derivative of the incomplete cylindrical function $E_0^+(w, z)$; that is,

$$\psi_{n/2}(w, z) = \frac{\pi z^{n/2}}{A_{n/2}} \frac{1}{i^n} \left(\frac{d}{dz}\right)^n E_0^+(w, z). \tag{1.37}$$

Explicit expressions in terms of the functions $E_0^+(w, z)$ and $E_1^+(w, z)$ for $n = 0; 2; 4$ are

$$\begin{aligned} &\psi_0(w, z) \equiv E_0^+(w, z); \ \psi_1(w, z) = zE_0^+(w, z) - E_1^+(w, z); \\ &\psi_2(w, z) = \left(\frac{z^2}{3} - 1\right) E_0^+(w, z) + \frac{2}{z}\left(1 - \frac{z^2}{3}\right) E_1^+(w, z) \\ &\qquad + \frac{z \sin w}{\pi}\left(\frac{\cos w}{z} + \frac{i \sin^2 w}{3}\right) e^{iz\cos w}. \end{aligned} \tag{1.38}$$

These formulae can be obtained in a simple way with the help of recurrence relations for the functions $E_\nu^+(w, z)$. These will be developed in the next section.

2. Recurrence Relations for the Incomplete Cylindrical Functions

We begin our investigation of properties of the incomplete cylindrical functions with the derivation of some recurrence relations. For this purpose we will use the definition of $E_\nu^+(w, z)$ as given by Eq. (1.4):

$$E_\nu^+(w, z) = \frac{2z^\nu}{A_\nu} \int_0^w e^{iz\cos t} \sin^{2\nu} t \, dt, \tag{2.1}$$

where, as in Eq. (1.3), $A_\nu = 2^\nu \Gamma(\nu + 1/2) \Gamma(1/2)$, and w is an arbitrary complex quantity. By analogy with ordinary cylindrical functions we will make use of the following operators provided $\mathrm{Re}\,(\nu - 1/2) > 0$

$$L_\nu E_\nu^+(w, z) \equiv E_{\nu-1}^+(w, z) + E_{\nu+1}^+(w, z) - \frac{2\nu}{z} E_\nu^+(w, z); \tag{2.2}$$

$$L'_\nu E_\nu^+(w, z) + E_{\nu-1}^+(w, z) - E_{\nu+1}^+(w, z) - 2\frac{d}{dz} E_\nu^+(w, z). \tag{2.3}$$

The first operator gives

$$zL_\nu E_\nu^+(w, z) = 2\frac{z^\nu}{A_\nu}\left[(2\nu - 1)\int_0^w e^{iz\cos t} \sin^{2\nu-2} t \, dt \right.$$
$$\left. + \frac{z^2}{2\nu + 1}\int_0^w e^{iz\cos t} \sin^{2\nu+2} t \, dt - 2\nu \int_0^w e^{iz\cos t} \sin^{2\nu} t \, dt\right].$$

where the property of the Γ-function $\Gamma(1 + x) = x\Gamma(x)$ has been used. Integrating the second term by parts and taking into account that $\mathrm{Re}\,(\nu - 1/2) > 0$, we find

$$\frac{z^2}{2\nu + 1}\int_0^w e^{iz\cos t} \sin^{2\nu+2} t \, dt$$
$$= \frac{iz}{2\nu + 1}\left[\sin^{2\nu+1} w - \frac{i(2\nu + 1)}{z} \cos w \sin^{2\nu-1} w\right] e^{iz\cos w}$$
$$- \int_0^w e^{iz\cos t} \left[(2\nu - 1)\sin^{2\nu-2} t - 2\nu \sin^{2\nu} t\right] dt.$$

Finally application of the operator L_ν to the function $E_\nu^+(w, z)$ gives the following relation between functions of three consecutive orders:

$$L_\nu E_\nu^+(w, z) = 2z^\nu \sin^{2\nu-1} w \left[\frac{\cos w}{A_\nu z} + \frac{i \sin^2 \omega}{A_{\nu+1}}\right] e^{iz\cos w}. \tag{2.4}$$

Proceeding in an analogous manner with the second operator L'_ν we find the second recurrence relation

$$L'_\nu E_\nu^+(w, z) = 2z^\nu \sin^{2\nu-1} w \left(\frac{\cos w}{A_\nu z} - i\frac{\sin^2 w}{A_{\nu+1}}\right) e^{iz\cos w}. \tag{2.5}$$

The formulae thus obtained may be transformed directly into the following recurrence relations for Hankel functions of the first kind if one assumes $\mathrm{Im}\, z > 0$ and lets w approach $-i\infty$;

$$\mathscr{H}_{\nu-1}^{(1)}(z) + \mathscr{H}_{\nu+1}^{(1)}(z) - \frac{2\nu}{z} \mathscr{H}_\nu^{(1)}(z) = 0; \tag{2.6}$$

$$\mathscr{H}_{\nu-1}^{(1)}(z) - \mathscr{H}_{\nu+1}^{(1)}(z) - 2\frac{d}{dz} \mathscr{H}_\nu^{(1)}(z) - 0. \tag{2.7}$$

In order to obtain recurrence relations for incomplete Bessel and Struve functions according to the rule given at the end of the preceeding section, it is sufficient to replace the function $E_\nu^+(w, z)$ on the left hand sides of Eqs. (2.4) and (2.5) by its value

$$E_\nu^+(w, z) = J_\nu(w, z) + iH_\nu(w, z),$$

and on the right hand side to set

$$e^{iz\cos w} = \cos(z\cos w) + i\sin(z\cos w),$$

so that for the incomplete Bessel function we have

$$J_{\nu-1}(w, z) + J_{\nu+1}(w, z) - \frac{2\nu}{z} J_\nu(w, z) = g_\nu^-(w, z); \tag{2.8}$$

$$J_{\nu-1}(w, z) - J_{\nu+1}(w, z) - 2\frac{d}{dz} J_\nu(w, z) = g_\nu^+(w, z), \tag{2.9}$$

where the following notations have been introduced

$$g_\nu^\pm(w, z) = 2z^\nu \sin^{2\nu-1} w \left[\frac{\cos w}{zA_\nu}\cos(z\cos w) \pm \frac{\sin^2 w}{A_{\nu+1}}\sin(z\cos w)\right]. \tag{2.10}$$

The corresponding formulae for the Struve function take the form

$$H_{\nu-1}(w, z) + H_{\nu+1}(w, z) - \frac{2\nu}{z} H_\nu(w, z) = f_\nu^+(w, z); \tag{2.11}$$

$$H_{\nu-1}(w, z) - H_{\nu+1}(w, z) - 2\frac{d}{dz} H_\nu(w, z) = f_\nu^-(w, z), \tag{2.12}$$

where

$$f_\nu^\pm(w, z) = 2z^\nu \sin^{2\nu-1} w \left[\frac{\cos w}{zA_\nu}\sin(z\cos w) \pm \frac{\sin^2 w}{A_{\nu+1}}\cos(z\cos w)\right]. \tag{2.13}$$

For $w = \pi/2$ we have $g_\nu^\pm(\pi/2, z) = 0$, and formulae (2.8), (2.9) reduce to the recurrence relations (1.15) and (1.16) of Chapter I for Bessel functions. In addition for $w = \pi/2$, $f_\nu^\pm(\pi/2, z) = \pm\frac{2z^\nu}{A_{\nu+1}}$, and formulae (2.11) and (2.12) reduce to the known recurrence relationships for the Struve function $H_\nu(z)$

$$H_{\nu-1}(z) + H_{\nu+1}(z) - \frac{2\nu}{z} H_\nu(z) = \frac{2z^\nu}{A_{\nu+1}}; \tag{2.14}$$

$$H_{\nu-1}(z) - H_{\nu+1}(z) - 2\frac{d}{dz} H_\nu(z) = -\frac{2z^\nu}{A_{\nu+1}}, \tag{2.15}$$

where as before

$$A_{\nu+1} = 2^{\nu+1}\Gamma(\nu + 3/2)\,\Gamma(1/2).$$

The relationship (2.4) therefore, makes it possible to determine the value of the function $E_\nu^+(w, z)$ for all values of (w, z), provided that $\mathrm{Re}\,(\nu - 1/2) > 0$, and that the values of any two functions with fixed indices are known. For the case of integral indicies, it is convenient to take

$$\left.\begin{aligned} E_0^+(w, z) &= \frac{2}{\pi} \int_0^w e^{iz\cos t}\, dt; \\ E_1^+(w, z) &= \frac{2z}{\pi} \int_0^w e^{iz\cos t} \sin^2 t\, dt, \end{aligned}\right\} \tag{2.16}$$

and for half-odd orders

$$\left.\begin{aligned} E_{1/2}^+(w, z) &= 2\sqrt{\frac{z}{2\pi}} \int_0^w e^{iz\cos t} \sin t\, dt; \\ E_{3/2}^+(w, z) &= 2\frac{z}{2}\sqrt{\frac{z}{2\pi}} \int_0^w e^{iz\cos t} \sin^3 t\, dt. \end{aligned}\right\} \tag{2.17}$$

Incomplete cylindrical functions $E_\nu^+(w, z)$ with half-odd indices

$$\nu = n + 1/2, \quad n = 0, 1, 2, \ldots$$

have, analogous to the Bessel functions $J_{n+1/2}(z)$, their own representation in terms of elementary functions. In particular, from Eq. (2.17)

$$\left.\begin{aligned} E_{1/2}^+(w, z) &= 2\sqrt{\frac{2}{\pi z}}\, \frac{e^{iz} - e^{iz\cos w}}{2i}; \\ E_{3/2}^+(w, z) &= \sqrt{\frac{2}{\pi z}} \left(\frac{e^{iz} - e^{iz\cos w}}{2iz} - \frac{e^{iz} - \cos w\, e^{iz\cos w}}{2} + \frac{iz}{4} \sin^2 w\, e^{iz\cos w}\right). \end{aligned}\right\} \tag{2.18}$$

For $w = \pi/2$ these functions reduce according to Eq. (1.6) to expressions for the Bessel and Struve functions with indices $\nu = 1/2$ and $\nu = 3/2$:

$$\left.\begin{aligned} J_{1/2}(\pi/2, z) &= J_{1/2}(z) = \sqrt{\frac{2}{\pi z}} \cdot \sin z; \\ J_{3/2}(\pi/2, z) &= J_{3/2}(z) = \sqrt{\frac{2}{\pi z}} \left(\frac{\sin z}{z} - \cos z\right); \end{aligned}\right\} \tag{2.19}$$

$$\left.\begin{aligned} H_{1/2}(\pi/2, z) &= H_{1/2}(z) = \sqrt{\frac{2}{\pi z}} (1 - \cos z); \\ H_{3/2}(\pi/2, z) &= H_{3/2}(z) = \sqrt{\frac{2}{\pi z}} \left(1 + \frac{2}{z^2} - \sin z - \frac{\cos z}{z}\right). \end{aligned}\right\} \tag{2.20}$$

The recurrence relations (2.4) and (2.5) represent a special case of the generalized functional equation determining the cylindrical functions (see § 2 Chapter I). If we denote the right hand sides of these equations, respectively, by $2g_\nu(w, z)/z$ and $2f_\nu(w, z)/z$, then, as one easily sees, the functions $g_\nu(w, z)$ and $f_\nu(w, z)$ actually satisfy the Nielson condition

(2.8), Chapter I. That is

$$\begin{aligned} & f_{\nu-1}(w, z) + f_{\nu+1}(w, z) - \frac{2\nu}{z} f_\nu(w, z) \\ & = g_{\nu-1}(w, z) - g_{\nu+1}(w, z) - 2 \frac{\partial}{\partial z} g_\nu(w, z). \end{aligned} \tag{2.21}$$

Therefore, Eqs. (2.4) and (2.5) may, in principle, serve as a definition for the incomplete cylindrical functions $E_\nu^+(w, z)$ for arbitrary indices. Moreover, the validity of Eq. (2.21) reduces the problem of determining solutions of Eqs. (2.4) and (2.5) to the problem of solving the inhomogeneous Bessel differential equation in which the right hand side is expressed as a linear combination of the functions f_ν and g_ν.

We introduce still another form of the recurrence relations (2.4), (2.5), by adding and subtracting these relations. We find

$$z \frac{\partial E_\nu^+}{\partial z} + \nu E_\nu^+ = z E_{\nu-1}^+ - \frac{2z^\nu}{A_\nu} \cos w \cdot \sin^{2\nu-1} w e^{iz\cos w}; \tag{2.22}$$

$$z \frac{\partial E_\nu^+}{\partial z} - \nu E_\nu^+ = -z E_{\nu+1}^+ + \frac{2iz^{\nu+1}}{A_{\nu+1}} \sin^{2\nu+1} w e^{iz\cos w}. \tag{2.23}$$

From the latter formulae it follows that, in particular,

$$\frac{\partial E_0^+}{\partial z} = -E_1^+ + \frac{2i \sin w}{\pi} e^{iz\cos w}, \tag{2.24}$$

which gives for the Bessel and Struve functions:

$$\begin{aligned} \frac{\partial J_0(w, z)}{\partial z} &= -J_1(w, z) - \frac{2 \sin w}{\pi} \sin(z \cos w); \\ \frac{\partial H_0(w, z)}{\partial z} &= -H_1(w, z) + \frac{2 \sin w}{\pi} \cos(z \cos w). \end{aligned} \tag{2.25}$$

All of the above relations were derived for the function $E_\nu^+(w, z)$. In order to obtain the analogous relations for the function

$$E_\nu^-(w, z) = \frac{2z^\nu}{A_\nu} \int_0^w e^{-iz\cos\theta} \sin^{2\nu} \theta \, d\theta, \tag{2.26}$$

it is sufficient to take the complex conjugate of the previously mentioned formulae formally, regarding the variables ν, w and z as real.

In conclusion it may be noted that the well known Struve function $H_\nu(z) \equiv H_\nu(\pi/2, z)$ can also be related to a class of incomplete cylindrical functions of Poisson form. This can be seen from the fact that these functions satisfy a non-homogeneous form of the recurrence relations (2.14) and (2.15).

3. Some General Properties of Incomplete Cylindrical Functions

In the preceeding paragraphs integral representations were introduced for the incomplete cylindrical functions of Poisson form (see for example the expressions (1.3) and (1.4)). Each of these may be taken as definition of these functions. These definitions are equivalent only in that region for which there exists a one to one correspondence between the variables w and c. Thus, for example the definition (1.4)

$$E_\nu^\pm(w, z) = \frac{2z^\nu}{A_\nu} \int_0^w e^{\pm iz\cos\theta} \sin^{2\nu}\theta \, d\theta \tag{3.1}$$

which represents the incomplete cylindrical function of Poisson form, is equivalent to other definitions, [for example, the expression (1.3)] only in that range of the variable w for which $w = \cos^{-1} c$ is a single valued function, i.e. $0 \leq \operatorname{Re} w \leq \pi$.

Within the above mentioned domain, the function $z^{-\nu} E_\nu^\pm(w, z)$ is continuous and single valued and, as we shall show later, can be expanded into an absolutely convergent power series in z:

$$E_\nu^\pm(w, z) = \frac{2z^\nu}{A_\nu} \sum_{m=0}^{\infty} C_{m,\nu}(w) \frac{(\pm iz)^m}{m!}. \tag{3.2}$$

However, in many practical examples, one may find integrals of the type (3.1) in which the variable w is outside the interval $0 \leq \operatorname{Re} w \leq \pi$, or where the argument of z changes by a multiple of π. Obviously these functions, shall also satisfy differential equations and recurrence relationships for incomplete cylindrical functions of Poisson form.

We shall now derive formulae for the evaluation of the incomplete cylindrical functions for such values of the above mentioned variables. First, from the expansion (3.2) it follows that

$$E_\nu^+(w, ze^{ik\pi}) = e^{ik\nu\pi} E_\nu^\pm(w, z), \tag{3.3}$$

where k is an arbitrary integer and the positive sign on the right hand side corresponds to an even k and the negative sign to an odd k. Analogously, there is the relationship

$$E_\nu^-(w, ze^{ik\pi}) = e^{ik\nu\pi} E_\nu^\mp(w, z). \tag{3.4}$$

Here, however, the negative sign corresponds to even k and the positive one to odd k. The formulae obtained in this way are identical with the analogous formulae for Bessel functions, and allow one to evaluate the function $E^\pm(w, z)$ for any value of arg (z) such that $-\pi \leq \arg z \leq \pi$.

We now proceed to the evaluation of $E_\nu^\pm(w, z)$, assuming that the variable w lies autside the strip $0 \leq \operatorname{Re} w \leq \pi$, and the variable z is restricted by $|\arg z| < \pi$. The right side of Eq. (3.1) has a branch point at $\theta = k\pi$, $k = 0, \pm 1, \pm 2, \ldots$ and is consequently a multivalued function. Thus, we will consider $E_\nu^\pm(w, z)$ in the complex w-plane with a cut along the real axis from $(-\infty, 0)$ and $\pi, \infty)$.

Let us now assume that w lies within the interval $k\pi \leq \operatorname{Re} w \leq (k+1)\pi$. Then one has

$$\begin{aligned} E_\nu^\pm(w, z) &= \frac{2z^\nu}{A_\nu} \int_0^w e^{\pm iz\cos\theta} \sin^{2\nu}\theta\, d\theta \\ &= \frac{2z^\nu}{A_\nu} \int_0^{k\pi} e^{\pm iz\cos\theta} \sin^{2\nu}\theta\, d\theta \\ &\quad + \frac{2z^\nu}{A_\nu} \int_{k\pi}^{w} e^{\pm iz\cos\theta} \sin^{2\nu}\theta\, d\theta, \end{aligned} \tag{3.5}$$

where the paths of integration depend on the sign of $\operatorname{Im} w$. The first term can be written in the equivalent form.

$$\frac{2z^\nu}{A_\nu} \int_0^{k\pi} e^{\pm iz\cos\theta} \sin^{2\nu}\theta\, d\theta = \sum_{l=0}^{k-1} \frac{2z^\nu}{A_\nu} \int_{l\pi}^{(l+1)\pi} e^{\pm iz\cos\theta} \sin^{2\nu}\theta\, d\theta. \tag{3.6}$$

Substituting $\theta - l\pi = u$ and (to avoid ambiguity) using the following rule

$$\begin{aligned} \sin(\theta \pm l\pi) &= e^{\mp il\pi + i\arg\theta} |\sin\theta| \cdots \theta = \alpha + i \cdot 0 \\ \sin(\theta \pm l\pi) &= e^{\pm il\pi + i\arg\theta} |\sin\theta| \cdots \theta = \alpha - i \cdot 0 \end{aligned} \tag{3.7}$$

we find that the expression (3.6) takes the following form

$$\frac{2z^\nu}{A_\nu} \int_0^{k\pi} e^{\pm iz\cos\theta} \sin^{2\nu}\theta\, d\theta = \sum_{l=0}^{k-1} \frac{2z^\nu}{A_\nu} e^{2i\nu\pi l} \int_0^{\pi} \sin^{2\nu} u \exp(\pm ize^{il\pi}\cos u)\, du.$$

From the definition of the Bessel function it follows that

$$\frac{z^\nu}{A_\nu} \int_0^{\pi} \sin^{2\nu} u \exp(\pm izc^{il\pi}\cos u)\, du = J_\nu(z).$$

Therefore, after summing over l we find

$$\frac{2z^\nu}{A_\nu} \int_0^{k\pi} e^{\pm iz\cos\theta} \sin^{2\nu}\theta\, d\theta = 2J_\nu(z) \frac{\sin k\nu\pi}{\sin \nu\pi} \begin{cases} e^{-i(k-1)\nu\pi} \cdots \theta = \alpha + i \cdot 0 \\ e^{+i(k-1)\nu\pi} \cdots \theta = \alpha - i \cdot 0. \end{cases} \tag{3.8}$$

The second term in Eq. (3.5) can be brought into the interval $k\pi \leq \operatorname{Re}\omega \leq (k+1)\pi$ by the change of variable $u = \theta - k\pi$.

$$\begin{aligned} &\frac{2z^\nu}{A_\nu} \int_{k\pi}^{w} e^{\pm iz\cos\theta} \sin^{2\nu}\theta\, d\theta \\ &= \frac{2z^\nu}{A_\nu} e^{2ik\nu\pi} \int_0^{w-k\pi} \sin^{2\nu} u \exp(\pm ize^{ik\pi}\cos u)\, du \\ &= e^{ik\nu\pi} E_\nu^\pm(w - k\pi, ze^{ik\pi}) \begin{cases} e^{-3ik\nu\pi}, & \operatorname{Im} w > 0 \\ e^{ik\nu\pi}, & \operatorname{Im} w < 0. \end{cases} \end{aligned} \tag{3.9}$$

Now the R. H. S. of this formula, according to Eqs. (3.3) and (3.4) can be expressed by $E_\nu^\pm(w - k\pi, z)$. Therefore, on the basis of the expression (3.8) and (3.9) we have finally

$$E_\nu^+(w, z) = 2\frac{\sin k\nu\pi}{\sin \nu\pi} J_\nu(z) \begin{cases} e^{-i(k-1)\nu\pi} \\ e^{+i(k-1)\nu\pi} \end{cases} + E_\nu^\pm(w - k\pi, z) \begin{cases} e^{-2ik\nu\pi}, \ \operatorname{Im} w > 0 \\ e^{+2ik\nu\pi}, \ \operatorname{Im} w < 0 \end{cases} \tag{3.10}$$

where $0 \leq \operatorname{Re}(w - k\pi) \leq \pi$ and the positive sign on the R. H. S. is to be taken for even k and the negative for odd k.

Analogously, we find

$$E_\nu^-(w, z) = 2\frac{\sin k\nu\pi}{\sin \nu\pi} J_\nu(z) \begin{cases} e^{-i(k-1)\nu\pi} \\ e^{+i(k-1)\nu\pi} \end{cases} + E^\mp(w - k\pi, z) \begin{cases} e^{-2ik\nu\pi}, \ \operatorname{Im} w > 0 \\ e^{+2ik\nu\pi}, \ \operatorname{Im} w < 0 \end{cases} \tag{3.11}$$

where also $0 \leq \operatorname{Re}(w - k\pi) \leq \pi$, but where now the negative sign corresponds to even k and the positive one to odd k. The formulae (3.10) and (3.11) retain their meaning for integer $\nu = n$, if we pass to the limit as $\nu - n = \delta \to 0$. In this case we have

$$\lim_{\delta \to 0} \frac{\sin k(n + \delta)\pi}{\sin(n + \delta)\pi} = k\frac{\cos kn\pi}{\cos n\pi} = k(-1)^{(k-1)n}.$$

The formula (3.10) then takes the form

$$E_n^+(w, z) \doteq 2kJ_n(z) + E_n^\pm(w - k\pi, z), \tag{3.12}$$

where $0 \leq \operatorname{Re}(w - k\pi) \leq \pi$ and the positive sign corresponds to even k and the negative sign to odd. Analogously, the expression (3.11) gives

$$E_n^-(w, z) = 2kJ_n(z) + E_n^\mp(w - k\pi, z), \tag{3.13}$$

where the negative sign is for even k and the positive is for odd.

In some cases we also need to evaluate the function $E_\nu^\pm(-w, z)$. However, for arbitrary ν the factor $(\sin\theta)^{2\nu}$ in the integrand can lead to lack of uniqueness. Therefore, we must define $-w = w\exp(-i\pi)$, in which case

$$E_\nu^\pm(-w, z) = E_\nu^\pm(e^{-i\pi}w, z) = e^{-i(2\nu+1)\pi} E_\nu^\pm(w, z). \tag{3.14}$$

In conclusion it may be mentioned that in practical cases the variable w can be limited to the strip $0 \leq \operatorname{Re} w \leq \pi/2$. In this way, if $\pi/2 < \operatorname{Re} w < \pi$. The expression (3.1) becomes

$$E_\nu^\pm(w, z) = \frac{2z^\nu}{A_\nu} \int_0^w e^{\pm iz\cos\theta} \sin^{2\nu}\theta \, d\theta = \frac{2z^\nu}{A_\nu} \int_0^\pi e^{\pm iz\cos\theta} \sin^{2\nu}\theta \, d\theta$$

$$+ \frac{2z^\nu}{A_\nu} \int_\pi^w e^{\pm iz\cos\theta} \sin^{2\nu}\theta \, d\theta.$$

The first term on the right can be expressed as a Bessel function while the second, after making the substitution $u = \pi - \theta$ can be reduced to the interval $0 \leq \mathrm{Re}\,(\pi - w) \leq \pi/2$. Therfore

$$E_\nu^{\pm}(w, z) = 2J_\nu(z) - E_\nu^{\mp}(\pi - w, z). \tag{3.15}$$

In this way the formulae obtained in this chapter allow one to define the incomplete cylindrical functions in the form (3.1) for arbitrary values of the variables w and z, provided that these functions are defined for values of these variables in the intervals $0 \leq \mathrm{Re}\, w \leq \pi/2$ and $|\arg z| \leq \pi$.

4. Differential Equations for Incomplete Cylindrical Functions

The incomplete cylindrical functions $E_\nu^+(w, z)$ have been defined previously by integrals of the Poisson type. As we shall now show, there exist other integral representations for these functions. In this section we shall consider a representation which can be obtained directly from the differential equation satisfied by the functions

$$E_\nu^+(w, z) = \frac{2z^\nu}{A_\nu} \int_0^w e^{iz\cos t} \sin^{2\nu} t\, dt. \tag{4.1}$$

Considering w as a parameter, we find this differential equation by applying to $E_\nu^+(w, z)$ the Bessel operator in the form

$$\frac{\nabla_\nu}{z^2} = \frac{\partial^2}{\partial z^2} + \frac{1}{z}\frac{\partial}{\partial z} + \left(1 - \frac{\nu^2}{z^2}\right).$$

We first have

$$\frac{\partial E_\nu^+}{\partial z} = \frac{2}{A_\nu}\left(\nu z^{\nu-1} \int_0^w e^{iz\cos t} \sin^{2\nu} t\, dt + i z^\nu \int_0^w \cos t\, e^{iz\cos t} \sin^{2\nu} t\, dt\right),$$

where $A_\mu = 2^\mu \Gamma(\mu + 1/2)\, \Gamma(1/2)$.

Integration of the second term by parts, subject to the condition $\mathrm{Re}\,(\nu + 1/2) > 0$ gives

$$\frac{\partial E_\nu^+}{\partial z} = \frac{2}{A_\nu}\left(\nu z^{\nu-1} \int_0^w e^{iz\cos t} \sin^{2\nu} t\, dt \right.$$

$$\left. + \frac{i z^\nu}{2\nu + 1} \sin^{2\nu+1} w e^{iz\cos w} - \frac{z^{\nu+1}}{2\nu + 1} \int_0^w e^{iz\cos t} \sin^{2\nu+2} t\, dt\right).$$

Also by an analogous method we obtain

$$\frac{\partial^2 E_\nu^+}{\partial z^2} = \frac{2}{A_\nu}\left(\nu(\nu - 1)\, z^{\nu-2} \int_0^w e^{iz\cos t} \sin^{2\nu} t\, dt \right.$$

$$+ i \frac{2\nu}{2\nu + 1} z^{\nu-1} \sin^{2\nu+1} w e^{iz\cos w} - \frac{2\nu z^\nu}{2\nu + 1} \int_0^w e^{iz\cos t} \sin^{2\nu+2} t\, dt$$

$$\left. - z^\nu \int_0^w e^{iz\cos t} \sin^{2\nu} t \cos^2 t\, dt\right).$$

Now, applying the Bessel operator ∇_ν / z^2, we have

$$\begin{aligned} \nabla_\nu E_\nu^+ &= z^2 \left[\frac{\partial^2 E_\nu^+}{\partial z^2} + \frac{1}{z} \frac{\partial E_\nu^+}{\partial z} + \left(1 - \frac{\nu^2}{z^2}\right) E_\nu^+ \right] \\ &= 2i \frac{z^{\nu+1} \sin^{2\nu+1} w}{A_\nu} e^{iz\cos w}. \end{aligned} \tag{4.2}$$

In other words, the function $E_\nu^+(w, z)$ satisfies a non-homogeneous Bessel equation whose inhomogeneous term is

$$\Psi_\nu(w, z) = 2i \frac{z^{\nu+1} \sin^{2\nu+1} w}{A_\nu} e^{iz\cos w}.$$

As is well known the general solution of such an equation can be expressed in terms of a fundamental system of solutions of the corresponding homogeneous equation. If we choose as such a fundamental system the Bessel function $J_\nu(z)$ and the Neumann function $N_\nu(z)$, the general solution of (4.2) may be written as

$$\begin{aligned} E_\nu^+(w, z) = J_\nu(z) &\left[C_\nu(w) - \frac{\pi}{2} \int_0^z N_\nu(t)\, \Psi_\nu(w, t) \frac{dt}{t} \right] \\ + N_\nu(z) &\left[D_\nu(w) + \frac{\pi}{2} \int_0^z J_\nu(t)\, \Psi_\nu(w, t) \frac{dt}{t} \right]. \end{aligned} \tag{4.3}$$

If however, this function is to satisfy the functional equations (2.4), (2.5), then according to the general theory (see § 2, Chapter I), one can obtain recurrence relations for the coefficients C_ν and D_ν. For this purpose, it is necessary to let a and b tend to zero in formulae (2.13) and (2.15), Chapter I, after first expressing β_ν and α_ν in terms of the right sides of the recurrence relations (2.4) and (2.5).

We can derive these same relations in a somewhat simpler way by considering the conditions imposed on the function $E_\nu^+(w, z)$, Eq. (4.3), by its integral representation, as $z \to 0$. From these considerations, we obtain not only recurrence relationships between the coefficients C_ν and D_ν, but also explicit expressions for them. For this purpose, we consider the dominant terms in the expressions (4.3) and (4.1) for $|z| \ll 1$. It should be mentioned first that in order for the solution (4.3) to be bounded at $z = 0$, it is necessary that $D_\nu(w) \equiv 0$. Now from the definition (4.1) for small z we have

$$E_\nu^+(w, z) = \frac{2z^\nu}{A_\nu} \int_0^w \sin^{2\nu} t\, dt + 0(z^{\nu+1}). \tag{4.4}$$

On the other hand, taking into account the expansions (1.11) and (1.12) of Chapter I for $N_\nu(z)$ for small z:

$$\left.\begin{aligned} N_0(z) &\underset{z\to 0}{\approx} \frac{2}{\pi}\ln z; \\ N_\nu(z) &\underset{z\to 0}{\approx} -\frac{\Gamma(\nu)}{\pi}\left(\frac{2}{z}\right)^\nu, \qquad \nu > 0, \end{aligned}\right\} \tag{4.5}$$

and the expansion for $J_\nu(z)$:

$$J_\nu(z) \simeq \frac{1}{\Gamma(\nu+1)}\left(\frac{z}{2}\right)^\nu, \qquad z \to 0 \tag{4.6}$$

we find from Eq. (4.3) that

$$E_\nu^+(w, z) = \frac{z^\nu}{2^\nu \Gamma(\nu+1)}\, C_\nu(w) + 0(z^{\nu+1}). \tag{4.7}$$

By equating the expressions (4.4) and (4.7) we obtain

$$\left.\begin{aligned} C_\nu(w) &= 2\frac{\Gamma(\nu+1)}{\Gamma\left(\nu+\frac{1}{2}\right)\Gamma\left(\frac{1}{2}\right)}\int_0^w \sin^{2\nu} t\, dt; \\ D_\nu(w) &= 0. \end{aligned}\right\} \tag{4.8}$$

Integrating the expression $C_\nu(w)$ for $\operatorname{Re}\nu > 1$ by parts, we obtain the following recurrence formula for the coefficients C_ν:

$$C_\nu - C_{\nu-1} = -\frac{\Gamma(\nu)}{\Gamma\left(\nu+\frac{1}{2}\right)\Gamma\left(\frac{1}{2}\right)}\sin^{2\nu-1} w\cos w. \tag{4.9}$$

Now substituting the coefficient $C_\nu(w)$ and the function $\Psi_\nu(w, z)$ into the expression (4.3) for $E_\nu^+(w, z)$, we find, after some easy transformations, a new integral representation for the study of the incomplete cylindrical functions

$$\begin{aligned} E_\nu^+(w, z) &= 2J_\nu(z)\frac{\Gamma(\nu+1)}{\Gamma\left(\nu+\frac{1}{2}\right)\Gamma\left(\frac{1}{2}\right)}\int_0^w \sin^{2\nu} t\, dt \\ &+ 2i\pi\frac{\sin^{2\nu+1} w}{2^{\nu+1}\Gamma\left(\nu+\frac{1}{2}\right)\Gamma\left(\frac{1}{2}\right)} \\ &\times\left[N_\nu(z)\int_0^z J_\nu(t)\, e^{it\cos w}\, t^\nu\, dt - J_\nu(z)\int_0^z N_\nu(t)\, e^{it\cos w}\, t^\nu\, dt\right]. \end{aligned} \tag{4.10}$$

We can write $E_\nu^+(w, z)$ with $w = \pi$ as a Bessel function $2J_\nu(z)$ if we use the known relationship for the β-function (Euler integral of the second kind) (see [11, p. 962]):

$$\int_0^\pi \sin^{2\nu} t\, dt = B\left(\nu+\frac{1}{2}, \frac{1}{2}\right) = \frac{\Gamma\left(\nu+\frac{1}{2}\right)\Gamma\left(\frac{1}{2}\right)}{\Gamma(\nu+1)}. \tag{4.11}$$

In the following sections we shall consider the functions $E_0^+(w, z)$ and $E_1^+(w, z)$ writing them explicitly as

$$E_0^+(w, z) = \frac{2w}{\pi} J_0(z) + i \sin w \left[N_0(z) \int_0^z J_0(t)\, e^{it\cos w}\, dt \right. \\ \left. - J_0(z) \int_0^z N_0(t)\, e^{it\cos w}\, dt \right]; \tag{4.12}$$

and

$$E_1^+(w, z) = \frac{2w}{\pi}\left(1 - \frac{\sin 2w}{2w}\right) J_1(z) \\ + i \sin^3 w \left[N_1(z) \int_0^z J_1(t)\, e^{it\cos w}\, t\, dt - J_1(z) \int_0^z N_1(t)\, e^{it\cos w}\, t\, dt \right]. \tag{4.13}$$

To obtain the corresponding expressions for $E_\nu^-(w, z)$ it is sufficient, as before, to take the complex conjugate, formally regarding ν, w and z as real. Replacing z by iz in the differential equation (4.2) satisfied by $E_\nu^+(w, z)$ which is valid for all z, we obtain the differential equation for the function $E_\nu^+(w, iz)$

$$\left[z^2 \frac{\partial^2}{\partial z^2} + z \frac{\partial}{\partial z} - (z^2 + \nu^2) \right] E_\nu^+(w, iz) = 2i\,(iz)^{\nu+1} \frac{\sin^{2\nu+1}}{A_\nu} e^{-z\cos w}. \tag{4.14}$$

The solution of the corresponding homogeneous equation in this case will consist of a linear combination of the modified Bessel function $I_\nu(z)$ and the MacDonald function $K_\nu(z)$. Since the Wronskian of this system equals $(-1/z)$, the solution of Eq. (4.14) can be written in the form:

$$E_\nu^+(w, iz) = I_\nu(z) \left[\tilde{C}_\nu(w) + \int_0^z K_\nu(t)\, \Psi_\nu(w, it) \frac{dt}{t} \right] \\ + K_\nu(z) \left[\tilde{D}_\nu(w) - \int_0^z I_\nu(t)\, \Psi_\nu(w, it) \frac{dt}{t} \right]; \tag{4.15}$$

where

$$\Psi_\nu(w, it) = 2 e^{i\nu\pi/2} z^{\nu+1} \frac{\sin^{2\nu+1} w}{A_\nu} e^{-z\cos w} \tag{4.16}$$

and the coefficients $\tilde{C}_\nu(w)$ and $\tilde{D}_\nu(w)$ are to be determined from the condition that $E_\nu^+(w, iz)$ and its derivatives be bounded at $z = 0$. From this condition we find that $\tilde{D}_\nu(w) \equiv 0$. In order to determine the coefficients $\tilde{C}_\nu(w)$ we proceed as before. From the expression (4.1), and assuming that $|z| \ll 1$ we have

$$E_\nu^+(w, iz) = \frac{2\,(iz)^\nu}{A_\nu} \int_0^w \sin^{2\nu} t\, dt + 0(z^{\nu+1}). \tag{4.17}$$

On the other hand, from the limiting behavior of the functions $K_\nu(z)$ and $I_\nu(z)$ as $z \to 0$, we find the approximation formulae

$$\left.\begin{aligned} K_0(z) &\underset{z\to 0}{\approx} -\ln z; \\ K_\nu(z) &\underset{z\to 0}{\approx} \frac{1}{2}\Gamma(\nu)\left(\frac{2}{z}\right)^\nu, \quad \nu > 0; \\ I_\nu(z) &\underset{z\to 0}{\approx} \left(\frac{z}{2}\right)^\nu \frac{1}{\Gamma(\nu+1)}, \end{aligned}\right\} \tag{4.18}$$

which arise by considering the relations (1.7) and (1.10) through (1.13) of Chapter I, for small z. Thus Eqs. (4.15) and (4.17) give

$$\tilde{C}_\nu(w) = 2(i)^\nu \frac{\Gamma(\nu+1)}{\Gamma\left(\nu+\frac{1}{2}\right)\Gamma\left(\frac{1}{2}\right)} \int_0^w \sin^{2\nu} t\, dt, \tag{4.19}$$

and by substituting these coefficients into Eq. (4.15) we obtain

$$\begin{aligned} E_\nu^+(w, iz) &= 2I_\nu(z) \frac{(i)^\nu \Gamma(\nu+1)}{\Gamma\left(\nu+\frac{1}{2}\right)\Gamma\left(\frac{1}{2}\right)} \int_0^w \sin^{2\nu} t\, dt \\ &+ \frac{2(i)^\nu \sin^{2\nu+1} w}{2^\nu \Gamma\left(\nu+\frac{1}{2}\right)\Gamma\left(\frac{1}{2}\right)} \left[K_\nu(z) \int_0^z I_\nu(t)\, t^\nu\, e^{-t\cos w}\, dt \right. \\ &\left. - I_\nu(z) \int_0^z K_\nu(t)\, t^\nu\, e^{-t\cos w}\, dt\right]. \end{aligned} \tag{4.20}$$

It should be mentioned that one can also obtain the same representation for $E_\nu^+(w, iz)$ directly from Eq. (4.10) by replacing z by iz, if one utilizes the relation:

$$N_\nu(iz) = -i\mathscr{H}_\nu^{(1)}(iz) + iJ_\nu(iz) = -\frac{2}{\pi} e^{-i\nu\pi/2} K_\nu(z) + ie^{i\nu\pi/2} I_\nu(z), \tag{4.21}$$

which follows from Eqs. (1.7) and (1.13), Chapter I.

For $\nu = 0$ and $\nu = 1$, formula (4.20) gives

$$\begin{aligned} E_0^+(w, iz) = \frac{2w}{\pi} I_0(z) + \frac{2\sin w}{\pi} &\left[K_0(z) \int_0^z I_0(t)\, e^{-t\cos w}\, dt \right. \\ &\left. - I_0(z) \int_0^z K_0(t)\, e^{-t\cos w}\, dt\right]; \end{aligned} \tag{4.22}$$

$$\begin{aligned} E_1^+(w, iz) &= \frac{2iw}{\pi}\left(1 - \frac{\sin 2w}{2w}\right) I_1(z) + \frac{2i\sin^3 w}{\pi} \\ \times &\left[K_1(z) \int_0^z I_1(t)\, e^{-t\cos w}\, t\, dt - I_1(z) \int_0^z K_1(t)\, e^{-t\cos w}\, t\, dt\right]. \end{aligned} \tag{4.23}$$

Now, considering $E_\nu^+(w, z)$ as a function of two variables one can easily derive a partial differential equation satisfied by this function by utilizing the definition (4.1), and by writing the right side of Eq. (4.2) in the form

$$\frac{2iz^{\nu+1}\sin^{2\nu+1}w}{A_\nu}\,\mathrm{e}^{iz\cos w} = 2\nu\cot w\,\frac{\partial E_\nu^+}{\partial w} - \frac{\partial^2 E_\nu^+}{\partial w^2}. \tag{4.24}$$

In this way we obtain the following partial differential equation

$$\frac{\partial^2 E_\nu^+}{\partial z^2} + \frac{1}{z}\,\frac{\partial E_\nu^+}{\partial z} + \left(1 - \frac{\nu^2}{z^2}\right)E_\nu^+ + \frac{1}{z^2}\,\frac{\partial^2 E_\nu^+}{\partial w^2} - \frac{2\nu\cot w}{z^2}\cdot\frac{\partial E_\nu^+}{\partial w} = 0. \tag{4.25}$$

We shall now look for incomplete cylindrical functions which satisfy this equation and which also fulfill the following boundary conditions with respect to w:

$$\left.\begin{array}{l} E_\nu^+(0, z) = 0; \quad E_\nu^+(\pi, z) = 2J_\nu(z) \\ \text{or} \qquad E_\nu^+(-i\infty, z) = \mathscr{H}_\nu^{(1)}(z). \end{array}\right\} \tag{4.26}$$

We first bring out the distinction between the integral representation (4.1) of Poisson form which has been established as a basic definition for the incomplete cylindrical functions, and the integral representation (4.10), (4.20) just derived. The function $E_\nu^+(w, z)$ normally appears for $\nu = \text{const}$ as a function of the variables w and z. However, there are instances in which one or the other of the variables w or z can be considered as a parameter. The representation (4.1) can be considered as a definition of the function $E_\nu^+(w, z)$ in the form of an integral with variable limit w and z as a parameter. On the other hand (4.10) and (4.20) can be regarded as a definition of the same function $E_\nu^+(w, z)$ in the form of integrals with variable limit z and w as parameter. Therefore, in accordance with the conditions of particular problems it may be better to consider w as variable and z as a parameter, and to utilize the representation (4.1). On the other hand if the conditions of a problem so dictate, it may be preferable to consider z as the variable and w as parameter, utilizing formulae (4.10) and (4.20). We introduce for this purpose the following notation for the integrals with upper limit to z which appear in (4.10)

$$Je_\nu(a, z) = \int_0^z \mathrm{e}^{-at} J_\nu(t)\, t^\nu\, dt; \tag{4.27}$$

$$Ne_\nu(a, z) = \int_0^z \mathrm{e}^{-at} N_\nu(t)\, t^\nu\, dt, \tag{4.28}$$

thus obtaining

$$E_\nu^+(w, z) = 2J_\nu(z) \frac{\Gamma(\nu+1)}{\Gamma\left(\nu+\frac{1}{2}\right)\Gamma\left(\frac{1}{2}\right)} \int_0^w \sin^{2\nu} t\, dt$$

$$+ \frac{2i\pi \sin^{2\nu+1} w}{2^{\nu+1}\Gamma\left(\nu+\frac{1}{2}\right)\Gamma\left(\frac{1}{2}\right)} [N_\nu(z)\, Je_\nu(-i\cos w, z) \tag{4.29}$$

$$- J_\nu(z)\, Ne_\nu(-i\cos w, z)].$$

In this way, knowledge of the functions $Je_\nu(a, z)$ and $Ne_\nu(a, z)$ completely determines the incomplete cylindrical functions. From this point of view they are important in describing the general properties of these functions. In addition, each one of them is of special interest *per se* as they appear frequently as solutions of many applied problems. Various particular cases of these functions have been studied by many authors, particularly in [4] where a detailed study of the functions $Je_0(a, z)$ and $Ne_0(a, z)$ was made. In this work series expansions, asymptotic representations, and relations to some other indefinite integrals are given.

A more detailed study of the functions $Je_\nu(a, z)$ and $Ne_\nu(a, z)$ and their relation to incomplete cylindrical functions will be found in later sections and in other chapters where their applications to various applied problems are investigated.

5. Incomplete Lipschitz-Hankel Integrals

Improper integrals of the form

$$\int_0^\infty e^{-at} J_\nu(t)\, t^{\mu-1}\, dt; \tag{5.1}$$

$$\int_0^\infty e^{-at} N_\nu(t)\, t^{\mu-1}\, dt. \tag{5.2}$$

(c.f. [10, p. 420]) are commonly called Lipschitz-Hankel integrals.

To assure convergence of these integrals, it is necessary that Re $(a) > 0$ and Re $(\mu + \nu) > 0$. These integrals have been investigated, and in general they can be expressed in terms of associated Legendre functions which, in turn, can be represented as hypergeometric functions. Thus, for the integral (5.1) the following formula can be written (see [10, p. 420] and [11, p. 725]):

$$\int_0^\infty e^{-at} J_\nu(t)\, t^{\mu-1}\, dt$$

$$= \frac{\Gamma(\nu+\mu)}{2^\nu \Gamma(\nu+1)\,(a^2+1)^{(\nu+\mu)/2}} F\left(\frac{\nu+\mu}{2}, \frac{1-\mu+\nu}{2}; \nu+1; \frac{1}{a^2+1}\right) \tag{5.3}$$

$$= \frac{\Gamma(\nu+\mu)}{(a^2+1)^{\mu/2}} P_{\mu-1}^{-\nu}\left(\frac{a}{\sqrt{a^2+1}}\right),$$

where $P_{\mu-1}^{-\nu}(z)$ is the associated Legrendre function of first kind. The corresponding expression for (5.2) is

$$\int_0^\infty e^{-at} N_\nu(t)\, t^{\mu-1}\, dt = -\frac{2e^{-i\nu\pi/2}}{\sqrt{\pi}} \cdot \frac{\Gamma(\nu+\mu)\,\Gamma(\mu-\nu)}{2^\mu \Gamma\left(\mu+\frac{1}{2}\right)} a^{(\nu-\mu)/2}$$

$$\times F\left(\frac{\mu-\nu+1}{2}, \frac{\mu-\nu}{2}; \mu+\frac{1}{2}; \frac{a^2+1}{a^2}\right) \tag{5.4}$$

$$= -\frac{2}{\pi}\Gamma(\mu+\nu)\,(a^2+1)^{-\mu/2}\, Q_{\mu-1}^{-\nu}\left(\frac{a}{\sqrt{a^2+1}}\right),$$

where $Q_{\mu-1}^{-\nu}(z)$ is the associated Legendre function of the second kind. The last formula holds for $\operatorname{Re}(\mu) > |\operatorname{Re}(\nu)|$ and $\operatorname{Re}(a) > 0$. Analogous relationships exist for integrals of Lipschitz-Hankel type involving the modified cylindrical functions (see e.g. [11, p. 727]).

For $\operatorname{Re}(a) > 1$

$$\int_0^\infty e^{-at} I_\nu(t)\, t^{\mu-1}\, dt = \frac{\Gamma(\nu+\mu)\cdot 2^{-\nu}}{(a^2-1)^{(\nu+\mu)/2}\,\Gamma(\nu+1)}$$

$$\times F\left(\frac{\nu+\mu}{2}, \frac{1-\mu+\nu}{2}; \nu+1; -\frac{1}{a^2-1}\right) \tag{5.5}$$

$$= -\frac{2\Gamma(\nu+\mu)}{\pi(a^2-1)^{\mu/2}} P_{\mu-1}^{-\nu}\left(\frac{a}{\sqrt{a^2-1}}\right);$$

For $\operatorname{Re}(a) > -1$

$$\int_0^\infty e^{-at} K_\nu(t)\, t^{\mu-1}\, dt = \frac{\Gamma(\nu+\mu)}{(a^2-1)^{\mu/2}} e^{i\nu\pi} \cdot Q_{\mu-1}^{-\nu}\left(\frac{a}{\sqrt{a^2-1}}\right). \tag{5.6}$$

We shall be interested in the special case where $\mu = \nu + 1$ and $\operatorname{Re}(\nu + 1/2) > 0$. In this case the hypergeometric functions which appear on the right hand side of (5.3) and (5.5) can be expressed by elementary functions and the corresponding integrals assume the form:

For $\operatorname{Re}(a) > 0$

$$\int_0^\infty e^{-at} J_\nu(t)\, t^\nu\, dt = \frac{2^\nu \Gamma\left(\nu+\frac{1}{2}\right)}{\sqrt{\pi}\,(a^2+1)^{\nu+1/2}};$$

For $\operatorname{Re}(a) > 1$ (5.7)

$$\int_0^\infty e^{-at} I_\nu(t)\, t^\nu\, dt = \frac{2^\nu \Gamma\left(\nu+\frac{1}{2}\right)}{\sqrt{\pi}\,(a^2-1)^{\nu+1/2}}.$$

For $\mu = \nu + 1$ the formulae (5.4) and (5.6) take the form:

For $\operatorname{Re}(a) > 0$

$$\int_0^\infty e^{-at} N_\nu(t)\, t^\nu\, dt = -\frac{2}{\pi}\frac{\Gamma(2\nu+1)}{(a^2+1)^{\nu+1/2}} Q_\nu^{-\nu}\left(\frac{a}{\sqrt{a^2+1}}\right). \tag{5.8}$$

For Re $(a) > -1$

$$\int_0^\infty e^{-at} K_\nu(t)\, t^\nu\, dt = \frac{\Gamma(2\nu+1)\, e^{i\nu\pi}}{(a^2-1)^{\nu+1/2}}\, Q_\nu^{-\nu}\left(\frac{a}{\sqrt{a^2-1}}\right). \tag{5.9}$$

The expressions for these integrals assume a simpler form if ν is an integer, since then the associated Legendre functions $Q_n^{-n}(z)$ can be expressed in terms of elementary functions. In particular (see [5, p. 308]),

$$\left.\begin{aligned} Q_0^0 &= \frac{1}{2}\ln\frac{z+1}{z-1}; \\ Q_1^{-1}(z) = \frac{1}{2} Q_1^1 &= \frac{\sqrt{z^2-1}}{2}\left(\frac{z}{z^2-1} - \frac{1}{2}\ln\frac{z+1}{z-1}\right). \end{aligned}\right\} \tag{5.10}$$

It is evident from the above formulae that the points $z = \pm 1$ are branch points in the complex z-plane for the function $Q_\nu^{-\nu}(z)$. Therefore, when evaluating these integrals it is customary to make a cut in the complex z-plane along the segment of the real axis $(-1, +1)$. For $-1 \le x \le 1$ the function $Q_\nu^{-\nu}(z)$ is defined (see e.g. [11, p. 1014]):

$$Q_\nu^{-\nu}(x) = \frac{1}{2} e^{-i\nu\pi} \lim_{\varepsilon\to 0} [e^{i\nu\pi/2} Q_\nu^{-\nu}(x+i\varepsilon) + e^{-i\nu\pi/2} Q_\nu^{-\nu}(x-i\varepsilon)]. \tag{5.11}$$

In this case (5.10) for $|x| \le 1$ becomes

$$\left.\begin{aligned} Q_0^0(x) &= \frac{1}{2}\ln\frac{1+x}{1-x}; \\ Q_1^{-1}(x) &= \frac{\sqrt{1-x^2}}{2}\left(\frac{x}{x^2-1} - \frac{1}{2}\ln\frac{1+x}{1-x}\right). \end{aligned}\right\} \tag{5.10'}$$

For the case $a = i\varrho$ the improper integrals considered here are discontinuous. We give below explicit expressions for some of them (see, e.g. [10, p. 444] and [11, p. 744 and 762/3]):

$$\int_0^\infty t^\nu \sin \varrho t J_\nu(t)\, dt = \begin{cases} \dfrac{\sqrt{\pi}\, 2^\nu (\varrho^2-1)^{-\nu-1/2}}{\Gamma\left(\frac{1}{2}-\nu\right)}, & \varrho > 1; \\ 0, & \varrho < 1; \end{cases} \tag{5.12}$$

$$\int_0^\infty t^\nu \cos \varrho t J_\nu(t)\, dt = \begin{cases} -\dfrac{2^\nu \sin \nu\pi}{\sqrt{\pi}} \cdot \dfrac{\Gamma\left(\nu+\frac{1}{2}\right)}{(\varrho^2-1)^{\nu+1/2}}, & \varrho > 1; \\ \dfrac{2^\nu}{\sqrt{\pi}} \cdot \dfrac{\Gamma\left(\nu+\frac{1}{2}\right)}{(1-\varrho^2)^{\nu+1/2}}, & \varrho < 1; \end{cases} \tag{5.13}$$

$$\int_0^\infty t^\nu \cos \varrho t N_\nu(t)\, dt = \begin{cases} -\dfrac{2^\nu \sqrt{\pi}\, (\varrho^2-1)^{-\nu-1/2}}{\Gamma\left(\frac{1}{2}-\nu\right)}, & \varrho > 1; \\ 0, & \varrho < 1; \end{cases} \tag{5.14}$$

$$\int_0^\infty t^\nu \cos \varrho t K_\nu(t)\, dt = \frac{\sqrt{\pi}\, 2^\nu}{2} \frac{\Gamma\left(\nu + \frac{1}{2}\right)}{(1+\varrho^2)^{\nu+1/2}} \left(\text{for } \operatorname{Re}\left(\nu + \frac{1}{2}\right) > 0\right); \tag{5.15}$$

$$\int_0^\infty \sin \varrho t N_0(t)\, dt = \begin{cases} \dfrac{2}{\pi} \dfrac{\ln(\varrho - \sqrt{\varrho^2 - 1})}{\sqrt{\varrho^2 - 1}}, & \varrho > 1; \\ \dfrac{2}{\pi} \cdot \dfrac{\sin^{-1} \varrho}{\sqrt{1 - \varrho^2}}, & \varrho < 1; \end{cases} \tag{5.16}$$

$$\int_0^\infty \sin \varrho t K_0(t)\, dt = \frac{1}{2\sqrt{\varrho^2 + 1}} \ln \frac{\sqrt{\varrho^2 + 1} + \varrho}{\sqrt{\varrho^2 + 1} - \varrho}. \tag{5.17}$$

In this way we see that many of the Lipschitz-Hankel integrals can be expressed in terms of elementary functions. In the subsequent discussion we shall define as incomplete Lipschitz-Hankel integrals the following functions of two variables:

$$Je_\nu(a, z) = \int_0^z e^{-at}\, t^\nu J_\nu(t)\, dt; \tag{5.18}$$

$$Ne_\nu(a, z) = \int_0^z e^{-at}\, t^\nu N_\nu(t)\, dt; \tag{5.19}$$

$$Ie_\nu(a, z) = \int_0^z e^{-at}\, t^\nu I_\nu(t)\, dt; \tag{5.20}$$

$$Ke_\nu(a, z) = \int_0^z e^{-at}\, t^\nu K_\nu(t)\, dt. \tag{5.21}$$

The symbol e in the above functions denotes the presence of the exponential function in the integrals. Analogously, we shall also encounter integrals which contain the functions cosh (at) and sinh (at) in place of exp $(-at)$. Such functions will be denoted by the symbols c and s. For example

$$Jc_\nu(a, z) = \int_0^z \cosh at\, t^\nu J_\nu(t)\, dt = \frac{1}{2} [Je_\nu(a, z) + Je_\nu(-a, z)]; \tag{5.22}$$

$$Js_\nu(a, z) = \int_0^z \sinh at\, t^\nu J_\nu(t)\, dt = \frac{1}{2} [Je_\nu(a, z) - Je_\nu(-a, z)]. \tag{5.23}$$

The incomplete Lipschitz-Hankel integral is also defined for arbitrary complex values of a and z and because of (5.22) and (5.23), it is sufficient to study their properties in the forms (5.18)—(5.21). We mention that some of these properties have been studied in [13]. Such integrals have already been considered in the preceding section [Eqs. (4.10) and (4.20)] for the cases $a = -i \cos w$ and $a = \cos w$. We shall study in more detail the relation of these integrals to incomplete cylindrical functions of Poisson form. First we will find some explicit expressions for them for particular values of their variables. It has already been mentioned that for $\mathrm{Re}(a) > 0$ and $z \to \infty$ the incomplete Lipschitz-Hankel integrals can be expressed by known functions. For $a = 0$ and any value of z, they can also be expressed by known functions. For example from the relationship [11, p. 697]

$$\int_0^1 x^\nu J_\nu(zx)\,dx = 2^{\nu-1} z^{-\nu} \sqrt{\pi}\,\Gamma\left(\nu + \frac{1}{2}\right)[J_\nu(z)\,H_{\nu-1}(z) - H_\nu(z)\,J_{\nu-1}(z)] \tag{5.24}$$

and using the recurrence relations for the Bessel functions $J_\nu(z)$ and the Struve function $H_\nu(z)$ which appear in this formula (see Chapter I, Eqs. (1.15) and (2.15)), it is easy to obtain the formula

$$\begin{aligned} Je_\nu(0, z) &= \int_0^z t^\nu J_\nu(t)\,dt = \frac{z^{\nu+1}}{2\nu+1} J_\nu(z) \\ &+ 2^{\nu-1}\sqrt{\pi}\,\Gamma\left(\nu + \frac{1}{2}\right) z\,[J_{\nu+1}(z)\,H_\nu(z) - J_\nu(z)\,H_{\nu+1}(z)]. \end{aligned} \tag{5.25}$$

From relations analogous to the expression (5.24), and with the aid of similar transformations, we find the following expressions for the remaining incomplete Lipschitz-Hankel integrals for $a = 0$:

$$\begin{aligned} Ne_\nu(0, z) &= \int_0^z t^\nu N_\nu(t)\,dt = \frac{z^{\nu+1}}{2\nu+1} N_\nu(z) \\ &+ 2^{\nu-1}\sqrt{\pi}\,\Gamma\left(\nu + \frac{1}{2}\right) z\,[N_{\nu+1}(z)\,H_\nu(z) - N_\nu(z)\,H_{\nu+1}(z)]; \end{aligned} \tag{5.26}$$

$$\begin{aligned} Ie_\nu(0, z) &= \int_0^z t^\nu I_\nu(t)\,dt = \frac{z^{\nu+1}}{2\nu+1} I_\nu(z) \\ &+ 2^{\nu-1}\sqrt{\pi}\,\Gamma\left(\nu + \frac{1}{2}\right) z\,[I_\nu(z)\,L_{\nu+1}(z) - L_\nu(z)\,I_{\nu+1}(z)]; \end{aligned} \tag{5.27}$$

$$\begin{aligned} Ke_\nu(0, z) &= \int_0^z t^\nu K_\nu(t)\,dt = \frac{z^{\nu+1}}{2\nu+1} K_\nu(z) \\ &+ 2^{\nu-1}\sqrt{\pi}\,\Gamma\left(\nu + \frac{1}{2}\right) z\,[K_\nu(z)\,L_{\nu+1}(z) + L_\nu(z)\,K_{\nu+1}(z)]. \end{aligned} \tag{5.28}$$

In the last two formulae $L_\nu(z) = -iH_\nu(iz)\exp\left(-\frac{i\nu\pi}{2}\right)$ denotes the modified Struve function. We shall now show that the incomplete Lipschitz-Hankel integrals $Je_\nu(a, z)$ and $Ne_\nu(a, z)$ can be expressed respectively by Bessel and Neumann functions for $a = \pm i$ and any complex value of z. As an example, let us consider the integral

$$Je_\nu(i, z) = \int_0^z \mathrm{e}^{-it}\,t^\nu J_\nu(t)\,dt. \tag{5.29}$$

If we introduce the auxiliary function

$$u = \int_0^z \cos(z - t)\, t^\nu J_\nu(t)\, dt, \tag{5.30}$$

it is easy to show that $Je_\nu(i, z)$ is related to u by

$$Je_\nu(i, z) = e^{-iz}\left[u + iz^\nu J_\nu(z) - i\frac{du}{dz}\right]. \tag{5.31}$$

Now, by applying the differential operator $d^2/dz^2 + 1$ to the auxiliary function we easily find:

$$\frac{d^2u}{dz^2} + u = \frac{d}{dz}\left[z^\nu J_\nu(z)\right]. \tag{5.32}$$

This equation can be solved by the method of variation of parameters and leads to Eq. (5.30). However, it can be shown that the expression

$$u_1 = C z^{\nu+1} J_\nu(z) \tag{5.33}$$

is a particular solution of the non-homogeneous equation, and therefore by substituting u_1 into the left hand side of (5.32) we have

$$\frac{d^2u_1}{dz^2} + u_1 = C\left[\nu(\nu + 1)\, z^{\nu-1} J_\nu(z) + 2(\nu + 1)\, z^\nu \frac{dJ_\nu(z)}{dz} + z^{\nu+1}\frac{d^2J_\nu(z)}{dz^2} + z^{\nu+1} J_\nu(z)\right].$$

Also, from the Bessel equation we have

$$z^2 \frac{d^2J_\nu(z)}{dz^2} = -z\frac{dJ_\nu(z)}{dz} - (z^2 - \nu^2)\, J_\nu(z).$$

Therefore

$$\frac{d^2u_1}{dz^2} + u_1 = (2\nu + 1)\, C\frac{d}{dz}\left[z^\nu J_\nu(z)\right].$$

From this equation it is evident that if $C = 1/(2\nu + 1)$ the expression (5.33) will be a particular solution of the non-homogeneous equation (5.32). In this way we can express the auxiliary function u in the form

$$u = A\cos z + B\sin z + \frac{z^{\nu+1}}{2\nu + 1} J_\nu(z), \tag{5.34}$$

where A and B are arbitrary constants. Now, by substituting (5.34) into (5.31), and taking advantage of the recurrence relations (1.18) of Chapter I for the Bessel functions we find after several simple transformations,

$$\begin{aligned} Je_\nu(i, z) &= e^{-iz}\left\{iz^\nu J_\nu(z) + \frac{z^{\nu+1}}{2\nu + 1} J_\nu(z) - \frac{i}{2\nu + 1}\frac{d}{dz}\left[z^{\nu+1} J_\nu(z)\right]\right\} + C_1 \\ &= C_1 + \frac{z^{\nu+1}}{2\nu + 1}\left[J_\nu(z) + iJ_{\nu+1}(z)\right] e^{-iz}. \end{aligned} \tag{5.35}$$

Noting that $Je_\nu(i, z) = 0$ for $z = 0$, it is seen that $C_1 = 0$. Therefore

$$Je_\nu(i, z) = \int_0^z e^{-it} t^\nu J_\nu(t)\, dt = \frac{e^{-iz} z^{\nu+1}}{2\nu + 1} [J_\nu(z) + iJ_{\nu+1}(z)]. \quad (5.36)$$

Since the relations used to obtain (5.35), (i. e. the differential equation and recurrence relations) remain unchanged if the Bessel function $J_\nu(t)$ is replaced by the Neumann function $N_\nu(t)$ it follows also that

$$Ne_\nu(i, z) = C_1 + \frac{e^{-iz} z^{\nu+1}}{2\nu + 1} [N_\nu(z) + iN_{\nu+1}(z)]. \quad (5.37)$$

In this case, since by Eq. (5.19), $Ne_\nu(i, 0) = 0$, for $z = 0$ and since

$$N_\nu(z) \underset{z\to 0}{\approx} -\frac{\Gamma(\nu)}{\pi} \left(\frac{2}{z}\right)^\nu,$$

we find

$$C_1 = i\,\frac{2^{\nu+1}\Gamma(\nu + 1)}{\pi(2\nu + 1)}$$

and finally

$$\begin{aligned} Ne_\nu(i, z) = \int_0^z e^{-it} t^\nu N_\nu(t)\, dt = i\,\frac{2^{\nu+1}\Gamma(\nu + 1)}{\pi(2\nu + 1)} \\ + \frac{e^{-iz} z^{\nu+1}}{2\nu + 1} [N_\nu(z) + iN_{\nu+1}(z)]. \end{aligned} \quad (5.38)$$

Corresponding expressions for the functions $Je_\nu(-i, z)$ and $Ne_\nu(-i, z)$ are

$$Je_\nu(-i, z) = \int_0^z e^{it} t^\nu J_\nu(t)\, dt = \frac{e^{iz} z^{\nu+1}}{2\nu + 1} [J_\nu(z) - iJ_{\nu+1}(z)]; \quad (5.39)$$

$$\begin{aligned} Ne_\nu(-i, z) = \int_0^z e^{it} t^\nu N_\nu(t)\, dt = -i\,\frac{2^{\nu+1}\Gamma(\nu + 1)}{\pi(2\nu + 1)} \\ + \frac{e^{iz} z^{\nu+1}}{2\nu + 1} [N_\nu(z) - iN_{\nu+1}(z)]. \end{aligned} \quad (5.40)$$

We note in passing that the incomplete Lipschitz-Hankel integrals $Je_\nu(a, z)$ and $Ne_\nu(a, z)$ for $a = \pm i$ can also be represented by the generalized trigonometric integrals of Kapteyn (see [10, p. 415]).

Replacing z by $z \exp(\pm i\pi/2)$ in Eq. (5.36) and taking into account the relation

$$I_\nu(z) = J_\nu(\pm iz) \exp(\mp i\nu\pi/2),$$

we can obtain the following expression for the incomplete integrals of Lipschitz-Hankel involving the modified Bessel functions

$$Ie_\nu(\pm 1, z) = \int_0^z e^{\mp t} t^\nu I_\nu(t)\, dt = \frac{e^{\mp z} z^{\nu+1}}{2\nu + 1} [I_\nu(z) \pm I_{\nu+1}(z)]. \quad (5.41)$$

The corresponding incomplete Lipschitz-Hankel integrals involving the MacDonald function for $a = \pm 1$, obtained from (5.36), (5.38), (5.39) and (5.40) are

$$Je_\nu(\pm i, z) + iNe_\nu(\pm i, z) = \int_0^z e^{\mp it} t^\nu \mathscr{H}_\nu^{(1)}(t)\,dt = \mp \frac{2^{\nu+1}\Gamma(\nu+1)}{\pi(2\nu+1)} + \frac{e^{\mp iz} z^{\nu+1}}{2\nu+1}\left[\mathscr{H}_\nu^{(1)}(z) \pm i\mathscr{H}_{\nu+1}^{(1)}(z)\right], \tag{5.42}$$

where $\mathscr{H}_\nu^{(1)}(z) = J_\nu(z) + iN_\nu(z)$ is the Hankel function of the first kind. Replacing z by $z \exp(i\pi/2)$ and using the relation

$$\mathscr{H}_\nu^{(1)}\left(z \exp i\frac{\pi}{2}\right) = \frac{2}{\pi i} K_\nu(z) \exp\left(-i\nu\frac{\pi}{2}\right),$$

we find after some simple transformations

$$Ke_\nu(\mp 1, z) = \int_0^z e^{\pm t} t^\nu K_\nu(t)\,dt = \mp \frac{2^\nu \Gamma(\nu+1)}{2\nu+1} + \frac{e^{\pm z} z^{\nu+1}}{2\nu+1}\left[K_\nu(z) \pm K_{\nu+1}(z)\right]. \tag{5.43}$$

The formulae just obtained were derived by other methods in [13]. Particular forms of these integrals for $\nu = 0$ and real a and z have been studied in a series of works treating applied problems. These will be presented in detail in the chapter dealing with applications.

6. The Relation between the Incomplete Lipschitz-Hankel Integrals and Incomplete Cylindrical Functions of Poisson Form

In the previous paragraph it was shown that the incomplete Lipschitz-Hankel integrals could, for certain specific values of the parameters, be expressed in terms of known functions. In this section it will be shown that, in the general case, these integrals are expressible in terms of incomplete cylindrical functions of Poisson form.

The incomplete cylindrical functions $E_\nu^+(w, z)$ which were defined in § 4 of this chapter satisfy the non-homogeneous Bessel differential equation, and can be represented in the form

$$E_\nu^+(w, z) = \frac{2}{\sqrt{\pi}} J_\nu(z) \frac{\Gamma(\nu+1)}{\Gamma\left(\nu+\frac{1}{2}\right)} \int_0^w \sin^{2\nu} t\,dt + \frac{i\sqrt{\pi}\sin^{2\nu+1} w}{2^\nu \Gamma\left(\nu+\frac{1}{2}\right)}\left[N_\nu(z)\int_0^z J_\nu(t)\, e^{it\cos w} t^\nu\,dt - J_\nu(z)\int_0^z N_\nu(t)\, e^{it\cos w} t^\nu\,dt\right]. \tag{6.1}$$

or, according to the definition of the incomplete Lipschitz-Hankel integral (see Eqs. (5.10)—(5.21)) we have

$$\begin{aligned} E_\nu^+(w, z) = c_\nu J_\nu(z) + i d_\nu [N_\nu(z)\, Je_\nu(-i \cos w, z) \\ - J_\nu(z)\, Ne_\nu(-i \cos w, z)], \end{aligned} \tag{6.2}$$

where for the sake of conciseness we have introduced the notation

$$\left.\begin{aligned} c_\nu &= \frac{2}{\sqrt{\pi}} \frac{\Gamma(\nu+1)}{\Gamma\left(\nu+\frac{1}{2}\right)} \int_0^w \sin^{2\nu} t\, dt; \\ d_\nu &= \sqrt{\pi} \frac{\sin^{2\nu+1} w}{2^\nu \Gamma\left(\nu+\frac{1}{2}\right)}. \end{aligned}\right\} \tag{6.3}$$

Analogously, formula (4.20) for $E_\nu^+(w, iz)$ can be written in the form

$$\begin{aligned} e^{-i\nu\pi/2} E_\nu^+(w, iz) = c_\nu I_\nu(z) + \frac{2}{\pi} d_\nu [K_\nu(z)\, Ie_\nu(\cos w, z) \\ - I_\nu(z)\, Ke_\nu(\cos w, z)]. \end{aligned} \tag{6.4}$$

From these formulae it can be seen immediately that the incomplete cylindrical function of Poisson form can be expressed in terms of the incomplete Lipschitz-Hankel integrals. We shall now show that, conversely, these integrals can be uniquely expressed in terms of the functions $E_\nu^+(w, z)$. By differentiating equation (6.2) with respect to z, we have

$$\begin{aligned} \frac{d}{dz} E_\nu^+(w, z) = c_\nu J_\nu'(z) + i d_\nu \Big[Je_\nu(-i \cos w, z)\, N_\nu'(z) \\ - Ne_\nu(-i \cos w, z)\, J_\nu'(z) + N_\nu(z) \frac{d}{dz} Je_\nu(-i \cos w, z) \\ - J_\nu(z) \frac{d}{dz} Ne_\nu(-i \cos w, z) \Big]. \end{aligned} \tag{6.5}$$

Or, taking into account that

$$\frac{d}{dz} Je_\nu(a, z) = \frac{d}{dz} \int_0^z e^{-at} t^\nu J_\nu(t)\, dt = e^{-az} z^\nu J_\nu(z);$$

$$\frac{d}{dz} Ne_\nu(a, z) = \frac{d}{dz} \int_0^z e^{-at} t^\nu N_\nu(t)\, dt = e^{-az} z^\nu N_\nu(z),$$

we find that

$$\begin{aligned} \frac{d}{dz} E_\nu^+(w, z) = c_\nu J_\nu'(z) + i d_\nu [Je_\nu(-i \cos w, z)\, N_\nu'(z) \\ - Ne_\nu(-i \cos w, z)\, J_\nu'(z)]. \end{aligned}$$

This expression can be transformed somewhat by taking advantage of the following relation for the function $E_\nu^+(w, z)$:

$$z \frac{d}{dz} E_\nu^+ = \nu E_\nu^+ - z E_{\nu+1}^+ + \frac{2 i d_\nu z^{\nu+1}}{\pi(2\nu+1)} e^{iz\cos w} \tag{6.6}$$

and of the analogous ones for the Bessel and Neumann functions

$$z J_\nu' = \nu J_\nu - z J_{\nu+1}; \quad z N_\nu' = \nu N_\nu - z N_{\nu+1}.$$

We thus obtain

$$\begin{aligned} E_{\nu+1}^+(w, z) - \frac{2 i d_\nu z^\nu}{\pi(2\nu+1)} e^{iz\cos w} \\ = c_\nu J_{\nu+1}(z) + i d_\nu [N_{\nu+1}(z)\, Je_\nu(-i\cos w, z) \\ - J_{\nu+1}(z)\, Ne_\nu(-i\cos w, z)]. \end{aligned} \tag{6.7}$$

Eqs. (6.7) and (6.2) can be considered as a system of two simultaneous linear equations in the unknowns Je_ν and Ne_ν. The determinant of this system

$$J_\nu(z)\, N_{\nu+1}(z) - J_{\nu+1}(z)\, N_\nu(z) = -\frac{2}{\pi z} \tag{6.8}$$

is different from zero and we obtain the pair of solutions

$$\begin{aligned} Je_\nu(-i\cos w, z) = \frac{z^{\nu+1}}{2\nu+1} e^{iz\cos w} J_\nu(z) \\ + \frac{\pi z}{2 i d_\nu} [E_\nu^+(w, z)\, J_{\nu+1}(z) - E_{\nu+1}^+(w, z)\, J_\nu(z)]; \end{aligned} \tag{6.9}$$

$$\begin{aligned} Ne_\nu(-i\cos w, z) = \frac{c_\nu}{i d_\nu} + \frac{z^{\nu+1}}{2\nu+1} e^{iz\cos w} N_\nu(z) \\ + \frac{\pi z}{2 i d_\nu} [E_\nu^+(w, z)\, N_{\nu+1}(z) - E_{\nu+1}^+(w, z)\, N_\nu(z)], \end{aligned} \tag{6.10}$$

where c_ν and d_ν are given in (6.3). It is evident that for $\nu = 0$, the above expressions become

$$\begin{aligned} Je_0(-i\cos w, z) = z e^{iz\cos w} J_0(z) \\ + \frac{\pi z}{2i \sin w} [E_0^+(w, z)\, J_1(z) - E_1^+(w, z)\, J_0(z)]; \end{aligned} \tag{6.11}$$

$$\begin{aligned} Ne_0(-i\cos w, z) = \frac{2w}{i\pi \sin w} + z e^{iz\cos w} N_0(z) \\ + \frac{\pi z}{2i \sin w} [E_0^+(w, z)\, N_1(z) - E_1^+(w, z)\, N_0(z)]. \end{aligned} \tag{6.12}$$

The expression for the incomplete Lipschitz-Hankel integrals involving the modified Bessel functions can be obtained in an analogous way from Eq. (6.4). However, we employ here a somewhat different approach. Replacing z by iz, we obtain after a simple transformation:

$$Ie_\nu(\cos w, z) = \int_0^z e^{-t\cos w} t^\nu I_\nu(t)\, dt = \frac{z^{\nu+1}}{2\nu+1} e^{-z\cos w} I_\nu(z)$$
$$+ \frac{\pi z}{d_\nu} [F_\nu^-(w, z) \cdot I_{\nu+1}(z) - F_{\nu+1}^-(w, z)\, I_\nu(z)], \tag{6.13}$$

where use has been made of the relation $I_\nu(z) = J_\nu(iz) \exp(-i\nu\pi/2)$ and the fact that

$$F_\nu^-(w, z) = \frac{z^\nu}{2^\nu \Gamma\left(\nu + \frac{1}{2}\right) \Gamma\left(\frac{1}{2}\right)} \int_0^w e^{-z\cos t} \sin^{2\nu} t\, dt$$
$$= \frac{1}{2} E_\nu^+(w, iz) \exp(-i\nu\pi/2) \tag{6.14}$$

(see § 1 of this chapter). Further, multiplying both sides of Eq. (6.10) by i, adding the result to (6.9), and replacing z everywhere by iz, we find

$$Ke_\nu(\cos w, z) = \int_0^z e^{-t\cos w} t^\nu K_\nu(t)\, dt$$
$$= \frac{\pi c_\nu}{2 d_\nu} + \frac{z^{\nu+1}}{2\nu+1} e^{-z\cos w} K_\nu(z) \tag{6.15}$$
$$- \frac{\pi z}{d_\nu} [F_\nu^-(w, z)\, K_{\nu+1}(z) + F_{\nu+1}^-(w, z)\, K_\nu(z)].$$

For $\nu = 0$ Eqs. (6.13) and (6.15) become

$$Ie_0(\cos w, z) = z e^{-z\cos w} I_0(z) + \frac{\pi z}{\sin w} [F_0^-(w, z)\, I_1(z) - F_1^-(w, z)\, I_0(z)]; \tag{6.16}$$

$$Ke_0(\cos w, z) = \frac{w}{\sin w} + z e^{-z\cos w} K_0(z)$$
$$- \frac{\pi z}{\sin w} [F_0^-(w, z)\, K_1(z) + F_1^-(w, z)\, K_0(z)]. \tag{6.17}$$

We have thus obtained explicit formulae for all of the incomplete Lipschitz-Hankel integrals in terms of incomplete cylindrical functions of Poisson form. From these formulae, as mentioned in the previous paragraph, special results for particular values of the variables can be obtained. Thus, taking for example $w = \pi/2$, we immediately obtain Eqs. (5.25)—(5.28). When w approaches 0 or π the expressions are inderterminate but their limiting values can be obtained from an expansion in w of the functions $E_\nu^+(w, z)$ in the neighborhood of these points with the aid of the coefficients $c_\nu(w)$ and $d_\nu(w)$.

Incomplete Lipschitz-Hankel integrals are frequently encountered in theoretical as well as in applied problems. The relations just established enable one to study them with the aid of the general theory of incomplete cylindrical functions.

On the other hand, from the above relations it is clear that all Lipschitz-Hankel integrals can be expressed in terms of incomplete cylindrical functions of the Poisson form.

7. Series Representations for Incomplete Cylindrical Functions

We shall first find a generating function for the incomplete cylindrical functions of Poisson form

$$E_\nu^+(w, z) = \frac{2z^\nu}{A_\nu} \int_0^w e^{iz\cos t} \sin^{2\nu} t \, dt, \tag{7.1}$$

where $A_\nu = 2^\nu \Gamma(\nu + 1/2)\, \Gamma(1/2)$ and $\mathrm{Re}\,(\nu + 1/2) > 0$. The generating function for a sequence of functions $\{f_n(z)\}$ with integer indices is defined as

$$\Psi(z, u) = \sum_{n=-\infty}^{\infty} u^n f_n(z).$$

Since the index in Eq. (7.1) is restricted to positive values, we define the generating function for the sequence $E_n^+(w, z)$ by

$$\Psi(w, z, u) = \sum_{n=0}^{\infty} u^n E_n^+(w, z). \tag{7.2}$$

Substituting $E_n^+(w, z)$ from Eq. (7.1) and interchanging the order of integration and summation which in the present case is valid due to the absolute convergence of the power series, we have

$$\Psi(w, z, u) = \frac{2}{\sqrt{\pi}} \int_0^w e^{iz\cos t} \left[\sum_{n=0}^{\infty} \left(\frac{zu \sin^2 t}{2} \right)^n \frac{1}{\Gamma\left(n + \frac{1}{2}\right)} \right] dt.$$

The inner series can be summed by taking advantage of the relation (4.11) for the beta-function with $n > 0$

$$\frac{1}{\Gamma\left(n + \frac{1}{2}\right)} = \frac{B\left(n, \frac{1}{2}\right)}{\Gamma(n)\sqrt{\pi}} = \frac{2}{\sqrt{\pi}\,\Gamma(n)} \int_0^{\pi/2} \sin^{2n-1} \varphi \, d\varphi. \tag{7.3}$$

Singling out the term for $n = 0$, gives

$$\sum_{n=0}^{\infty} \frac{x^n}{\Gamma\left(n + \frac{1}{2}\right)} = \frac{1}{\sqrt{\pi}} \left[1 + 2 \sum_{n=1}^{\infty} \frac{x^n}{\Gamma(n)} \int_0^{\pi/2} \sin^{2n-1} \varphi \, d\varphi \right],$$

where $x = (zu/2) \sin^2 t$. Changing again the order of summation and integration gives

$$\sum_{n=0}^{\infty} \frac{x^n}{\Gamma\left(n + \frac{1}{2}\right)} = \frac{1}{\sqrt{\pi}} \left[1 + 2x \int_0^{\pi/2} \sin\varphi \, d\varphi \sum_{k=0}^{\infty} \frac{(x \sin^2\varphi)^k}{\Gamma(k+1)}\right].$$

The last sum is obviously equal to $\exp(x \sin^2\varphi)$, so that the generating function $\Psi(w, z, u)$ takes the form

$$\Psi(w, z, u) = \frac{2}{\pi} \int_0^w e^{iz\cos t} \, dt + \frac{2zu}{\pi} \int_0^w \sin^2 t \, e^{iz\cos t} \, dt \times \int_0^{\pi/2} \sin\varphi \exp\left(\frac{zu}{2} \sin^2 t \sin^2\varphi\right) d\varphi. \tag{7.4}$$

The first term in this formula is equal to $E_0^+(w, z)$, while the inner integral can be cxpressed in terms of the normal probability integral. Formula (7.4) is taken as the generating function for incomplete cylindrical functions of Poisson form. The double integral in the second term occurs in many applied problems and the analysis of such problems is aided by the theory of incomplete cylindrical functions.

We shall now devclope series expansions for these functions. To obtain an expansion of the function $E_\nu^\pm(w, z)$ as a power series in z we shall employ its integral representation (7.1). Expanding the function $\exp(\pm iz \cos t)$ as a power series in $(iz \cos t)$ and integrating term by term we have

$$E_\nu^\pm(w, z) = \frac{z^\nu}{A_\nu} \sum_{m=0}^{\infty} C_{m,\nu}(w) \frac{(\pm iz)^m}{m!}. \tag{7.5}$$

We now introduce the notation

$$C_{m,\nu}(w) = 2 \int_0^w \cos^m t \sin^{2\nu} t \, dt. \tag{7.6}$$

For real w the coefficients $C_{m,\nu}(w)$ can be expressed as incomplete beta-functions, and for $w = \pi/2$ they transform directly into the well known complete beta-functions:

$$C_{m,\nu}(\pi/2) = 2 \int_0^{\pi/2} \cos^m t \sin^{2\nu} t \, dt = B\left(\frac{m+1}{2}, \nu + \frac{1}{2}\right).$$

The series (7.5) converges rapidly and is thus suitable for calculating the function $E_\nu^\pm(w, z)$ for $|z| < 1$.
To facilitate these calculations, we make use of the following recurrence relation for the coefficients $C_{m,\nu}(w)$:

$$C_{m,\nu}(w) = \frac{m-1}{2\nu + m} C_{m-2,\nu}(w) + \frac{2}{2\nu + m} \sin^{2\nu+1} w \cos^{m-1} w, \tag{7.7}$$

which can be obtained by integrating Eq. (7.6) by parts. The expressions for the first two coefficients $C_{0\nu}$ and $C_{1\nu}$ are

$$\left.\begin{aligned} C_{0,\nu}(w) &= C_\nu = 2\int_0^w \sin^{2\nu} t\,dt; \\ C_{1,\nu}(w) &= \frac{2}{2\nu+1}\sin^{2\nu+1} w. \end{aligned}\right\} \tag{7.8}$$

In this way all of the coefficients can be expressed for a fixed ν in terms of the integral $C_{0,\nu} = C_\nu$ and elementary functions. When $\nu = n$ all of the coefficients are expressible in terms of simple elementary functions. Thus, for example, the first five coefficients for $E_0^{\pm}(w, z)$ are

$$\left.\begin{aligned} C_{0,0} &= 2w;\ C_{1,0}(w) = 2\sin w;\ C_{2,0}(w) = w + \sin w \cos w; \\ C_{3,0}(w) &= \frac{2}{3}\sin w \cos^2 w + \frac{4}{3}\sin w; \\ C_{4,0}(w) &= \frac{3}{4}(\sin w \cos w + w) + \frac{1}{2}\sin w \cos^3 w. \end{aligned}\right\} \tag{7.9}$$

The corresponding coefficients for $E_1^{\pm}(w, z)$ are

$$\left.\begin{aligned} C_{0,1}(w) &= w - \sin w \cdot \cos w;\ C_{1,1}(w) = \frac{2}{3}\sin^3 w; \\ C_{2,1}(w) &= \frac{1}{4}\left(w - \frac{1}{4}\sin 4w\right); \\ C_{3,1}(w) &= \frac{2}{5}\sin^3 w\left(\frac{5}{3} - \sin^2 w\right). \end{aligned}\right\} \tag{7.10}$$

In addition to having an expansion (7.5) as a power series in z, the functions $E_n^{\pm}(w, z)$ also possesses an expansion as a series of Bessel functions $J_n(z)$. It is sufficient to expand the exponential $\exp(\pm iz\cos t)$ in terms of Bessel functions:

$$\exp(\pm iz\cos t) = J_0(z) + 2\sum_{n=1}^{\infty}(\pm i)^n J_n(z)\cos nt \tag{7.11}$$

and obtain from this expansion an explicit expression for $E_0^{\pm}(w, z)$ and $E_1^{\pm}(w, z)$. Substituting the expansion (7.11) into (7.1) we find, after an elementary transformation

$$E_0^{\pm}(w, z) = \frac{2w}{\pi}\left\{J_0(z) + 2\sum_{n=1}^{\infty}(\pm i)^n J_n(z)\frac{\sin(wn)}{wn}\right\}; \tag{7.12}$$

$$\begin{aligned} E_1^{\pm}(w, z) = \frac{zw}{\pi}\left(1 - \frac{\sin 2w}{2w}\right)J_0(z) + \frac{zw}{\pi}\sum_{n=1}^{\infty}(\pm i)^n J_n(z) \\ \times\left\{\frac{2\sin(wn)}{wn} - \frac{\sin(n+2)w}{(n+2)w} - \frac{\sin(n-2)w}{(n-2)w}\right\}. \end{aligned} \tag{7.13}$$

Setting $w = \pi$ in the above formulae and making use of the recurrence relations for Bessel functions $J_0 + J_2 = \frac{2}{z} J_1$, we obtain

$$E_0^{\pm}(\pi, z) = 2J_0(z); \quad E_1^{\pm}(\pi, z) = 2J_1(z).$$

Expansions of the functions $E_0^{\pm}(w, z)$ and $E_1^{\pm}(w, z)$ in the form (7.12) and (7.13) are convenient in practical applications. These expansions can also be obtained by solving the partial differential equation (4.25) by the method of separation of variables. The series (7.12) and (7.13) are actually Fourier series in the complete set of functions $\sin(mw)$ and $\cos(mw)$. In addition to the aforementioned developments, it is possible to obtain other expansions for the function $E_\nu^{\pm}(w, z)$. They can be readily obtained by writing formula (7.1) in the form

$$\begin{aligned} E_\nu^{+}(w, z) &= \frac{2z^\nu}{A_\nu} \int\limits_{\cos w}^{1} e^{izt}(1 - t^2)^{\nu - 1/2}\, dt \\ &= J_\nu(z) + iH_\nu(z) - \frac{2z^\nu}{A_\nu} \int\limits_0^{\cos w} e^{izt}(1 - t^2)^{\nu - 1/2}\, dt \end{aligned} \tag{7.14}$$

and taking advantage of a suitable representation of the function $(1 - t^2)^{\nu - 1/2}$. For $\nu = 0$, the Fourier expansion (see [10, p. 694])

$$\frac{1}{\sqrt{1 - t^2}} = \frac{\pi}{2} + \pi \sum_{m=1}^{\infty} J_0(m\pi) \cos m\pi t, \tag{7.15}$$

may be used, being uniformly convergent for $-1 < t < 1$. Substituting this into (7.14) and integrating term by term we obtain

$$\begin{aligned} E_0^{+}(w, z) &= J_0(z) + iH_0(z) + ie^{iz\cos w} \\ &\times \left[\frac{1}{z} + 2\sum_{m=1}^{\infty} J_0(m\pi) \frac{z\cos(m\pi\cos w) - im\pi\sin(m\pi\cos w)}{z^2 - m^2\pi^2}\right] \\ &- i\left[\frac{1}{z} + 2\sum_{m=1}^{\infty} J_0(m\pi) \frac{z}{z^2 - m^2\pi^2}\right], \end{aligned} \tag{7.16}$$

$|\cos w| < 1$.

An expansion of a similar type was obtained by Nagaoka [2]. It is actually an expansion of Schlömilch type [see (10, Chapter XIX)]. A similar expansion for the function $E_1^{+}(w, z)$ can be obtained by making use of formula (2.24):

$$E_1^{+}(w, z) = -\frac{\partial E_0^{+}(w, z)}{\partial z} + \frac{2i}{\pi} \sin w e^{iz\cos w}.$$

Calculation of $E_\nu^{+}(w, z)$ for $\nu > 1$ can be accomplished by the recurrence relation for higher orders. Repeated application of the recurrence formula can be avoided by means of the following formula:

$$E^{+}_{\nu+k}(w,z) = E^{+}_{\nu}(w,z)\, R_{k,\nu}(z) - E^{+}_{\nu-1}(w,z)\, R_{k-1,\nu+1}(z)$$
$$+ 2e^{iz\cos w} \sum_{m=0}^{k-1} \sin^{2(\nu+m)-1} w \left(\frac{\cos w}{A_{\nu+m}} + \frac{i \sin^2 w}{A_{\nu+1+m}}\right) z^{\nu+m} R_{k-m+1,\nu+m+1}(z), \tag{7.17}$$

where, as before $A_\mu = 2^\mu \Gamma(\mu + 1/2)\, \Gamma(1/2)$ and

$$R_{m,\mu}(z) = \sum_{n=0}^{[m/2]} \frac{(-1)^n (m-n)!\, \Gamma(\mu+m-n) \left(\frac{z}{2}\right)^{-m+2n}}{n!\,(m-2n)!\,\Gamma(\mu+n)} \tag{7.18}$$

are Lommel's Polynomials [10, p. 324] with $[m/2]$ equal to the nearest integer less than or equal to $m/2$.

In conclusion we shall obtain an expansion of the function $E^{\pm}_{\nu}(w,z)$ with respect to the variable w, in the form

$$E^{\pm}_{\nu}(w,z) = \frac{2z^\nu}{A_\nu} e^{iz} w^{2\nu+1} \sum_{k=0}^{\infty} \frac{w^{2k}}{2\nu+2k+1} \sum_{m=0}^{k} a_m(\nu)\, b^{\pm}_{k-m}(z), \tag{7.19}$$

where the coefficients $a_m(\nu)$ and $b^{\pm}_{\mu}(z)$ for arbitrary ν and z are found in an elementary way from the following expansion

$$(w^{-1}\sin w)^{2\nu} = \left(1 - \frac{w^2}{3!} + \frac{w^4}{5!} - \cdots\right)^{2\nu} = \sum_{m=0}^{\infty} a^m(\nu)\, w^{2m},$$
$$\exp[\pm iz(\cos w - 1)] = \exp\left[\pm iz\left(-\frac{w^2}{2!} + \frac{w^4}{4!} - \cdots\right)\right] = \sum_{\mu=0}^{\infty} b^{\pm}_{\mu}(z)\, w^{2\mu}. \tag{7.20}$$

The series (7.19) is converges rapidly for $|zw^2| < 1$. If the variable w is such that $|z(\pi/2 - w)| < 1$, it is advantageous to use the following expansion

$$E^{\pm}_{\nu}(w,z) = J_\nu(z) \pm iH_\nu(z) - \frac{2z^\nu}{A_\nu} \sum_{k=0}^{\infty} \frac{\left(\frac{\pi}{2} - w\right)^{k+1}}{k+1} \sum_{m=0}^{[k/2]} c_m(\nu) \cdot d^{\pm}_{k-2m}(z). \tag{7.21}$$

Here the symbol $[k/2]$ denotes the integral part of $k/2$ and the coefficients $c_m(\nu)$ and $d^{\pm}_{\mu}(z)$ found from the following expansions:

$$\cos^{2\nu} u = \left(1 - \frac{u^2}{2!} + \frac{u^4}{4!} - \cdots\right)^{2\nu} = \sum_{m=0}^{\infty} c_m(\nu)\, u^{2m},$$
$$\exp(\pm iz \sin u) = \exp\left[\pm iz\left(u - \frac{u^3}{3!} + \cdots\right)\right] = \sum_{\mu=0}^{\infty} d^{\pm}_{\mu}(z)\, u^{\mu}. \tag{7.22}$$

The expansion (7.21) is easily obtained by taking advantage of relations (1.35), (1.32) and (1.6) of Section 1.

8. Incomplete Beta and Gamma Functions and their Relation to Hypergeometric Functions

In the previous section, while developing series for the incomplete cylindrical functions, we encountered the following integral

$$C_{m,\nu}(w) = 2\int_0^w \cos^m t \cdot \sin^{2\nu} t\, dt. \tag{8.1}$$

Later the following integrals will also be needed:

$$\gamma(p, y) = \int_0^y e^{-t} t^{p-1}\, dt; \tag{8.2}$$

$$\tilde{\gamma}(p, y) = \int_0^y e^{t} t^{p-1}\, dt. \tag{8.3}$$

For $w = \pi/2$, $C_{m,\nu}(w)$ becomes the Euler integral of the first kind, or beta function while the function $\gamma(p, y)$ for $y \to \infty$ becomes the Euler integral of the second kind or gamma function. For other values of the variables w or y, they are expressed as incomplete beta and gamma functions. Analysis of these functions leads naturally to their expression in terms of hypergeometric series, the properties of which have been well studied. As is known, the generalized hypergeometric series is defined by

$$\sum_{m=0}^{\infty} \frac{(\alpha_1)_m (\alpha_2)_m \cdots (\alpha_p)_m}{(\gamma_1)_m (\gamma_2)_m \cdots (\gamma_q)_m} \cdot \frac{z^m}{m!},$$

where the symbol $(\alpha)_m$ is defined by

$$(\alpha)_m = \alpha(\alpha+1)\cdots(\alpha+m-1) = \frac{\Gamma(m+\alpha)}{\Gamma(\alpha)}. \tag{8.4}$$

This series can be regarded as a function of z with $(p+q)$ parameters $\alpha_1, \alpha_2, \ldots, \alpha_p$ and $\gamma_1, \gamma_2, \ldots, \gamma_q$. Such a series can be conveniently represented by the Pochhammer symbol

$${}_pF_q(\alpha_1, \alpha_2, \ldots, a_p; \gamma_1, \gamma_2, \ldots, \gamma_q; z) = \sum_{m=0}^{\infty} \frac{(\alpha_1)_m (\alpha_2)_m \cdots (\alpha_p)_m}{(\gamma_1)_m (\gamma_2)_m \cdots (\gamma_q)_m} \cdot \frac{z^m}{m!}. \tag{8.5}$$

The left index p of the symbol ${}_pF_q$ denotes the number of parameters in the numerator while the right index indicates the number of denominator parameters. The most commonly used of these series are the hypergeometric series

$$\left.\begin{aligned} {}_2F_1(\alpha, \beta; \gamma; z) &\equiv F(\alpha, \beta; \gamma; z); \\ F(\alpha, \beta; \gamma; z) &= 1 + \frac{\alpha\beta}{\gamma!} z + \frac{\alpha(\alpha+1)\beta(\beta+1)}{\gamma(\gamma+1)\,2!} z^2 + \cdots \end{aligned}\right\} \tag{8.6}$$

and the confluent hypergeometric series

$$ {}_1F_1(\alpha;\gamma;z) = F(\alpha;\gamma;z) = 1 + \frac{\alpha}{\gamma!} z + \frac{\alpha(\alpha+1)}{\gamma(\gamma+1)\,2!} z^2 + \cdots. \quad (8.7)$$

Both of these series evidently terminate if either α or β is a negative integer, and are not defined if γ is a negative integer. For all other cases, the hypergeometric series (8.6) is convergent within the unit circle, while (8.7) is convergent for all values of z.

One of the integral representations of the hypergeometric function $F(\alpha,\beta;\gamma;z)$ for $\operatorname{Re}\gamma > \operatorname{Re}\beta > 0$ (see [11, p. 1054]) is

$$F(\alpha,\beta;\gamma;z) = \frac{\Gamma(\gamma)}{\Gamma(\beta)\,\Gamma(\gamma-\beta)} \int_0^1 t^{\beta-1}(1-t)^{\gamma-\beta-1}(1-tz)^{-\alpha}\,dt. \quad (8.8)$$

These functions posses a number of interesting properties. Some of the more important of these are the following

$$\begin{aligned} F(\alpha,\beta;\gamma;z) &= (1-z)^{-\alpha} F\left(\alpha,\gamma-\beta;\gamma;\frac{z}{z-1}\right) \\ &= (1-z)^{-\beta} F\left(\beta,\gamma-\alpha;\gamma;\frac{z}{z-1}\right) \\ &= (1-z)^{\gamma-\alpha-\beta} F(\gamma-\alpha,\gamma-\beta;\gamma;z); \end{aligned} \quad (8.9)$$

$$\begin{aligned} F(\alpha,\beta;\gamma;z) &= \frac{\Gamma(\gamma)\,\Gamma(\gamma-\alpha-\beta)}{\Gamma(\gamma-\alpha)\,\Gamma(\gamma-\beta)} F(\alpha,\beta;\alpha+\beta-\gamma+1;1-z) \\ &\quad + (1-z)^{\gamma-\alpha-\beta} \frac{\Gamma(\gamma)\,\Gamma(\alpha+\beta-\gamma)}{\Gamma(\alpha)\,\Gamma(\beta)} \\ &\quad \times F(\gamma-\alpha,\gamma-\beta;\gamma-\alpha-\beta+1;1-z); \end{aligned} \quad (8.10)$$

$$\begin{aligned} &F(\alpha,;\beta;\gamma;z) \\ &= \frac{\Gamma(\gamma)\,\Gamma(\beta-\alpha)}{\Gamma(\beta)\,\Gamma(\gamma-\alpha)} (-1)^{\alpha} z^{-\alpha} F\left(\alpha,\alpha+1-\gamma;\alpha+1-\beta;\frac{1}{z}\right) \\ &\quad + \frac{\Gamma(\gamma)\,\Gamma(\alpha-\beta)}{\Gamma(\alpha)\,\Gamma(\gamma-\beta)} (-1)^{\beta} z^{-\beta} F\left(\beta,\beta+1-\gamma;\beta+1-\alpha;\frac{1}{z}\right); \end{aligned} \quad (8.11)$$

$$\begin{aligned} F\left(2\alpha,2\beta;\alpha+\beta+\frac{1}{2};\frac{1-\sqrt{z}}{2}\right) &= \frac{\Gamma\left(\alpha+\beta+\frac{1}{2}\right)\Gamma\left(\frac{1}{2}\right)}{\Gamma\left(\alpha+\frac{1}{2}\right)\Gamma\left(\beta+\frac{1}{2}\right)} F\left(\alpha,\beta;\frac{1}{2};z\right) \\ &\quad + \sqrt{z}\,\frac{\Gamma\left(\alpha+\beta+\frac{1}{2}\right)\Gamma\left(-\frac{1}{2}\right)}{\Gamma(\alpha)\,\Gamma(\beta)} F\left(\alpha+\frac{1}{2},\beta+\frac{1}{2};\frac{3}{2};z\right). \end{aligned} \quad (8.12)$$

These formulae are necessary for the analytical continuation of the hypergeometric function to the region $|z| > 1$. In addition (8.11) can be used to obtain asymptotic estimates for $F(\alpha;\beta;\gamma;z)$ when $|z| \gg 1$.

Along with the above, we can also deduce the corresponding relations for the confluent hypergeometric function $F(\alpha;\gamma;z)$. One of the integral representations of this function for $\operatorname{Re}\gamma > \operatorname{Re}\alpha > 0$ is

$$F(\alpha,\gamma,z) = \frac{\Gamma(\gamma)\, z^{1-\gamma}}{\Gamma(\alpha)\,\Gamma(\gamma-\alpha)} \int_0^z e^t\, t^{\alpha-1} (z-t)^{\gamma-\alpha-1}\, dt. \tag{8.13}$$

The confluent hypergeometric function satisfies a number of relations among which are

$$\left.\begin{aligned} F(\alpha,\gamma;z) &= e^z F(\gamma-\alpha,\gamma;-z); \\ F(\alpha,\alpha;z) &= e^z \end{aligned}\right\} \tag{8.14}$$

while for $|z| \gg 1$ they possess the following asymptotic property (c.f. [5, p. 573])

$$F(\alpha,\gamma;z) \approx \frac{\Gamma(\gamma)}{\Gamma(\alpha)} z^{\alpha-\gamma} e^z + \frac{\Gamma(\gamma)}{\Gamma(\gamma-\alpha)} (z\, e^{\pm i\pi})^{-\alpha}, \tag{8.15}$$

where the positive sign is taken for $\operatorname{Im} z < 0$ and the negative sign for $\operatorname{Im} z > 0$.

We shall now show that the incomplete beta function $C_{m,\nu}(w)$ can be expressed by means of the hypergeometric function $F(\alpha,\beta;\gamma;z)$. In fact, by making the change of the variable $\cos^2 t = u$, in Eq. (8.1) we obtain

$$C_{m,\nu}(w) = \int_{\cos^2 w}^{1} u^{(m-1)/2} (1-u)^{\nu-1/2}\, du$$

or

$$C_{m,\nu}(w) = \int_0^1 u^{(m-1)/2} (1-u)^{\nu-1/2}\, du - \int_0^{\cos^2 w} u^{(m-1)/2} (1-u)^{\nu-1/2}\, du.$$

The first term in this expression is the beta-function $B\left(\frac{m+1}{2}, \nu+1/2\right)$, while the second term, after the change of variable $u = t\cos^2 w$ becomes

$$\int_0^{\cos^2 w} u^{(m-1)/2} (1-u)^{\nu-1/2}\, du = \cos^{m+1} w \int_0^1 t^{(m-1)/2} (1-t\cos^2 w)^{\nu-1/2}\, dt.$$

Comparison of the last integral with formula (8.8) shows that it can be expressed as a hypergeometric function $F(\alpha,\beta;\gamma;z)$ by setting:

$$\beta - 1 = \frac{m-1}{2}; \quad \gamma = \beta + 1; \quad -\alpha = \frac{1}{2} - \nu; \quad z = \cos^2 w.$$

For the final result we obtain

$$\begin{aligned} C_{m,\nu}(w) = B\left(\frac{m+1}{2}, \nu+1/2\right) &- \cos^{m+1} w B\left(\frac{m+1}{2}, 1\right) \\ &\times F\left(-\nu+1/2, \frac{m+1}{2}; \frac{m+3}{2}; \cos^2 w\right). \end{aligned}$$

Or, by making use of the well known relation between the beta- and gamma-functions

$$B(x, y) = \frac{\Gamma(x)\,\Gamma(y)}{\Gamma(x+y)}\,,$$

the latter formula can be written

$$C_{m,\nu}(w) = \frac{\Gamma\left(\frac{m+1}{2}\right)\Gamma\left(\nu+\frac{1}{2}\right)}{\Gamma\left(\nu+\frac{m}{2}+1\right)} - \frac{\Gamma\left(\frac{m+1}{2}\right)}{\Gamma\left(\frac{m+3}{2}\right)}\cos^{m+1} w \tag{8.16}$$
$$\times F\left(-\nu+1/2, \frac{m+1}{2}; \frac{m+3}{2}; \cos^2 w\right).$$

The expression for $C_{m,\nu}(w)$ just obtained, although somewhat complicated in form, is nevertheless useful especially when the argument $w = \alpha$ is real and near $\pi/2$. In addition, formula (8.16) offers the possibility of extending all of the known properties of the hypergeometric series to the incomplete β function.

We now pass to the study of the incomplete gamma function (8.2), (8.3). Comparing (8.3) with the integral representation of the confluent hypergeometric series (8.13), we find directly by taking $\gamma - \alpha - 1 = 0$, $\alpha = p$,

$$\tilde{\gamma}(p, y) = B(p, 1)\, y^p F(p; p+1; y) \tag{8.17}$$

or

$$\tilde{\gamma}(p, y) = \frac{y^p}{p} F(p; p+1; y) \tag{8.18}$$

In order to represent formula (8.2) as a hypergeometric series, we change the variable of integration from t to $-t$ and obtain

$$\gamma(p, y) = (-1)^p \int_0^{-y} e^t\, t^{p-1}\, dt.$$

Comparison of the above with Eq. (8.13) gives

$$\gamma(p, y) = \frac{y^p}{p} F(p; p+1; -y), \tag{8.19}$$

or, using the relation (8.14), we obtain for the incomplete gamma function

$$\gamma(p, y) = \frac{y^p}{p} e^{-y} F(1; 1+p; y). \tag{8.20}$$

Expressions (8.18) and (8.19) are well suited for analysis of the incomplete gamma function since the confluent hypergeometric series are well studied and possess many properties which facilitate practical calculations. These formulae, according to Eq. (8.7), define the incomplete gamma function as a power series in y. One can immediately see that they may be written in the form

$$\tilde{\gamma}(p, y) = y^p\left(\frac{1}{p} + \frac{1}{p+1}\frac{y}{1!} + \frac{1}{p+2}\frac{y^2}{2!} + \cdots\right) = y^p \sum_{k=0}^{\infty} \frac{y^k}{(k+p)\,k!}; \quad (8.21)$$

$$\begin{aligned}\gamma(p, y) &= y^p\left\{\frac{1}{p} - \frac{1}{p+1}\frac{y}{1!} + \frac{1}{p+2}\frac{y^2}{2!} + \cdots\right\}\\ &= y^p \sum_{k=0}^{\infty} (-1)^k \frac{1}{k+p}\frac{y^k}{k!}, \quad p > 0.\end{aligned} \quad (8.22)$$

Asymptotic expansions for the incomplete gamma and beta functions can be obtained from their relations to the corresponding hypergeometric functions. In particular, the asymptotic expansion for the incomplete gamma function $\gamma(p, y)$ for $|y| \gg 1$ is

$$\gamma(p, y) \approx \Gamma(p) - \frac{e^{-y}}{y^{p-1}} \sum_{k=0}^{\infty} \frac{(-1)^k\, \Gamma(1-p+k)}{\Gamma(1-p)\, y^k}. \quad (8.23)$$

9. Asymptotic Series and Methods for their Construction

Along with expansions of incomplete cylindrical functions as power series in w and z, which are defined for $|z \sin w| < 1$, it is also necessary to have expansions for these functions when $|z \sin w| > 1$. Such expansions are asymptotic and, as a rule, non-convergent. Here we establish basic properties of such series and methods for their construction.

To study of the behavior of certain functions $F(z)$ for large values of $|z|$ it is frequently expedient to express them as series in inverse powers of z

$$F(z) = f(z)\left(A_0 + \frac{A_1}{z} + \frac{A_2}{z^2} + \cdots\right), \quad (9.1)$$

where $f(z)$ is a function which describes the behavior for large $|z|$. An example of an expansion of this type is the series (8.23) for the incomplete gamma function.

When the ratio $F(z)/f(z)$ has an essential singularity at infinity, these series will be divergent. Nevertheless they can prove useful not only for qualitative analysis of the behavior of functions, but also for direct evaluation for large $|z|$. For this purpose it is necessary that the difference between $F(z)/f(z)$ and the sum of first $n+1$ terms of the series in (9.1) should be of the order of z^{-n-1}. In this case it is said that the function $F(z)$ is expressed asymptotically by the series (9.1). In other words

$$F(z) \approx f(z) \sum_{k=0}^{\infty} \frac{A_k}{z^k}, \quad (9.2)$$

provided

$$\lim_{z\to\infty}\left\{z^n\left[F(z) - f(z)\sum_{k=0}^{n}\frac{A_k}{z^k}\right]\right\} = 0. \tag{9.3}$$

In this form, for a given n, the sum of the first $n+1$ terms of the series for sufficiently large values of $|z|$ becomes arbitrarily close to $F(z)$. For fixed n, the deviation will be of the order of $|z|^{-n-1}$. Since the series (9.2) diverges, it follows that to a given z there corresponds a fixed n, for which the partial sum of the first $n+1$ terms approximates the given function most closely. Such an optimum value of n determines the inherent error in the expression (9.2) for a given z. As z increases, the optimum n also increases, while the inherent error decreases.

In most cases the coefficients A_k decrease as k increases until a value $k = m$ is reached after which they start to increase. Therefore, for practical calculations one restricts himself to partial sums up to the order m. Different methods exist for the construction of asymptotic series (see, e.g. [14, 15]). For functions with an integral representation of the form:

$$F(z) = \int_a^b e^{zh(t)}\varphi(t)\,dt, \tag{9.4}$$

(which include the incomplete cylindrical functions) a very fruitful procedure proves to be the saddle point method, or as it is sometimes called, the method of stationary phase (see also [5, p. 415], [10, p. 262], and [12]). We shall explain here the method in a somewhat simplified form, including only those details necessary for the present considerations.

For sufficiently large z and Re $[zh(t)] \neq 0$, the greatest contribution to the integral (9.4) evidently occurs either near the endpoints of the interval $[a, b]$ or near points $t = c$ in $[a, b]$ where $h(t)$ attains an extremal; that is, where

$$h'(c) = 0. \tag{9.5}$$

These contributions can be of different orders of magnitude depending on the relations between the quantities Re $[zh(a)]$, Re $[zh(b)]$ and Re $[zh(c)]$. If, for example, the first of these quantities is greater than the others, then the basic contribution to the value Eq. (9.4) comes from the neighborhood of the lower limit. On the other hand, for Re$[zh(c)]$ $\gg$ Re $[zh(a)]$ and Re $[zh(b)]$, the main contribution in (9.4) comes from values in the interval of integration near c. The procedure for determining the asymptotic expansion for $F(z)$ in $[a, b]$ consists of estimating the separate contributions mentioned above. We will follow mainly the procedures outlined in [14] and [15].

Examining first the contribution from one end of the interval, say a, we shall assume that at this point $h(t)$ assumes its greatest value in $[a, b]$ but that a is not an extremal point, i.e. $h'(a) \neq 0$. In this case the function $h(t)$ can be represented in the form

$$h(t) = h(a) - u(t - a), \tag{9.6}$$

where $u(t - a)$ represents a function which tends linearly to zero as $(t - a) \to 0$. We shall further denote by $\Psi(u)$ the function inverse to $u(t - a)$,

$$\left.\begin{aligned} \Psi(u) &= t - a; \\ \Psi(0) &= 0, \end{aligned}\right\} \tag{9.7}$$

Equation (9.6) then assumes the form

$$h[a + \Psi(u)] = h(a) - u. \tag{9.8}$$

Since, by hypothesis $h'(a) \neq 0$, the implicit function theorem assures the unique existence of the function $\Psi(u)$, which approaches 0 for $u \to 0$. Substituting $h(t)$ from (9.6) in the expression (9.4), and changing the variable of integration from t to u according to Eq. (9.7) with $dt = \Psi'(u)\, du$, we obtain the following expression for the contribution in the vicinity of the point a for $|z| \gg 1$:

$$\int\limits_a^{a+\varepsilon} \varphi(t)\, e^{zh(t)}\, dt = e^{zh(a)} \int\limits_0^{\varepsilon_1} \varphi[a + \Psi(u)]\, \Psi'(u)\, e^{-uz}\, du. \tag{9.9}$$

Now we expand $\varphi[a + \Psi(u)]\, \Psi'(u)$ as a power series in u:

$$\varphi[a + \Psi(u)]\, \Psi'(u) = \sum_{k=0}^{\infty} A_k(a)\, u^k. \tag{9.10}$$

Substituting this into Eq. (9.9), and integrating term by term gives

$$\begin{aligned} \int\limits_a^{a+\varepsilon} \varphi(t)\, e^{zh(t)}\, dt &= e^{zh(a)} \sum_{k=0}^{\infty} A_k \int\limits_0^{\varepsilon_1} e^{-uz}\, u^k\, du \\ &= e^{zh(a)} \sum_{k=0}^{\infty} \frac{A_k}{z^{k+1}}\, \gamma(k + 1, z\varepsilon_1), \end{aligned} \tag{9.11}$$

where $\gamma(p, y)$ is the incomplete gamma function which for $|y| \gg 1$ can be approximated by $\Gamma(p)$ according to (8.23). Therefore, if $|z| \gg 1$, formula (9.11) can be written in the following form:

$$\int\limits_a^{a+\varepsilon} \varphi(t)\, e^{zh(t)}\, dt \approx e^{zh(a)} \sum_{k=0}^{\infty} \frac{\Gamma(k + 1)\, A_k(a)}{z^{k+1}}. \tag{9.12}$$

This formula characterises the contribution to (9.4) from values in the integration interval near a. The contribution from values near to the other end of the interval is given by

$$\int_{b-\varepsilon}^{b} \varphi(t)\, e^{zh(t)}\, dt \simeq -e^{zh(b)} \sum_{k=0}^{\infty} \frac{\Gamma(k+1)\, A_k(b)}{z^{k+1}}, \tag{9.13}$$

where the coefficients $A_k(b)$ are found from the system of equations

$$\left.\begin{aligned} \varphi[\alpha + \Psi(u)]\, \Psi'(u) &= \sum_{k=0}^{\infty} A_k(\alpha)\, u^k; \\ h[\alpha + \Psi(u)] &= h(\alpha) - u. \end{aligned}\right\} \tag{9.14}$$

Thus for the coefficient $A_0(\alpha)$, the above system, together with the condition $\Psi(0) = 0$, gives

$$\varphi(\alpha)\, \Psi'(0) = A_0(\alpha);$$

$$h'(\alpha)\, \Psi'(0) = -1,$$

so that

$$A_0(\alpha) = -\varphi(\alpha)/h'(\alpha). \tag{9.15}$$

For the contribution from the interval $(+\varepsilon, -\varepsilon)$ near the extremal point c, where $h'(c) = 0$, it is necessary, according to the saddle point method ([14], [15]) to replace the system (9.14) by the system

$$\left.\begin{aligned} \varphi[c + \Psi(u)]\, \Psi'(u) &= \sum_{k=0}^{\infty} B_k(c)\, u^k; \\ h[c + \Psi(u)] &= h(c) - u^2. \end{aligned}\right\} \tag{9.16}$$

In this case we obtain the analogous result

$$\begin{aligned} \int_{c-\varepsilon}^{c+\varepsilon} \varphi(t)\, e^{zh(t)}\, dt &= e^{zh(c)} \int_{-\delta}^{\delta} e^{-zu^2} \varphi[c + \Psi(u)]\, \Psi'(u)\, du \\ &= e^{zh(c)} \sum_{k=0}^{\infty} B_k(c) \int_{-\delta}^{\delta} e^{-zu^2} u^k\, du, \end{aligned} \tag{9.17}$$

where $\delta = \varepsilon \sqrt{-h''(c)/2}$, and the coefficients $B_k(c)$ are found from the system (9.16). Thus for B_0 and B_1, we have

$$B_0(c) = \varphi(c) \sqrt{-\frac{2}{h''(c)}}; \quad B_1(c) = \frac{2}{h''(c)} \left[\varphi(c) \frac{h'''(c)}{3h''(c)} - \varphi'(c)\right]. \tag{9.18}$$

Formula (9.17) was written under the assumption that the extremal point c did not coincide with either of the limits of integration. When this occurs, the corresponding integration limit on the right side of (9.17) must be taken as zero. Thus if $c = a$ the lower limit is equal to zero, while if $c = b$, the upper limit is taken to be zero. To include all possible cases, formula (9.17) may be recast in the following form

$$\int_{c-\varepsilon}^{c+\varepsilon} \varphi(t)\, e^{zh(t)}\, dt = e^{zh(c)} \sum_{k=0}^{\infty} \Pi_k(c) \int_{0}^{\delta} e^{-zu^2} u^k\, du, \tag{9.19}$$

where

$$\Pi_k(c) = B_k(c) \begin{cases} 1, & \text{for } c = a; \\ (-1)^k, & \text{for } c = b; \\ 1 + (-1)^k, & \text{for } c \neq a, b. \end{cases}$$

For an estimate of the contribution from an extremal point c for large $|z|$ the contour of integration $(c - \varepsilon, c + \varepsilon)$ in the neighborhood of this point is chosen so that, along it, the integrand decreases rapidly. For this it is sufficient to consider

$$\operatorname{Re}(z\delta^2) > 0, \qquad \delta = \varepsilon\sqrt{-h''(c)/2}, \tag{9.20}$$

which also insures convergence of the integral on the right hand side of (9.19) for $|z| \to \infty$.

In this case the right hand side of (9.19) can be expressed by incomplete gamma functions and the basic equation (8.23) becomes

$$\int_0^\delta e^{-zu^2} u^k\, du \approx \frac{1}{2} z^{-(k+1)/2}\, \Gamma\left(\frac{k+1}{2}\right). \tag{9.21}$$

Consequently formula (9.19) for large z can be written

$$\int_{c-\varepsilon}^{c+\varepsilon} \varphi(t)\, e^{zh(t)}\, dt \approx \frac{1}{2} e^{zh(c)} \sum_{k=0}^{\infty} \Pi_k(c)\, z^{-(k+1)/2}\, \Gamma\left(\frac{k+1}{2}\right). \tag{9.22}$$

We can now write an asymptotic representation for the given integral (9.4) which takes into account the contributions from the end points of the interval and the extremal points. If no stationary point coincides with either end point, we have the following asymptotic approximation:

$$\begin{aligned} \underset{c\neq a,b}{F(z)} = \int_a^b \varphi(t) e^{zh(t)}\, dt &\approx e^{zh(a)} \sum_{k=0}^{\infty} \frac{A_k(a)\, \Gamma(1+k)}{z^{k+1}} \\ &- e^{zh(b)} \sum_{k=0}^{\infty} \frac{A_k(b)\, \Gamma(k+1)}{z^{k+1}} + \frac{1}{\sqrt{z}} e^{zh(c)} \sum_{k=0}^{\infty} \frac{B_{2k}(c)\, \Gamma\left(k+\frac{1}{2}\right)}{z^k}. \end{aligned} \tag{9.23}$$

If an extremal point c coincides with the end point a of the interval of integration, the expression (9.23) becomes

$$\begin{aligned} \underset{c=a}{F(z)} \approx \frac{1}{2} e^{zh(a)} \sum_{k=0}^{\infty} \frac{B_{2k+1}(a)\, \Gamma(k+1)}{z^{k+1}} &- e^{zh(b)} \sum_{k=0}^{\infty} \frac{A_k(b)\, \Gamma(k+1)}{z^{k+1}} \\ &+ \frac{1}{2\sqrt{z}} e^{zh(a)} \sum_{k=0}^{\infty} \frac{B_{2k}(a)\, \Gamma\left(k+\frac{1}{2}\right)}{z^k}. \end{aligned} \tag{9.24}$$

Finally if an extremal point coincides with the other end points $c = b$, the asymptotic expression assumes the following form

$$\begin{aligned} F_{c=b}(z) \approx e^{zh(a)} \sum_{k=0}^{\infty} \frac{A_k(a)\,\Gamma(k+1)}{z^{k+1}} - \frac{1}{2} e^{zh(b)} \sum_{k=0}^{\infty} \frac{B_{2k+1}(b)\,\Gamma(k+1)}{z^{k+1}} \\ + \frac{1}{2\sqrt{z}} e^{zh(b)} \sum_{k=0}^{\infty} \frac{B_{2k}(b)\,\Gamma\left(k+\frac{1}{2}\right)}{z^k}. \end{aligned} \tag{9.25}$$

The coefficients $A_m(\alpha)$ and $B_m(\alpha)$ in these formulae are determined from the relations (9.14) and (9.16) respectively. These results were obtained with the assumption of one extremal point but can obviously be generalized to the case of several.

We will now give explicit expressions for the dominant terms in the expansions (9.23)—(9.25).

$$\begin{aligned} F_{c\neq a,b}(z) \approx \sqrt{2\pi}\varphi(c)\, e^{zh(c)} [-zh''(c)]^{-1/2} \\ + \frac{1}{z}\left[\frac{\varphi(b)}{h'(b)} e^{zh(b)} - \frac{\varphi(a)}{h'(a)} e^{zh(a)}\right]; \end{aligned} \tag{9.26}$$

$$\begin{aligned} F_{c=a}(z) \approx \frac{1}{2}\sqrt{2\pi}\varphi(a)\, e^{zh(a)} [-zh''(a)]^{-1/2} \\ + \frac{1}{z}\left\{\frac{1}{h''(a)}\left[\varphi(a)\frac{h'''(a)}{3h''(a)} - \varphi'(a)\right] e^{zh(a)} + \frac{\varphi(b)}{h'(b)} e^{zh(b)}\right\}; \end{aligned} \tag{9.27}$$

$$\begin{aligned} F_{c=b}(z) \approx \frac{1}{2}\sqrt{2\pi}\varphi(b)\, e^{zh(b)} [-zh''(b)]^{-1/2} \\ - \frac{1}{z}\left\{\frac{1}{h''(b)}\left[\varphi(b)\frac{h'''(b)}{3h''(b)} - \varphi'(b)\right] e^{zh(b)} + \frac{\varphi(a)}{h'(a)} e^{zh(a)}\right\}. \end{aligned} \tag{9.28}$$

In the term $[-zh''(c)]^{-1/2}$ above the principal value of the square root is understood.

10. Asymptotic Expansion for Incomplete Cylindrical Functions of Poisson Form

We shall now apply the results of the previous section to determine the asymptotic behavior of the incomplete cylindrical functions:

$$E_\nu^+(w, z) = \frac{2z^\nu}{A_\nu} \int_0^w e^{iz\cos t} \sin^{2\nu} t\, dt, \qquad \mathrm{Re}\left(\nu + \frac{1}{2}\right) > 0. \tag{10.1}$$

The integral under consideration is of the type just considered and therefore, according to the procedure outlined, its asymptotic evaluation amounts determining the contributions from neighboorhods of the limits of integration and of extremal points. The extremal points are given by the equation $h'(c) = -i \sin c = 0$, or $c_k = k\pi$. However, by restricting ourselves to the region $0 < \operatorname{Re} w < \pi$ in the complex w-plane, we have but one extremum at $c = 0$ — the lower limit. The asymptotic expansion in this case can then be obtained from formulae (9.24) and (9.27). However, because of the factor z^ν and the fact that the function $\varphi(t) = \sin^{2\nu} t$ is multiple-valued, the process of determining the coefficients for arbitrary ν is rather difficult. On the other hand, the determination is facilitated in the present case because of the special character of the functions which, for $w \to \infty \exp i\lambda$, according to (12.5), tend to Hankel function of the first kind. Since for large z the contribution to Eq. (10.1), from the extremal point $c = 0$ comes from a small neighborhood of this point, the integration contour can always be chosen so that it coincides with the contour for the Hankel functions.

In other words, the contribution to (10.1) from the extremal point is completely characterized by the behavior of the Hankel function for large $|z|$. Therefore, in finding asymptotic expansions of incomplete cylindrical functions we shall separate the portion converging to the Hankel function from the remaining expression. The asymptotic behavior of the latter are determined from the contribution at the upper limit in (10.1) which corresponds to the third term in formula (9.24). Substituting $u = iz(1 - \cos t)$, we can write (10.1) in the form:

$$E_\nu^+(w, z) = \sqrt{\frac{2}{\pi z}} \frac{1}{\Gamma\left(\nu + \frac{1}{2}\right)} e^{i(z - \nu\pi/2 - \pi/4)}$$

$$\times \left[\int_0^{\infty \exp i\lambda} e^{-u} u^{\nu - 1/2} \left(1 - \frac{u}{2iz}\right)^{\nu - 1/2} du \right. \tag{10.2}$$

$$\left. - \int_y^{\infty \exp i\lambda} e^{-u} u^{\nu - 1/2} \left(1 - \frac{u}{2iz}\right)^{\nu - 1/2} du \right],$$

where

$$y = iz(1 - \cos w) = 2iz \sin^2(w/2). \tag{10.3}$$

In the first term of (10.2) the integration is along a ray through the origin with slope λ, while the path of integration for the second term is taken along a parallel ray through the point y. The convergence of both of these integrals is guaranteed if the angle λ is chosen so that $|\lambda| < \pi/2$. The first term in this formula is precisely the Hankel function $\mathscr{H}_\nu^{(1)}(z)$, (see Eq. (3.31), Chapter I). The second can be written in a somewhat different form by a change of variable $u - y = t$. The formula (10.2), on account of (10.3), becomes

$$E_\nu^+(w, z) = \mathscr{H}_\nu^{(1)}(z) + \frac{2iz^{\nu - 1}}{A_\nu} \sin^{2\nu - 1} w e^{iz\cos w} \Psi_\nu(w, z), \tag{10.4}$$

where $\Psi_\nu(w, z)$ is defined by

$$\Psi_\nu(w, z) = \int_0^\infty e^{-t}\left(1 - 2i \cot w \frac{t}{z \sin w} + \frac{t^2}{z^2 \sin^2 w}\right)^{\nu-1/2} dt. \qquad (10.5)$$

In the above expression, the upper limit is actually equal to $-y+\infty \exp i\lambda$. However, since $|\lambda| < \pi/2$, the path of integration can be taken as the positive real axis provided the integrand in (10.5) has no other singularities in the first quadrant of the t-plane.

In this way an asymptotic expansion for the function $E_\nu^+(w, z)$ for large values of $|z|$ can be deduced from the asymptotic expansion of the function $\Psi_\nu(w, z)$, together with the following well known asymptotic expansion of the Hankel function:

$$\mathscr{H}_\nu^{(1)}(z) \approx \left(\frac{2}{\pi z}\right)^{1/2} e^{i(z-\nu\pi/2-\pi/4)} \sum_{m=0}^\infty \left(-\frac{1}{2iz}\right)^m \frac{\Gamma\left(\nu + m + \frac{1}{2}\right)}{\Gamma\left(\nu - m + \frac{1}{2}\right)}. \qquad (10.6)$$

We shall therefore consider the asymptotic expansion for the function $\Psi_\nu(w, z)$. We note first that the integrand without the exponential can be represented by the following power series in $t/z \sin w$

$$\left[1 - 2i \cot\left(\frac{t}{z \sin w}\right) + \left(\frac{t}{z \sin w}\right)^2\right]^{\nu-1/2}$$
$$= \frac{2^\nu \sqrt{\pi}}{\Gamma\left(\frac{1}{2} - \nu\right)} \sum_{m=0}^\infty \left(\frac{t}{z \sin w}\right)^m T_m^{-\nu}(i \cot w), \qquad (10.7)$$

where $T_m^{-\nu}(x)$ are the well known Gegenbauer polynomials (see [5, pp. 726—727] and [11, pp. 1043—1045]) which are related to the hypergeometric series

$$T_m^{-\nu}(x) = \frac{2^\nu \Gamma(1 + m - 2\nu)}{\Gamma(m + 1)\,\Gamma(1 - \nu)} F\left(-m, m + 1 - 2\nu; 1 - \nu; \frac{1 - x}{2}\right) \qquad (10.8)$$

This can be written in a more suitable form if we take advantage of Eq. (8.12) of this chapter for the hypergeometric series:

$$F\left(2\alpha, 2\beta; \beta + \alpha + \frac{1}{2}; \frac{1 - x}{2}\right) = \frac{\sqrt{\pi}\,\Gamma\left(\alpha + \beta + \frac{1}{2}\right)}{\Gamma\left(\alpha + \frac{1}{2}\right)\Gamma\left(\beta + \frac{1}{2}\right)} F\left(\alpha, \beta; \frac{1}{2}; x^2\right)$$
$$-2x \frac{\sqrt{\pi}\,\Gamma\left(\alpha + \beta + \frac{1}{2}\right)}{\Gamma(\alpha)\,\Gamma(\beta)} F\left(\alpha + \frac{1}{2}, \beta + \frac{1}{2}; \frac{3}{2}; x^2\right). \qquad (10.9)$$

For even $m = 2k$ (10.8) can be reduced to

$$T_{2k}^{-\nu}(x) = \frac{2^{2k-\nu}(-1)^k \Gamma\left(k + \frac{1}{2}\right)\Gamma\left(k + \frac{1}{2} - \nu\right)}{\pi \Gamma(2k + 1)} F\left(-k, k + \frac{1}{2} - \nu; \frac{1}{2}; x^2\right) \qquad (10.10)$$

while for odd $m = 2k + 1$, $k = 0, 1, 2, \ldots$ it can be written as

$$T_{2k+1}^{-\nu}(x) = \frac{x 2^{2k-\nu+2}(-1)^k \Gamma\left(k+\frac{3}{2}\right) \Gamma\left(k+\frac{3}{2}-\nu\right)}{\pi \Gamma(2k+2)} \times F\left(-k, k+\frac{3}{2}-\nu; \frac{3}{2}; x^2\right). \tag{10.11}$$

where use has been made of the following two relations for the gamma function:

$$\Gamma(2y) = \frac{2^{2y-1}}{\sqrt{\pi}} \Gamma(y)\, \Gamma\left(y+\frac{1}{2}\right); \tag{10.12}$$

$$\Gamma(1-y)\, \Gamma(y) = \frac{\pi}{\sin \pi y}. \tag{10.13}$$

We return now to the problem of finding an asymptotic expansion for $\Psi_\nu(w, z)$ for large z. If the expansion (10.7) is restricted to a finite sum of $2N - 1$ terms, and the sum is substituted into (10.5), then after changing the order of summation and integration, we find

$$\begin{aligned} \Psi_\nu(w, z) = {} & \frac{1}{\sqrt{\pi}} \sum_{k=0}^{N} \frac{(-1)^k 2^{2k}}{(z \sin w)^{2k}} \frac{\Gamma\left(k+\frac{1}{2}\right)\Gamma\left(k-\nu+\frac{1}{2}\right)}{\Gamma\left(\frac{1}{2}-\nu\right)} \\ & \times F\left(-k, k-\nu+\frac{1}{2}; \frac{1}{2}; -\cot^2 w\right) \\ & + \frac{2i \cot w}{\sqrt{\pi}} \sum_{k=0}^{N-1} \frac{(-1)^k 2^{2k+1}}{(z \sin w)^{2k+1}} \frac{\Gamma\left(k+\frac{3}{2}\right)\Gamma\left(k-\nu+\frac{3}{2}\right)}{\Gamma\left(\frac{1}{2}-\nu\right)} \\ & \times F\left(-k, k-\nu+\frac{3}{2}; \frac{3}{2}; -\cot^2 w\right) + O(|z \sin w|^{-2N-1}]. \end{aligned} \tag{10.14}$$

By combining the above expression with the asymptotic expansion (10.6) we obtain a formula which completely describes the characteristic behavior of $E_\nu^+(w, z)$ for large values of $|z \sin w|$. The principal terms of this expansion are of the form

$$E_\nu^+(w, z) \approx \sqrt{\frac{2}{\pi z}}\, \mathrm{e}^{i(z-\nu\pi/2-\pi/4)} + \frac{2iz^{\nu-1} \sin^{2\nu-1} w}{A_\nu} \mathrm{e}^{iz\cos w}. \tag{10.15}$$

Analogously, one can obtain a corresponding asymptotic expansion for the function

$$E_\nu^-(w, z) = \frac{2z^\nu}{A_\nu} \int_0^w \mathrm{e}^{-iz\cos\theta} \sin^{2\nu}\theta\, d\theta.$$

by formally replacing i everywhere by $-i$. For example, formula (10.4) becomes under this transformation

$$E_\nu^-(w, z) = \mathscr{H}_\nu^{(2)}(z) - \frac{2iz^{\nu-1}}{A_\nu} \sin^{2\nu-1} w \mathrm{e}^{-iz\cos w}\, \overline{\Psi_\nu(w, z)}, \tag{10.16}$$

where $\mathscr{H}_\nu^{(2)}(z)$ is the Hankel function of the second kind (see Eq. (3.32), Chapter I) and

$$\overline{\Psi_\nu(w, z)} = \int_0^\infty e^{-t}\left[1 + 2i \cot w \left(\frac{t}{z \sin w}\right) + \left(\frac{t}{z \sin w}\right)^2\right]^{\nu-1/2} dt, \quad (10.17)$$

is defined by Eq. (10.5) with i replaced by $-i$. In this way we find the principal terms in the asymptotic expansion of the function $E_\nu^-(w, z)$ for $|z \sin w| \gg 1$ to be

$$E_\nu^-(w, z) \approx \sqrt{\frac{2}{\pi z}}\, e^{-i(z-\nu\pi/2-\pi/4)} - \frac{2i}{A_\nu} z^{\nu-1} \sin^{2\nu-1} w e^{-iz\cos w}. \quad (10.18)$$

The asymptotic expansions (10.15) and (10.18) obtained above contain terms of two different types. The first terms of these formulae are identical to the dominant terms in the asymptotic expansions of the complete cylindrical functions, while the second term characterizes the incomplete nature of the functions. For certain values of w and z the latter terms can be shown to dominate the first ones, and actually define the asymptotic behavior of the incomplete functions. A similar situation occurs in the determination of the asymptotic behavior of other functions, for example the confluent hypergeometric function (Eq. (8.15), this Chapter). Because of this behavior, the formulae just obtained must be used with certain precautions. For example for real $w = \alpha$ the basic contribution to the asymptotic behavior of $E_\nu^+(\alpha, z)$ is given by the first term if $\operatorname{Im} z < 0$ and conversely by the second term if $\operatorname{Im} z > 0$. The reverse situation holds for $E_\nu^-(\alpha, z)$.

Knowledge of asymptotic expansions for $E_\nu^+(w, z)$ and $E_\nu^-(w, z)$ provides a means of obtaining the corresponding asymptotic expansions for the incomplete Bessel, Struve and Hankel functions and their modifications if use is made of the following relations among these functions (see § 1 of this Chapter):

$$\left.\begin{aligned}
J_\nu(w, z) &= \frac{E_\nu^+(w, z) + E_\nu^-(w, z)}{2};\\
H_\nu(w, z) &= \frac{E_\nu^+(w, z) - E_\nu^-(w, z)}{2i};\\
I_\nu(w, z) &= \frac{1}{2} e^{i\nu\pi/2} E_\nu^+(w, -iz) + \frac{1}{2} e^{-i\nu\pi/2} E_\nu^+(w, iz);\\
L_\nu(w, z) &= \frac{1}{2} e^{i\nu\pi/2} E_\nu^+(w, -iz) - \frac{1}{2} e^{-i\nu\pi/2} E_\nu^+(w, iz);\\
\mathscr{H}_\nu(w, z) &= E_\nu^+(-iw, z);\\
K_\nu(w, z) &= i\,\frac{\pi}{2} e^{i\nu\pi/2} E_\nu^+(-iw, iz).
\end{aligned}\right\} \quad (10.19)$$

Thus the first relation gives the following asymptotic estimate for the incomplete Bessel functions:

$$J_\nu(w, z) \approx J_\nu(z) - \frac{z^{\nu-1}\sin^{2\nu-1} w}{2^{\nu-1}\pi}\Bigg[\sin(z\cos w)\sum_{k=0}^{\infty}\left(\frac{2}{z\sin w}\right)^{2k}$$

$$\times \frac{\Gamma\left(k+\frac{1}{2}\right)}{\Gamma\left(\frac{1}{2}+\nu-k\right)} F\left(-k, k+\frac{1}{2}-\nu; \frac{1}{2}; -\cot^2 w\right) \tag{10.20}$$

$$-\frac{2\cos w\cos(z\cos w)}{z\sin^2 w}\sum_{k=0}^{\infty}\left(\frac{2}{z\sin w}\right)^{2k}$$

$$\times \frac{\Gamma\left(k+\frac{3}{2}\right)}{\Gamma\left(\nu-\frac{1}{2}-k\right)} F\left(-k, k+\frac{3}{2}-\nu; \frac{3}{2}; -\cot^2 w\right)\Bigg]$$

and for the incomplete Struve functions we have from the second

$$H_\nu(w, z) \approx N_\nu(z) + \frac{z^{\nu-1}\sin^{2\nu-1} w}{2^{\nu-1}\pi}$$

$$\times \Bigg[\cos(z\cos w)\sum_{k=0}^{\infty}\left(\frac{2}{z\sin w}\right)^{2k}\frac{\Gamma\left(k+\frac{1}{2}\right)}{\Gamma\left(\nu+\frac{1}{2}-k\right)}$$

$$+ F\left(-k, k+\frac{1}{2}-\nu; \frac{1}{2}; -\cot^2 w\right) \tag{10.21}$$

$$+\frac{2\cos w\sin(z\cos w)}{z\sin^2 w}\sum_{k=0}^{\infty}\left(\frac{2}{z\sin w}\right)^{2k}\frac{\Gamma\left(k+\frac{3}{2}\right)}{\Gamma\left(\nu-\frac{1}{2}-k\right)}$$

$$\times F\left(-k, k+\frac{3}{2}-\nu; \frac{3}{2}; -\cot^2 w\right)\Bigg].$$

where use has been made of the following relation for the gamma functions:

$$\Gamma\left(k-\nu+\frac{1}{2}\right)\Gamma\left(\nu-k+\frac{1}{2}\right) = (-1)^k\frac{\pi}{\cos\nu\pi}.$$

The expansions (10.20) and (10.21) are valid for $|z\sin w| > 1$. In addition, since $|z| > 1$, the functions $J_\nu(z)$ and $N_\nu(z)$ therein can be replaced by their appropriate asymptotic expansions. If one sets $w = \pi/2$ in (10.20), then, as would be expected, one obtains the known asymptotic expansion for the Bessel function, while (10.21) reduces to the asymptotic expansion for the Struve functions (see [10], p. 363)

$$H_\nu\left(\frac{\pi}{2}, z\right) \equiv H_\nu(z) \approx N_\nu(z) + \frac{1}{\pi}\sum_{k=0}^{\infty}\left(\frac{2}{z}\right)^{2k-\nu+1} \frac{\Gamma\left(k+\frac{1}{2}\right)}{\Gamma\left(\nu+\frac{1}{2}-k\right)}. \tag{10.22}$$

The principal terms in the expansions of the incomplete Bessel and Struve functions for real argument $w = \alpha$ and $z = x$ are given by

$$J_\nu(\alpha, x) \approx \sqrt{\frac{2}{\pi x}}\cos\left(x - \frac{\nu\pi}{2} - \frac{\pi}{4}\right) - \frac{x^{\nu-1}\sin^{2\nu-1}\alpha}{\sqrt{\pi}\,2^{\nu-1}\,\Gamma\left(\nu+\frac{1}{2}\right)}\sin(x\cos\alpha); \tag{10.23}$$

$$H_\nu(\alpha, x) \approx \sqrt{\frac{2}{\pi x}}\sin\left(x - \frac{\nu\pi}{2} - \frac{\pi}{4}\right) + \frac{x^{\nu-1}\sin^{2\nu-1}\alpha}{\sqrt{\pi}\,2^{\nu-1}\,\Gamma\left(\nu+\frac{1}{2}\right)}\cos(x\cos\alpha). \tag{10.24}$$

It follows from the above that for $x\sin\alpha \gg 2$ $(\alpha \neq \pi/2)$ the dominant asymptotic terms of these expansions are determined by the second terms for $\nu > 1/2$ and, conversely by the first terms when $\nu < 1/2$.

In an analogous manner, one can obtain asymptotic expansions for the remaining incomplete cylindrical function of Poisson form. We shall decuce only the principal terms. From the third and fourth relations (10.19) and from formulae (10.15) and (10.18) we find asymptotic expansions for the modified Bessel and Struve functions. For $|x\sin\alpha| \gg 1$ we obtain:

$$I_\nu(\alpha, x) \approx \frac{1}{\sqrt{2\pi x}}\left(e^x + e^{-x-i(\pi/2+\nu\pi)}\right) - \frac{2x^{\nu-1}\sin^{2\nu-1}\alpha}{\sqrt{\pi}\,2^\nu\,\Gamma\left(\nu+\frac{1}{2}\right)}\sinh(x\cos\alpha); \tag{10.25}$$

$$L_\nu(\alpha, x) \approx \frac{1}{\sqrt{2\pi x}}\left(e^x - e^{-x-i(\nu\pi+\pi/2)}\right) - \frac{2x^{\nu-1}\sin^{2\nu-1}\alpha}{\sqrt{\pi}\,2^\nu\,\Gamma\left(\nu+\frac{1}{2}\right)}\cosh(x\cos\alpha). \tag{10.26}$$

Thus for $|x|$ large, the asymptotic expansions for the functions $I_\nu(\alpha, x)$ and $L_\nu(\alpha, x)$ coincide almost exactly with the corresponding expansion for the modified Bessel and Struve functions. They are even closer if $\nu < 1/2$.

The principal terms of the asymptotic expansion for the incomplete Hankel functions, according to appropriate relation of (10.19), can be written

$$\mathscr{H}_\nu(\beta, z) \approx \sqrt{\frac{2}{\pi z}}\, e^{i(z-\nu\pi/2-\pi/4)} - \frac{2z^{\nu-1}\sinh^{2\nu-1}\beta}{\sqrt{\pi}\,2^\nu\,\Gamma\left(\nu+\frac{1}{2}\right)}\, e^{i(z\cosh\beta-\nu\pi)}. \tag{10.27}$$

For $\beta \to \infty$ and $\operatorname{Im} z > 0$ these formulae reduce to the corresponding asymptotic formulae for Hankel functions of the first kind when $\operatorname{Re} \nu > -1/2$. For the case $z = x$, the asymptotic form does not change. However, the principal contribution is from the second term if $\operatorname{Re} \nu > 1/2$. In order that these formulae reduce to the corresponding Hankel function for real $z = x$, when $\beta \to \infty$, it is necessary, in accordance with (3.22) of Chapter I, that ν be restricted to the interval $-1/2 < \operatorname{Re} \nu < 1/2$.

We shall now write down the principal terms of the asymptotic expansions for the incomplete MacDonald functions. From the appropriate relation of (10.19) we have

$$K_\nu(\beta, x) \approx \sqrt{\frac{\pi}{2x}}\, e^{-x} - \frac{\pi x^{\nu-1} \sinh^{2\nu-1} \beta}{\sqrt{\pi}\, 2^\nu \Gamma\left(\nu + \frac{1}{2}\right)} e^{-x\cosh\beta}. \tag{10.28}$$

and as is to be expected, this formula reduces for $\beta \to \infty$ to the asymptotic expression for the MacDonald function.

We shall now set forth a method which permits us to find the asymptotic expansion for the incomplete Hankel and MacDonald functions — not only for complex values of z, but also for arbitrary complex w providing their imaginary part is large i. e. $|\operatorname{Im} w| \gg 1$. These asymptotic developments characterize the difference between the incomplete cylindrical functions and the corresponding complete ones. In particular, for $w = -i\beta$ and $\beta \gg 1$ we find from formulae (10.4) and (10.16)

$$\begin{aligned} E_\nu^+(-i\beta, z) - \mathscr{H}_\nu^{(1)}(z) &\equiv \mathscr{H}_\nu(\beta, z) - \mathscr{H}_\nu^{(1)}(z) \\ &\approx -\frac{2z^{\nu-1} \sinh^{2\nu-1} \beta}{\sqrt{\pi}\, 2^\nu \Gamma\left(\nu + \frac{1}{2}\right)} e^{i(z\cosh\beta - \nu\pi)}; \end{aligned} \tag{10.29}$$

$$E_\nu^-(-i\beta, z) - \mathscr{H}_\nu^{(2)}(z) \approx \frac{2z^{\nu-1} \sinh^{2\nu-1} \beta}{\sqrt{\pi}\, 2^\nu \Gamma\left(\nu + \frac{1}{2}\right)} e^{-i(z\cosh\beta - \nu\pi)}. \tag{10.30}$$

The corresponding relations for the MacDonald functions for real $z = x$ can be written in the form

$$K_\nu(\beta, x) - K_\nu(x) \approx \frac{\pi x^{\nu-1} \sinh^{2\nu-1} \beta}{\sqrt{\pi}\, 2^\nu \Gamma\left(\nu + \frac{1}{2}\right)} e^{-x\cosh\beta}. \tag{10.31}$$

It can be seen from this formula that the difference, which characterizes the incompleteness of the functions under investigation, can vary over a wide range depending on the value of the variables z, β, ν. For example, the magnitude of the right sides of Eqs. (10.29) and (10.30) can be very large for real z if $2\nu - 1 > 0$. Conversely, this difference is small for $2\nu - 1 < 0$. Thus for direct application of the above asymptotic representations we must perform a supplementary analysis in each case. This also holds true for formula (10.31), although at first glance, it might appear that the right side is always small.

As noted above, all of the asymptotic representations obtained in this section were derived for the strip $0 \leq \operatorname{Re} w < \pi$ in the complex w-plane. Thus the $E_\nu^\pm(w, z)$ and $E_\nu^\pm(\overline{w}, z)$ are characterized for large values of $|z \sin w|$ for all values of w in this strip. Direct application of these expressions is not possible when investigating behavior outside this interval because in obtaining (10.2) the substitution $u = \cos\theta$ was used, and all the following investigations were carried out relative to $\cos w$. It is obvious that the formulae obtained cannot reflect the behaviour of $E^\pm(w, z)$ as the variable w changes by multiples of π. This could also be inferred from the general methods described in the previous section. If, for example, $m\pi < \operatorname{Re} w < (m+1)\pi$, it is necessary to take into account not only the contribution from the extremal point $(c = 0)$ but also that from the m extrema at $c = \pi, 2\pi, \ldots, m\pi$, defined by $\sin c = 0$, $0 \leq c < (m+1)\pi$. However, the contributions from the outer limits do not change.

Thus, in order to obtain the asymptotic expansions of the functions $E_\nu^\pm(w, z)$ outside of the strip $0 < \operatorname{Re} w < \pi$, we shall make use of the method of analytic continuation as described in Section 3 of this Chapter. Let $\operatorname{Re} w$ lie in the interval $k\pi < \operatorname{Re} w < (k+1)\pi$ $(k = 0;\ 1;\ 2;\ \ldots)$. Then according to Eq. (3.10) e. g. for $\operatorname{Im} w < 0$

$$E_\nu^+(w, z) = \frac{2 \sin k\nu\pi}{\sin \nu\pi} e^{i(k-1)\nu\pi} J_\nu(z) + e^{2ik\nu\pi} E_\nu^\pm(w - k\pi, z), \quad (10.32)$$

where the sign on the right side is taken positive when k is even and negative when k is odd. Analogously, from (3.11) we find for the same case $\operatorname{Im} w < 0$

$$E_\nu^-(w, z) = \frac{2 \sin k\nu\pi}{\sin \nu\pi} e^{i(k-i)\nu\pi} J_\nu(z) + e^{2ik\nu\pi} E_\nu^\pm(w - k\pi, z). \quad (10.33)$$

where, conversely, the positive sign corresponds to odd k and the negative one to even k. The right hand sides now provide formulae for $E_\nu^\pm(w - k\pi, z)$ in terms of the basic expressions (10.4) and (10.16), and taking into account the accepted convention of (3.7), we find, after some transformations that for $\operatorname{Im} w < 0$

$$\begin{aligned} E_\nu^+(w, z) &= \frac{2iz^{\nu-1} \sin^{2\nu-1} w}{A_\nu} e^{iz\cos w} \Psi_\nu(w, z) \\ &+ 2 \frac{\sin k\nu\pi}{\sin \nu\pi} e^{i(k-1)\nu\pi} J_\nu(z) \qquad (10.34) \\ &+ e^{2ik\nu\pi} \begin{Bmatrix} \mathscr{H}_\nu^{(1)}(z), & k - \text{even} \\ \mathscr{H}_\nu^{(2)}(z), & k - \text{odd} \end{Bmatrix} \end{aligned}$$

$$E_\nu^-(w, z) = -\frac{2iz^{\nu-1}\sin^{2\nu-1} w}{A_\nu} e^{iz\cos w}\,\overline{\Psi_\nu(w, z)}$$
$$+ 2\frac{\sin k\nu\pi}{\sin \nu\pi} e^{i(k-1)\nu\pi} J_\nu(z) \tag{10.35}$$
$$+ e^{2ik\nu\pi}\begin{Bmatrix} \mathscr{H}_\nu^{(2)}(z), & k - \text{even} \\ \mathscr{H}_\nu^{(1)}(z), & k - \text{odd} \end{Bmatrix}$$

where $k\pi < \text{Re}\, w < (k + 1)\pi$, $A_\nu = 2^\nu \Gamma(\nu + 1/2)\,\Gamma(1/2)$, and the functions $\Psi_\nu(w, z)$ and $\overline{\Psi_\nu(w, z)}$ are defined by Eqs. (10.5) and (10.17), the asymptotic expansions of which are given by (10.14). These formulae retain their meaning for integer values of ν. In this manner we obtain formulae which, together with (10.19) allow us to find asymptotic expansions for all of the incomplete cylindrical functions of Poisson form. These expansions are valid for all w in the right half-plane, including the imaginary axis. In establishing asymptotic expansions of the functions $E_\nu^\pm(w \exp i\pi, z)$ for w in the left half-plane, we must use caution. In fact, according to definition, the function

$$E_\nu^\pm(w \exp i\pi, z) = \frac{2z^\nu}{A_\nu}\int_0^{w\exp i\pi} e^{\pm iz\cos\theta} \sin^{2\nu}\theta\, d\theta \tag{10.36}$$

must differ from the functions $E_\nu^\pm(w, z)$, but since the transformation $\cos\theta = u$ used in obtaining the asymptotic expansions is unaffected by the sign of w, expressions identical to those found previously would be obtained in this case. Therefore, to avoid ambiguity it is necessary at the start to make the substitution $\theta = \psi \exp i\pi$ in Eq. (10.36). In other words

$$E_\nu^\pm(w \exp i\pi, z) = e^{(2\nu+1)i\pi}\frac{2z^\nu}{A_\nu}\int_0^{w} e^{\pm iz\cos\psi} \sin^{2\nu}\psi\, d\psi$$
$$= e^{(2\nu+1)i\pi} E_\nu^\pm(w, z). \tag{10.37}$$

This expression enables one to make use of the previous results for w in the right half-plane together with formulae (10.34) and (10.35) to define the analytic continuation of the functions $E_\nu^\pm(w, z)$ onto the left half-plane.

The asymptotic expansions discussed in this section were obtained under the assumption that $|\arg z| < \pi$. However, by considering the asymptotic representations for $E_\nu^+(w, z)$ and $E_\nu^-(w, z)$ separately, the range of z can be extended by writing $z = z_0 \exp ik\pi$ with $|\arg z_0| < \pi$, and making use of formulae (3.3), and (3.4).

Chapter III

Incomplete Cylindrical Functions of Bessel Form

1. Definitions of Incomplete Cylindrical Functions of Bessel Form

In the preceeding chapter we considered a class of functions expressed in the form of a Poisson integral with an arbitrary integration contour. In addition, the cylindrical functions may be also represented as Bessel-Schlaefli-Sonine integrals with fully determined integration contours. In the present chapter we shall consider another class of functions, defined by similar integrals, but with arbitrary contours of integration. Here, as before, these will be so constructed that for appropriately chosen contours they tend continuously to the well known cylindrical functions. In this connection we shall call them incomplete cylindrical functions of Bessel form and denote them by $\varepsilon_\nu(w, z)$.

In the first chapter fundamental integral representations were given for cylindrical functions in the form of Poisson integrals and Bessel-Schlaefli-Sonine integrals with definite integration contours. Demonstrating the equivalence of these representations is not trivial: it has been the subject of special investigations by various mathematicians and is of a profound character.

If the corresponding integrals are considered with arbitrary contours, then they represent functions of several variables. In the general case such functions, constructed from the Poisson integral, cannot be reduced to the analogous functions, derived from the Bessel-Sonine-Schlaefli integrals. It is therefore natural to consider the latter class of such functions in some detail, especially since they have wide application in a series of applied problems.

To construct the incomplete cylindrical functions of Bessel form, we could use any of the integral representations given in § 4 of Chapter I by varying the integration contour. For the sake of definitness we shall take the incomplete cylindrical function of the Bessel form, $\varepsilon_\nu(w, z)$, to be

$$\varepsilon_\nu(w, z) = \frac{1}{\pi i} \int_0^w e^{z\sinh t - \nu t}\, dt, \qquad (1.1)$$

where ν, w, and z are all complex quantities.

For integer $\nu = n$ and $w = 2\pi i$ this function reduces to the Bessel function, i.e.

$$\varepsilon_n(2\pi i, z) = 2J_n(z). \tag{1.2}$$

Definition (1.1) may also be written in other forms. For example, taking $e^t = u$ or $ze^t/2 = \xi$, we find, respectively,

$$\left.\begin{aligned} \varepsilon_\nu(w, z) &= \frac{1}{\pi i}\int_1^{\exp w} u^{-\nu-1} \exp\frac{z}{2}\left(u - \frac{1}{u}\right) du; \\ \varepsilon_\nu(w, z) &= \frac{\left(\frac{z}{2}\right)^\nu}{\pi i}\int_{z/2}^{z/2\exp w} \xi^{\nu-1} \exp\left(\xi - \frac{z^2}{4\xi}\right) d\xi. \end{aligned}\right\} \tag{1.3}$$

None of these forms for arbitrary ν reduces immediately to a cylindrical function even for $w = 2\pi i$ (c.f. formula (1.2)). Therefore, we will first of all formulate properly constructed incomplete cylindrical functions which do reduce to the corresponding cylindrical functions.

We consider the following identities

$$\frac{1}{2}\left[\varepsilon_\nu(w, z) - \varepsilon_\nu(\bar{w}, z)\right]$$

$$= \frac{1}{2\pi i}\int_0^{\bar{w}} e^{z\sinh t - \nu t}\, dt - \frac{1}{2\pi i}\int_0^{w} e^{z\sinh t - \nu t}\, dt$$

$$= \frac{1}{2\pi i}\int_{\bar{w}}^{w} e^{z\sinh t - \nu t}\, dt,$$

where $\bar{w}$ is the complex conjugate of w.

If $|\arg z| < \frac{\pi}{2}$ and w tends to $\infty + i\pi$ in the above then, by (4.7) of Chapter I we have the Bessel function $J_\nu(z)$. Therefore, the function

$$j_\nu(w, z) = \frac{1}{2}\left[\varepsilon_\nu(w, z) - \varepsilon_\nu(\bar{w}, z)\right] = \frac{1}{2\pi i}\int_{\bar{w}}^{w} e^{z\sinh t - \nu t}\, dt \tag{1.4}$$

will be called the incomplete Bessel function.

We consider further the function $\varepsilon_\nu(w - i/\pi 2, iz)$. By definition (1.1) we have

$$\varepsilon_\nu\left(w - i\frac{\pi}{2}, iz\right) = \frac{1}{\pi i}\int_0^{w - i\pi/2} e^{iz\sinh t - \nu t}\, dt.$$

Substituting $t + i\pi/2 = u$, we find

$$\varepsilon_\nu\left(w - i\frac{\pi}{2}, iz\right) = \frac{e^{i\nu\pi/2}}{\pi i}\int_{i\pi/2}^{w} e^{z\cosh u - \nu u}\, du. \tag{1.5}$$

We now construct the following identity:

$$\frac{1}{2}\,e^{-i\nu\pi/2}\left[\varepsilon_\nu\left(w - i\frac{\pi}{2}, iz\right) - \varepsilon_\nu\left(\bar{w} - i\frac{\pi}{2}, iz\right)\right] = \frac{1}{2\pi i}\int\limits_{\bar{w}}^{w} e^{z\cosh u - \nu u}\,du.$$

If in this identity $|\arg z| < \pi/2$ and w tends to $\infty + i\pi$, then, by (4.15) of Chapter I one obtains the modified Bessel function $I_\nu(z)$. Therefore, we agree to call

$$\begin{aligned} i_\nu(w, z) &= \frac{1}{2}\left[\varepsilon_\nu\left(w - i\frac{\pi}{2}, iz\right) - \varepsilon_\nu\left(\bar{w} - i\frac{\pi}{2}, iz\right)\right] e^{-i\nu\pi/2} \\ &= \frac{1}{2\pi i}\int\limits_{\bar{w}}^{w} e^{z\cosh\mu - \nu u}\,du \end{aligned} \tag{1.6}$$

the incomplete Bessel function of purely imaginary argument of the Bessel form.

Similarly the function

$$\begin{aligned} h_\nu(w, z) &= \varepsilon_\nu\left(w + i\frac{\pi}{2}, z\right) - \varepsilon_\nu\left(-w + i\frac{\pi}{2}, z\right) \\ &= \frac{1}{\pi i}\,e^{-i\nu\pi/2}\int\limits_{-w}^{w} e^{iz\cosh t - \nu t}\,dt, \end{aligned} \tag{1.7}$$

which for $|\arg z| < \pi/2$ and $w \to \infty + i\pi/2$ tends to the Hankel function, shall be called the incomplete Hankel function of the Bessel form. Finally, we shall call the function

$$\begin{aligned} k_\nu(w, z) &= i\frac{\pi}{2}\,e^{i\nu\pi/2}\left[\varepsilon_\nu\left(w + i\frac{\pi}{2}, iz\right) - \varepsilon_\nu\left(-w + i\frac{\pi}{2}, iz\right)\right] \\ &= \frac{1}{2}\int\limits_{-w}^{w} e^{-z\cosh t - \nu t}\,dt, \end{aligned} \tag{1.8}$$

which for $w \to \infty$ reduces to the MacDonald function $K_\nu(z)$, the incomplete MacDonald function of the Bessel form. For incomplete MacDonald functions and Hankel functions defined in this way the same relation holds as between MacDonald and Hankel functions, i.e.

$$k_\nu(w, z) = \frac{\pi i}{2}\exp\left(\frac{i\nu\pi}{2}\right)h_\nu(w, iz). \tag{1.9}$$

In practical applications, functions of this class often appear with purely imaginary values of the variable $w = i\beta$, i.e.

$$\varepsilon_\nu(i\beta, z) = \frac{1}{\pi i}\int\limits_{0}^{i\beta} e^{z\sinh t - \nu t}\,dt = \frac{1}{\pi}\int\limits_{0}^{\beta} e^{i(z\sin\theta - \nu\theta)}\,d\theta, \tag{1.10}$$

It is convenient to write this expression in the form

$$\varepsilon_\nu(i\beta, z) = A_\nu(\beta, z) - iB_\nu(\beta, z), \tag{1.11}$$

where we have put

$$A_\nu(\beta, z) = \frac{1}{\pi} \int_0^\beta \cos(z \sin\theta - \nu\theta)\, d\theta; \tag{1.12}$$

$$B_\nu(\beta, z) = \frac{1}{\pi} \int_0^\beta \sin(\nu\theta - z \sin\theta)\, d\theta. \tag{1.13}$$

For $\beta = \pi$, the right sides of these formulae represent well known expressions for the Anger and Weber functions (c.f. [10, p. 336]):

$$A_\nu(z) = \frac{1}{\pi} \int_0^\pi \cos(\nu\theta - z \sin\theta)\, d\theta; \tag{1.14}$$

$$B_\nu(z) = \frac{1}{\pi} \int_0^\pi \sin(\nu\theta - z \sin\theta)\, d\theta. \tag{1.15}$$

In other words, we have

$$A_\nu(\pi, z) \equiv A_\nu(z); \tag{1.16}$$

$$B_\nu(\pi, z) \equiv B_\nu(z). \tag{1.17}$$

The functions $A_\nu(\beta, z)$ and $B_\nu(\beta, z)$ will consequntly be called the incomplete Anger and Weber functions, respectively.

We have introduced the concepts of incomplete Bessel, Hankel, and MacDonald functions in the form of Bessel integrals, and also the incomplete Anger and Weber functions. All these functions reduce to the corresponding cylindrical functions for certain values of the variable w. They are therefore proper incomplete cylindrical functions of Bessel form, in accordance with our previous definitions. Since all of these incomplete functions can be simply expressed in terms of $\varepsilon_\nu(w, z)$, we shall study chiefly the properties of this function in the following sections. Here we remark that in order to obtain any relation (differential equation, recursion formula, etc.) for incomplete Anger and Weber functions from the corresponding relation for the incomplete cylindrical function $\varepsilon_\nu(w, z)$, it is only necessary to carry out the following formal operation on these relations: For the function $\varepsilon_\nu(w, z)$ substitute the expression $A_\nu(w, z) - iB_\nu(w, z)$, and subsequently separate real and imaginary parts, proceeding formally as if ν, w, and z were real quantities. The real part in the resulting expressions then represents the desired relation for the incomplete Anger function, and the imaginary part that for the incomplete Weber function.

2. Recursion Relations and Differential Equations for the Function $\varepsilon_\nu(w, z)$

We begin our study of the properties of the incomplete cylindrical function of Bessel form, i.e.

$$\varepsilon_\nu(w, z) = \frac{1}{\pi i} \int_0^w e^{z \sinh t - \nu t}\, dt, \tag{2.1}$$

by introducing the recursion relation which it satisfies. To this end we apply the operators

$$L_\nu \varepsilon_a(w, z) = \varepsilon_{\nu-1} + \varepsilon_{\nu+1} - 2\frac{\nu}{z}\varepsilon_\nu;$$

$$L_\nu \varepsilon_\nu(w, z) = \varepsilon_{\nu-1} - \varepsilon_{\nu+1} - 2\frac{\partial \varepsilon_\nu}{\partial z}.$$

The first of these gives

$$L_\nu \varepsilon_\nu = \frac{1}{\pi i}\left[\int_0^w e^{z\sinh t-(\nu-1)t}\,dt + \int_0^w e^{z\sinh t-(\nu+1)t}\,dt - \frac{2\nu}{z}\int_0^w e^{z\sinh t-\nu t}\,dt\right]$$

$$= \frac{2}{\pi i}\left(\int_0^w e^{z\sinh t-\nu t}\cosh t\,dt - \frac{\nu}{z}\int_0^w e^{z\sinh t-\nu t}\,dt\right).$$

After integrating the first term by parts:

$$\int_0^w e^{z\sinh t-\nu t}\cosh t\,dt = \frac{1}{z}e^{z\sinh t-\nu t}\Big|_0^w + \frac{\nu}{z}\int_0^w e^{z\sinh t-\nu t}\,dt$$

we find

$$\varepsilon_{\nu-1}(w, z) + \varepsilon_{\nu+1}(w, z) - \frac{2\nu}{z}\varepsilon_\nu(w, z) = \frac{2}{\pi z i}\left(e^{z\sinh w-\nu w} - 1\right). \quad (2.2)$$

Dealing similarly with the second operator, we have

$$\varepsilon_{\nu-1}(w, z) - \varepsilon_{\nu+1}(w, z) - 2\frac{\partial \varepsilon_\nu(w, z)}{\partial z} = 0. \quad (2.3)$$

The relation just obtained reduces for $w = i\pi$ to the recursion relation for Anger and Weber functions (c.f. [10, p. 340]), as one would expect. Actually, if we set $\varepsilon_\nu(i\pi, z) = A_\nu(z) - iB_\nu(z)$, separate real and imaginary parts, and take for the moment ν and z to be real quantities, we find:

$$A_{\nu-1}(z) + A_{\nu+1}(z) - 2\frac{\nu}{z}A_\nu(z) = -2\frac{\sin \pi\nu}{\pi z}; \quad (2.4)$$

$$A_{\nu-1}(z) - A_{\nu+1}(z) - 2\frac{\partial}{\partial z}A_\nu(z) = 0; \quad (2.5)$$

$$B_{\nu-1}(z) + B_{\nu+1}(z) - 2\frac{\nu}{z}B_\nu(z) = \frac{2(\cos \nu\pi - 1)}{\pi z}; \quad (2.6)$$

$$B_{\nu-1}(z) - B_{\nu+1}(z) - 2\frac{\partial}{\partial z}B_\nu(z) = 0. \quad (2.7)$$

We note, in distinction from the previously considered incomplete cylindrical function of the Poisson form, that the right side of the second recursion relation (2.3) for the function $\varepsilon_\nu(w, z)$ is identically zero. Functions of this type possess a series of additional properties some of which will be taken up below.

We turn now to the derivation of a differential equation which is satisfied by the function $\varepsilon_\nu(w, z)$. To this end we set up the Bessel operator

$$\nabla_\nu \varepsilon_\nu(w, z) = \left(z^2 \frac{\partial^2}{\partial z^2} + z \frac{\partial}{\partial z} + (z^2 - \nu^2)\right) \varepsilon_\nu(w, z).$$

Applying this to $\varepsilon_\nu(w, z)$ in expression (2.1), we have

$$\nabla \varepsilon_\nu(w, z) = \frac{1}{\pi i}\left[z^2 \int_0^w e^{z\sinh t - \nu t} (\sinh^2 t + 1)\, dt \right.$$
$$\left. + z \int_0^w e^{z\sinh t - \nu t} \sinh t\, dt - \nu^2 \int_0^w e^{z\sinh t - \nu t}\, dt \right].$$

Integrating the last term twice by parts, we find

$$\nu^2 \int_0^w e^{z\sinh t - \nu t}\, dt = -(\nu + z \cosh t)\, e^{z\sinh - \nu t}\Big|_0^w$$
$$+ z^2 \int_0^w e^{z\sinh t - \nu t} \cosh^2 t\, dt + z \int_0^w e^{z\sinh t - \nu t} \sinh t\, dt.$$

Consequently,

$$\begin{aligned} \nabla_\nu \varepsilon_\nu(w, z) &= z^2 \frac{\partial^2 \varepsilon_\nu}{\partial z^2} + z \frac{\partial \varepsilon_\nu}{\partial z} + (z^2 - \nu^2)\, \varepsilon_\nu \\ &= \frac{1}{\pi i}\left[(\nu + z \cosh w)\, e^{z\sinh w - \nu w} - (\nu + z)\right]. \end{aligned} \tag{2.8}$$

Thus the incomplete cylindrical function $\varepsilon_\nu(w, z)$ is the solution of an inhomogeneous Bessel differential equation and satisfies the boundary conditions

$$\begin{aligned} \varepsilon_\nu(w, 0) &= \frac{1}{\pi i \nu}(1 - e^{-\nu w}); \\ \frac{\partial}{\partial z} \varepsilon_\nu(w, z)\,\Big|_{z=0} &= \frac{1}{\pi i (1 - \nu^2}\left[-1 + e^{-\nu w}(\cosh w + \nu \sinh w)\right]. \end{aligned} \tag{2.9}$$

By the theory of differential equations, a solution of (2.8) can be written in the form

$$\begin{aligned} \varepsilon_\nu(w, z) &= \frac{\pi}{2} J_\nu(z) \int^z N_\nu(t) \frac{1}{\pi i}\left[(\nu + t \cosh w)\, e^{t\sinh w - \nu w} - (\nu + t)\right] dt \\ &- \frac{\pi}{2} N_\nu(z) \int^z J_\nu(t) \frac{1}{\pi i}\left[(\nu + t \sinh w)\, e^{t\sinh w - \nu w} - (\nu + t)\right] dt. \end{aligned} \tag{2.10}$$

It is not difficult to verify that this solution satisfies conditions (2.9). In fact, developing (2.10) as a series in z, and retaining only the dominant terms for small z, we have, for $\nu > 0$

$$\varepsilon_\nu(w, z) \approx \frac{1}{\pi} \frac{\left(\frac{z}{2}\right)^\nu}{\Gamma(\nu+1)} \cdot \frac{\pi}{2} \int^z \left(\frac{t}{2}\right)^{-\nu} \frac{\Gamma(\nu) \cdot \nu(e^{-\nu w} - 1)}{\pi i t} dt$$

$$- \frac{1}{\pi}\left(\frac{z}{2}\right)^{-\nu} \Gamma(\nu) \frac{\pi}{2} \int^z \left(\frac{t}{2}\right)^\nu \frac{\nu(e^{-\nu w} - 1)}{\pi i t \Gamma(\nu+1)} dt.$$

Letting $z \to 0$ we find, after some computation, the first of the conditions (2.9). This reasoning remains valid for $\nu = 0$. In order to verify the second condition of (2.9), we differentiate (2.10) with respect to z, and retain as before only the principal terms for small z; we have

$$\frac{d\varepsilon_\nu}{dt} = \frac{1}{2i}\left\{\frac{\nu z^{\nu-1}}{2^\nu \Gamma(\nu+1)} \int^z \frac{[(\nu + t\cosh w + t\sinh w)\, e^{-\nu w} - (\nu + t)]}{t} \frac{t^{-\nu}\, 2^\nu\, \Gamma(\nu)}{\pi} dt\right.$$

$$\left. + \frac{\nu\Gamma(\nu)}{\pi z^{\nu+1}} 2^\nu \int^z \frac{[(\nu + t\cosh w + t\sinh w)\, e^{-\nu w} - (\nu + t)]}{t} \frac{z^\nu}{2^\nu} \frac{dt}{\Gamma(\nu+1)}\right\} + 0(z).$$

Letting $z \to 0$, we find

$$\varepsilon_\nu'(w, 0) = \frac{1}{\pi i(1-\nu^2)} [e^{-\nu w}(\cosh w + \nu \sinh w) - 1].$$

A more detailed analysis of solution (2.10) will be given in the chapter devoted to incomplete cylindrical functions of Sonine-Schlaefli form.

3. Connection between Incomplete Cylindrical Functions of the Poisson and Bessel Forms

For arbitrary index ν, the functions

$$E_\nu^+(w, z) = 2\frac{z^\nu}{A^\nu} \int_0^w e^{iz\cos\theta} \sin^{2\nu}\theta\, d\theta \tag{3.1}$$

and

$$\varepsilon_\nu(w, z) = \frac{1}{\pi i} \int_0^w e^{z\sinh t - \nu t} dt \tag{3.2}$$

can be expressed in terms of each other only for special values of the variable w. This comes from the fact that for certain values of w these functions are expressible in terms of cylindrical functions, for which the Poisson and Bessel integral representations are equivalent.

$E_\nu^+(w, z)$ and $\varepsilon_\nu(w, z)$ belong in general to different classes of functions; only for integer indices $\nu = n$ does there exist between them a one-to-one relationship. To find this relation we introduce in to (3.1) the new variable of integration $\mu = \cos\theta$ and find

$$E_n^+(w, z) = \frac{2z^n}{A_n} \int_{\cos w}^1 e^{iz\mu} (1-\mu^2)^{n-1/2} d\mu.$$

Carrying out n integrations by parts, we find

$$E_n^+(w, z) = 2\frac{z^n}{A_n}\left[\sum_{k=0}^{n-1}\frac{(-1)^k e^{iz\mu}}{(iz)^{k+1}}\frac{d^k(1-\mu^2)^{n-1/2}}{d\mu^k}\Bigg|_{\cos w}^{1} + \frac{(-1)^n}{(iz)^n} J\right], \quad (3.3)$$

where we have introduced the notation

$$J = \int_{\cos w}^{1} e^{iz\mu}\frac{d^n}{d\mu^n}\sin^{2n-1}\theta\, d\mu.$$

With the aid of the well known Jacobi relation (c.f., e.g. [10, p. 36])

$$\frac{d^{n-1}}{d\mu^{n-1}}\sin^{2n-1}\theta = (-1)^{n-1}\frac{1\cdot 3\cdot 5\cdots(2n-1)}{n}\sin n\theta, \quad (3.4)$$

the integral J may be transformed into

$$J = \frac{(-1)^{n-1}\cdot 1\cdot 3\cdot 5\cdots(2n-1)}{n}\int_{\cos w}^{1} e^{iz\mu}\, d(\sin n\theta)$$

or, using the relation

$$1\cdot 3\cdot 5\cdots(2n-1) = \frac{1}{\pi}2^n\Gamma\left(n+\frac{1}{2}\right)\Gamma\left(\frac{1}{2}\right) = \frac{A_n}{\pi}$$

and re-introducing the parameter $\theta = \arccos\mu$, it is not difficult to obtain

$$\begin{aligned} J &= \frac{(-1)^n A_n}{\pi}\int_0^w e^{iz\cos\theta}\cos n\theta\, d\theta = \frac{(-1)^n A_n}{2\pi} \\ &\times\left(\int_0^w e^{iz\cos\theta+in\theta}\, d\theta + \int_0^w e^{iz\cos\theta-in\theta}\, d\theta\right). \end{aligned} \quad (3.5)$$

If now we introduce the new variable $\theta = \pi/2 + it$ into the first term, and $\theta = -(\pi/2 + it)$ in the second, we have

$$J = \frac{(-1)^n (i)^{n+1} A_n}{2\pi}\left(\int_{i\pi/2}^{i\pi/2-iw} e^{z\sinh t - nt}\, dt - \int_{i\pi/2}^{i\pi/2+iw} e^{z\sinh t - nt}\, dt\right)$$

or, by expression (3.2),

$$J = \frac{1}{2}(-1)^n (i)^{n+2} A_n\left[\varepsilon_n\left(i\frac{\pi}{2} - iw, z\right) - \varepsilon_n\left(\frac{i\pi}{2} + iw, z\right)\right]. \quad (3.6)$$

Returning to formula (3.3), we find the following relation between incomplete cylindrical functions of the Poisson and Bessel forms:

$$\begin{aligned} E_n^+(w, z) &= \left[\varepsilon^n\left(i\frac{\pi}{2} + iw, z\right) - \varepsilon_n\left(i\frac{\pi}{2} - iw, z\right)\right] \\ &- 2\frac{e^{iz\cos w}}{A_n}z^n\sum_{k=0}^{n-1}\frac{(-1)^k}{(iz)^{k+1}}\frac{d^k\sin^{2n-1}w}{(d\cos w)^k}. \end{aligned} \quad (3.7)$$

This formula remains valid for $n = 0$, with the summation taken as empty, i.e.

$$E_0^+(w, z) = \varepsilon_0\left(i\frac{\pi}{2} + iw, z\right) - \varepsilon_0\left(i\frac{\pi}{2} - iw, z\right). \tag{3.8}$$

For $n = 1$ (3.7) gives

$$E_1^+(w, z) = \varepsilon_1\left(i\frac{\pi}{2} + iw, z\right) - \varepsilon_1\left(i\frac{\pi}{2} - iw, z\right) + 2\frac{e^{iz\cos w}}{iz\pi}\sin w. \tag{3.9}$$

Conversely, we can derive an explicit expression from (3.7) for the functions $\varepsilon_n(w, z)$ in terms of the incomplete cylindrical functions of the Poisson form $E_n(w, z)$. However, these expressions for $\varepsilon_0(w, z)$ and $\varepsilon_1(w, z)$ are simple to construct directly from their definitions:

$$\left.\begin{aligned} \varepsilon_0(w, z) &= \frac{1}{\pi i}\int_0^w e^{z\sinh t}\,dt; \\ \varepsilon_1(w, z) &= \frac{1}{\pi i}\int_0^w e^{z\sinh t - t}\,dt. \end{aligned}\right\} \tag{3.10}$$

Thus, taking $t = i\pi/2 - i\theta$, we find

$$\varepsilon_0(w, z) = \frac{1}{\pi}\int_{\pi/2+iw}^{\pi/2} e^{iz\cos\theta}\,d\theta,$$

which, by definition (3.1) yields

$$\varepsilon_0(w, z) = \frac{1}{2}\left[E_0^+\left(\frac{\pi}{2}, z\right) - E_0^+\left(\frac{\pi}{2} + iw, z\right)\right]. \tag{3.11}$$

Similarly, carrying out the same transformation of the integration variable in expression (3.10) for $\varepsilon_1(w, z)$, we find

$$\varepsilon_1(w, z) = \frac{1}{\pi}\int_{\pi/2+iw}^{\pi/2} e^{iz\cos\theta}(\sin\theta - i\cos\theta)\,d\theta.$$

The first summand in the integrand is immediately integrable, and the second represents the derivative with respect to z of $E_0^+(w, z)$, i.e.

$$\varepsilon_1(w, z) = \frac{e^{z\sinh w} - 1}{\pi i z} - \frac{1}{2}\frac{\partial}{\partial z}\left[E_0^+\left(\frac{\pi}{2}, z\right) - E_0^+\left(\frac{\pi}{2} + iw, z\right)\right]. \tag{3.12}$$

Making use of (2.24) of Chapter II

$$\frac{\partial}{\partial z}E_0(w, z) = \frac{2i}{\pi}e^{iz\cos w}\sin w - E_1(w, z),$$

we obtain finally for $\varepsilon_1(w, z)$ the expression

$$\begin{aligned} \varepsilon_1(w, z) = {} & \frac{1}{\pi i}\left(1 - \frac{1}{z}\right) + \frac{1}{\pi i}\left(\frac{1}{z} - \cosh w\right)e^{z\sinh w} \\ & - \frac{1}{2}\left[E_1^+\left(\frac{\pi}{2} + iw, z\right) - E_1^+\left(\frac{\pi}{2}, z\right)\right]. \end{aligned} \tag{3.13}$$

Expressions for the functions $\varepsilon_n(w, z)$ for $n > 1$ may be obtained from the relevant recursion relations.

The formulae given above completely determine the connection between incomplete cylindrical functions of Poisson and Bessel forms for all integer values of the indices $\nu = n$. This allows one to carry out detailed analysis with tabulated values of only one of these of function classes.

4. Incomplete Cylindrical Functions of Bessel Form with Half-Odd Indices

We have seen above that incomplete cylindrical functions of the Poisson form with half-odd indices can be expressed in terms of elementary functions, (c.f., e.g. formula (2.18) of Chapter II). We show now that incomplete cylindrical functions of the Bessel form with half-odd index, $\varepsilon_{n+1/2}(w, z)$, can be expressed by means of known and tabulated functions. For this purpose, because of the recursion relation (2.2), we need only consider the functions $\varepsilon_{1/2}(w, z)$ and $\varepsilon_{-1/2}(w, z)$. For $\varepsilon_{1/2}(w, z)$ we have

$$\varepsilon_{1/2}(w, z) = \frac{1}{\pi i} \int_0^w e^{z \sinh t - 1/2 t} \, dt. \tag{4.1}$$

Introducing the new integration variable $t = 2i \times (u - \pi/4)$, we find after some transformations

$$\begin{aligned} \varepsilon_{1/2}(w, z) &= \frac{2}{\pi} e^{-iz + i\pi/4} \int_{\pi/4}^{\pi/4 - iw/2} e^{2iz \sin^2 u - iu} \, du \\ &= \frac{2}{\pi} e^{-iz + i\pi/4} \left[\int_{\pi/4}^{\pi/4 - iw/2} \cos u \exp(2iz \sin^2 u) \, du \right. \\ &\quad \left. - i e^{2iz} \int_{\pi/4}^{\pi/4 - iw/2} \sin u \exp(-2iz \cos^2 u) \, du \right]. \end{aligned}$$

Or, with the further changes of integration variable $\xi = \sqrt{2iz} \sin u$ in the first term and $\eta = \sqrt{-2iz} \cos u$ in the second, we obtain

$$\varepsilon_{1/2}(w, z) = \frac{2}{\pi} \frac{e^{-iz}}{\sqrt{2z}} \int_{x_0}^{x} e^{\xi^2} \, d\xi - \frac{2 e^{2iz}}{\pi \sqrt{2z}} \int_{y_0}^{y} e^{\eta^2} \, d\eta, \tag{4.2}$$

where for brevity we have introduced the notations

$$\left. \begin{aligned} x &= x_0 [\cosh(w/2) - i \sinh(w/2)], \quad & x_0 &= \sqrt{z} \exp(i\pi/4); \\ y &= y_0 [\cosh(w/2) + i \sinh(w/2)], \quad & y_0 &= \sqrt{z} \exp(-i\pi/4). \end{aligned} \right\} \tag{4.3}$$

To avoid ambiguity, we shall interpret $\sqrt{z}$ here and in the following by its principal value, i.e.

$$\sqrt{z} = \sqrt{|z|} \exp\left(\frac{i \arg z}{2}\right).$$

with $-\pi \leq \arg z \leq \pi$. The expression for $\varepsilon_{1/2}(w, z)$ is not difficult to represent now in terms of the probability integral, either in the form [16]

$$W(x) = e^{-x2} \int_0^x e^{t^2} dt, \tag{4.4}$$

or in the form [17]

$$w(x) = e^{-x^2}\left(1 + \frac{2i}{\sqrt{\pi}} \int_0^x e^{t^2} dt\right). \tag{4.4'}$$

For definiteness, we express $\varepsilon_{1/2}(w, z)$ by means of integral (4.4). After some transformations we obtain

$$\varepsilon_{1/2}(w, z) = \frac{1}{\pi}\sqrt{\frac{2}{z}}\{e^{z\sinh w}[W(x) - W(y)] + W(y_0) - W(x_0)\}. \tag{4.5}$$

In a similar way we find

$$\varepsilon_{-1/2}(w, z) = \frac{1}{\pi i}\int_0^w e^{z\sinh t + t/2} dt = \frac{1}{\pi i}\sqrt{\frac{2}{z}} \times \{e^{z\sinh w}[W(x) + W(y)] - W(x_0) - W(y_0)\}, \tag{4.6}$$

where x and y are defined by relations (4.3).

In view of the recursion relation, these results may be used to obtain expressions for all incomplete cylindrical functions of Bessel form with half-odd index in terms of the probability integral.

We analyze now the formulae just obtained in some limiting cases. First of all we note that for $w \to 0$ $(x \equiv x_0; y \equiv y_0)$,

$$\varepsilon_{\pm 1/2}(0, z) = 0, \text{ as we would expect.}$$

In order to obtain from these formulae the other well known results

$$\varepsilon_{1/2}(w, 0) = \frac{2}{\pi i}(1 - e^{-w/2}), \tag{4.7}$$

and

$$\varepsilon_{-1/2}(w, 0) = \frac{2}{\pi i}(-1 + e^{w/2}); \tag{4.8}$$

it is sufficient to make use of the limiting behavior of the probability integral for $t \sim \sqrt{z} \to 0$:

$$W(t) \approx t.$$

Considering, for example $\varepsilon_{1/2}(w, z)$ when $z \to 0$, we have taking account of (4.3)

$$\varepsilon_{1/2}(w, z)_{z\to 0} = \frac{1}{\pi}\sqrt{\frac{2}{z}}_{z\to 0}(x - y + y_0 - x_0) = \frac{2}{\pi i}(1 - e^{-w/2}).$$

Furthermore, taking $w = i\pi$ in (4.5) and (4.6) and using the odd function character of the probalitity integral, $W(-t) = -W(t)$, we find

$$\varepsilon_{1/2}(i\pi, z) = -i\varepsilon_{-1/2}(i\pi, z) = \frac{2}{\sqrt{\pi z}} W[\sqrt{z} \exp(-i\pi/4)]. \tag{4.9}$$

There is basic interest in the behavior of these functions for $w \to -\infty + i\lambda$ and $w \to \infty + i(\pi - \lambda)$, where the parameter λ is chosen so that for fixed values of the variable z, convergence of the relevant integrals is assured. From (4.1) we see that for this convergence to occur, it is sufficient that

$$|\arg z - \lambda| < \frac{\pi}{2}. \tag{4.10}$$

We return to (4.2) and write it in the form

$$\begin{aligned}\varepsilon_{1/2}(w, z) &= \frac{1}{\pi}\sqrt{\frac{2}{z}}\,[W(\sqrt{-iz}) - W(\sqrt{iz})] \\ &+ \frac{ie^{-iz}}{\pi}\sqrt{\frac{2}{z}}\int_0^a e^{-t^2}\,dt - \frac{ie^{iz}}{\pi}\sqrt{\frac{2}{z}}\int_0^b e^{-t^2}\,dt,\end{aligned} \tag{4.11}$$

where we have put

$$a = \sqrt{\frac{z}{2}}\left[\exp\left(-\frac{w}{2}\right) + \exp\frac{w - \pi i}{2}\right];$$

$$b = \sqrt{\frac{z}{2}}\left[-\exp\left(-\frac{w}{2}\right) + \exp\frac{w - \pi i}{2}\right].$$

If we now let w tend to $-\infty + i\lambda$, then because (4.10) $a \to \infty$, and $b \to -\infty$. Therefore, formula (4.11) yields

$$\varepsilon_{1/2}(-\infty + i\lambda, z) = \frac{1}{\pi}\sqrt{\frac{2}{z}}\,[W(\sqrt{-iz}) - W(\sqrt{iz})] + i\sqrt{\frac{2}{\pi z}}\cos z. \tag{4.12}$$

It $w \to +\infty + i(\pi - \lambda)$, then also because of (4.10), both a and b tend to $+\infty$, and from (4.11) we find

$$\varepsilon_{1/2}(\infty + i\pi - i\lambda, z) = \frac{1}{\pi}\sqrt{\frac{2}{z}}\,[W(\sqrt{-iz}) - W(\sqrt{iz})] + \sqrt{\frac{2}{\pi z}}\sin z. \tag{4.13}$$

Similarly we obtain limiting values for $\varepsilon_{-1/2}(w, z)$:

$$\varepsilon_{-1/2}(-\infty + i\lambda, z) = \frac{i}{\pi}\sqrt{\frac{2}{z}}\,[W(\sqrt{-iz}) - W(\sqrt{iz})] - i\sqrt{\frac{2}{\pi z}}\sin z; \tag{4.14}$$

$$\varepsilon_{-1/2}(\infty + i\pi - i\lambda, z) = \frac{i}{\pi}\sqrt{\frac{2}{z}}\,[W(\sqrt{-iz}) + W(\sqrt{iz})] + i\sqrt{\frac{2}{\pi z}}\cos z. \tag{4.15}$$

Subtracting (4.12) from (4.13) and (4.14) from (4.15), we obtain an expression for the Hankel function of first order

$$\begin{aligned} &\varepsilon_{\pm 1/2}(\infty + i\pi - i\lambda, z) - \varepsilon_{\pm 1/2}(-\infty + i\lambda, z) \\ &\quad = \frac{1}{\pi i} \int_{-\infty + i\lambda}^{\infty - i(\pi - \lambda)} e^{z \sinh t \pm t/2}\, dt = \mathscr{H}^{(1)}_{\pm 1/2}(z), \end{aligned} \tag{4.16}$$

where

$$\left.\begin{aligned} \mathscr{H}^{(1)}_{1/2}(z) &= \sqrt{\frac{2}{\pi z}} \frac{e^{iz}}{i}; \\ \mathscr{H}^{(1)}_{-1/2}(z) &= \sqrt{\frac{2}{\pi z}} e^{iz}. \end{aligned}\right\} \tag{4.17}$$

We now use the relations obtained here to illustrate the assertion, made in the preceeding section, that for non-integer index ν, the incomplete cylindrical functions of Bessel and Poisson forms represent different classes of functions.

As an example, the incomplete Hankel function of Bessel form $h_{1/2}(w, z)$ given by expression (1.7) may be represented in the form

$$h_{1/2}(w, z) = \varepsilon_{1/2}(w + i\pi/2, z) - \varepsilon_{1/2}(-w + i\pi/2, z). \tag{4.18}$$

In order to use formula (4.5), we first of all note that by (4.3), the values of x and y are given for this case by

$$\left.\begin{aligned} x(\pm w + i\pi/2) &= \sqrt{2}\, x_0 \cosh\frac{w}{2}, & x_0 &= \sqrt{z} \exp\frac{i\pi}{4}; \\ y(\pm w + i\pi/2) &= \pm i \sqrt{2}\, y_0 \sinh\frac{w}{2}, & y_0 &= \sqrt{z} \exp\left(-\frac{i\pi}{4}\right). \end{aligned}\right\} \tag{4.19}$$

Therefore from (4.18) and (4.5) we find the following expression for the incomplete Hankel function of the Bessel form:

$$h_{1/2}(w, z) = -\frac{2}{\pi} \sqrt{\frac{2}{z}}\, e^{iz \cos w}\, W\left(e^{i\pi/4} \sqrt{2z} \sinh\frac{w}{2}\right). \tag{4.20}$$

For this result, the odd character of $W(t)$ was used. Using the same substitutions one can express $h_{-1/2}(w, z)$ as well as the other incomplete cylindrical functions of the Bessel form, in terms of the probability integral.

From (4.20) and (4.16) we see immediately that, because of (4.10):

$$\lim_{w \to \infty - i\lambda + i\pi/2} h_{1/2}(w, z) = \mathscr{H}^{(1)}_{1/2}(z) = \sqrt{\frac{2}{\pi z}} \frac{e^{iz}}{i}. \tag{4.21}$$

On the other hand, from the fundamental formulae (1.25) and (2.18) of Chapter II, the incomplete Hankel function of the Poisson form can be written

$$\mathscr{H}_{1/2}(w, z) = E^{+}_{1/2}(-iw, z) = i \sqrt{\frac{2}{\pi z}} (e^{iz \cos w} - e^{iz}). \tag{4.22}$$

From this it follows, that for $|\arg z - \lambda| < \pi/2$,

$$\lim_{w \to \infty - i\lambda + i\pi/2} \mathscr{H}_{1/2}(w, z) = \sqrt{\frac{2}{\pi z}} \frac{e^{iz}}{i} = \mathscr{H}^{(1)}_{1/2}(z). \tag{4.23}$$

Consequently, for all z satisfying the condition $|\arg z - \lambda| < \pi/2$, the incomplete Hankel function in either the Poisson or Bessel form degenerates into the Hankel function $\mathscr{H}^{(1)}_{1/2}(z)$ as $w \to \infty - i\lambda + i\pi/2$. For all other fixed values of w these represent entirely different functions of z. The functions $\mathscr{H}_{1/2}(w, z)$ are expressible in terms of elementary functions. The functions $h_{1/2}(w, z)$, an the other hand, can be represented only in terms of the probability integral. It is not difficult to extend this result to other incomplete cylindrical functions of half-odd index. Thus, we see that the incomplete cylindrical functions of the Poisson and Bessel forms represent different classes of functions.

We close this section with the remark that the probability integral, which has been extensively studied and tabulated is related to the incomplete cylindrical functions of Bessel form.

5. A Generating Function, Addition Formula, and Series for the Function $\varepsilon_\nu(w, z)$

The generating function for an arbitrary set of functions $\Psi_m(z)$ is defined by

$$\Pi_1(t, z) = \sum_{m=-\infty}^{\infty} t^m \Psi_m(z). \tag{5.1}$$

If this series exist only for positive m then the generating function is determined by

$$\Pi(t, z) = \sum_{m=0}^{\infty} t^m \Psi_m(z). \tag{5.2}$$

We will find the generating function for the incomplete cylindrical functions $\varepsilon_m(w, z)$. We substitute for $\Psi_m(t)$ in (5.2) the function

$$\varepsilon_m(w, z) = \frac{1}{\pi i} \int_0^w e^{z \sinh\theta - m\theta}\, d\theta$$

and obtain

$$\Pi(t, w, z) = \frac{1}{\pi i} \sum_{m=0}^{\infty} t^m \int_0^w e^{z \sinh\theta - m\theta}\, d\theta. \tag{5.3}$$

Assuming for the present that $\operatorname{Re} w \geq 0$, it is permissible to interchange the order of summation and integration:

$$\Pi(t, w, z) = \frac{1}{\pi i} \int_0^w e^{z \sinh\theta} \left(\sum_{m=0}^{\infty} t^m e^{-m\theta}\, d\theta \right).$$

Introducing the notation $e^\theta = \xi$ and summing the series, formally regarding $|t| \leq 1$, we find

$$\Pi(t, w, z) = \frac{1}{\pi i} \int_1^{\exp w} e^{\frac{z}{2}\left(\xi - \frac{1}{\xi}\right)} \frac{d\xi}{\xi - t}. \tag{5.4}$$

For $w = 2\pi i$, $\varepsilon_n(w, z)$ can be immediately expressed in terms of the Bessel function, $\varepsilon_n(2\pi i, z) = 2J_n(z)$. The Bessel function, as is well known, has the generating function

$$\exp \frac{z}{2}(t - t^{-1}) = \sum_{n=-\infty}^{\infty} t^n J_n(z). \tag{5.5}$$

This same expression can be obtained formally from (5.4) if we set $w = 2\pi i$ and take $|t| < 1$. For this value of w, the integration in (5.4) can be regarded as integration around the unit circle, whence, by the residue theorem

$$\Pi(t, 2\pi i, z) = 2e^{z/2(t-t^{-1})}. \tag{5.6}$$

We shall not perform a detailed analysis of the generating function (5.4); we remark however, that in practice it may be convenient to use this formula in the converse sense, i.e. for deriving series representations in powers of t for integrals of the form

$$\int_1^{\exp w} e^{\frac{z}{2}\left(\xi - \frac{1}{\xi}\right)} \frac{d\xi}{\xi - t}.$$

In this sense the above restriction $\operatorname{Re} w \geq 0$ may be unessential.

We turn now to the introduction of addition formulae for the functions $\varepsilon_\nu(w, z)$. In § 2 of this Chapter, it was shown that the $\varepsilon_1(w, z)$ satisfy the recurrence

$$\varepsilon_{\nu-1}(w, z) - \varepsilon_{\nu+1}(w, z) - 2\frac{\partial}{\partial z}\varepsilon_\nu(w, z) = 0.$$

By a general theorem (c.f. [10, p. 386]), all functions satisfying such a recurrence obey the addition formula

$$\varepsilon_\nu(w, z + t) = \sum_{n=-\infty}^{\infty} J_n(t)\, \varepsilon_{\nu-n}(w, z). \tag{5.7}$$

We shall prove this same relation by methods somewhat different from [10], which we shall subsequently use for obtaining other addition formulae. We construct

$$\sum_{n=-\infty}^{\infty} J_n(t)\varepsilon_{\nu-n}(w, z) = \frac{1}{\pi i} \sum_{n=-\infty}^{\infty} J_n(t) \int_0^w e^{z \sinh u - (\nu - n)u}\, du,$$

and, interchanging in this the order of summation and integration, we have

$$\sum_{n=-\infty}^{\infty} J_n(t)\, \varepsilon_{\nu-n}(w, z) = \frac{1}{\pi i} \int_0^w e^{z\sinh u - \nu u} \left(\sum_{n=-\infty}^{\infty} J_n(t)\, e^{nu} \right) du.$$

Carrying out here the summation under the integral sign by means of (5.5), we obtain the addition formula (5.7):

$$\sum_{n=-\infty}^{\infty} J_n(t)\, \varepsilon_{\nu-n}(w, z) = \frac{1}{\pi i} \int_0^w e^{(z+t)\sinh u - \nu u}\, du \equiv \varepsilon_\nu(w, z + t).$$

The relation (5.7) for Bessel functions is usually called the Neumann addition formula. We shall therefore call (5.7) the Neumann addition formula for incomplete cylindrical functions.

By analogy with the generalized Neumann addition formula for Bessel functions (c.f. [10, p. 392])

$$\left.\begin{aligned} J_\nu(\omega) \left\{ \frac{z - \zeta e^{-i\varphi}}{z - \zeta e^{i\varphi}} \right\}^{\nu/2} &= \sum_{m=-\infty}^{\infty} J_{\nu+m}(z)\, J_m(\zeta)\, e^{im\varphi}; \\ \omega &= \sqrt{z^2 + \zeta^2 - 2\zeta z \cos\varphi}. \end{aligned}\right\} \tag{5.8}$$

we introduce now a more general addition formula for incomplete cylindrical functions of the Bessel form. For this purpose we consider an expression of the form

$$\sum_{m=-\infty}^{\infty} \varepsilon_{\nu+m}(w, z)\, J_m(\zeta)\, e^{im\varphi} = \frac{1}{\pi i} \sum_{m=-\infty}^{\infty} J_m(\zeta)\, e^{im\varphi} \int_0^w e^{z\sinh u - (\nu+m)u}\, du.$$

Once more changing the order of summation and integration and using (5.5), we find

$$\begin{aligned} \sum_{m-=\infty}^{\infty} \varepsilon_{\nu+m}(w, z)\, J_m(\zeta)\, e^{im\varphi} &= \frac{1}{\pi i} \int_0^w e^{zs\ inhu - \nu u + \zeta\sinh(i\varphi - u)}\, du \\ &= \frac{1}{\pi i} \int_0^w e^{(z - \zeta\cos\varphi)\sinh u + i\sin\varphi\cosh u - \nu u}\, du. \end{aligned}$$

We introduce now the auxilliary quantity Ψ defined by

$$\exp(2i\Psi) = \frac{z - \zeta e^{-i\varphi}}{z - \zeta e^{i\varphi}}, \tag{5.9}$$

so that

$$\left.\begin{aligned} z - \zeta\cos\varphi &= \omega\cos\Psi; \\ \zeta\sin\varphi &= \omega\sin\Psi. \end{aligned}\right\} \tag{5.10}$$

Then the substitution $u + i\Psi = t$ for the variable of integration gives

$$\sum_{m=-\infty}^{\infty} \varepsilon_{\nu+m}(w, z)\, J_m(\zeta)\, e^{im\varphi} = \frac{1}{\pi i} e^{i\nu\Psi} \int_{i\Psi}^{w+i\Psi} e^{\omega\sinh t - \nu t}\, dt.$$

The right side of this equality represents, by definition, the difference of two incomplete cylindrical functions. Thus, we have the analogue of the generalized addition theorem (5.8) for incomplete cylindrical functions in the form

$$\left(\frac{z-\zeta e^{-i\varphi}}{z-\zeta e^{i\varphi}}\right)^{\nu/2}[\varepsilon_\nu(w+i\Psi,\omega)-\varepsilon_\nu(i\Psi,\omega)] \\ =\sum_{m=-\infty}^{\infty}\varepsilon_{\nu+m}(w,z)J_m(\zeta)\,e^{im\varphi}, \tag{5.11}$$

where as before $\omega=\sqrt{z^2+\zeta^2-2z\zeta\cos\varphi}$, and Ψ is defined by (5.9). Evidently the addition formula obtained earlier, (5.7), is a particular case of (5.11) for $\varphi=\pi$, since for this case $\omega=z+\zeta$ and $\Psi=0$. On the other hand, this formula for integer $\nu=n$ and $w=2\pi i$ reduces to (5.8), since $\varepsilon_n(2\pi i,z)\equiv 2J_n(z)$.

We will now consider series expansions analogous to the Lommel expansions for Bessel functions, (c.f. [10, p. 154]), which may be written as

$$\frac{J_\nu(z\sqrt{1+k})}{(1+k)^{\nu/2}}=\sum_{m=0}^{\infty}\frac{(-1)^m}{m!}\left(\frac{1}{2}kz\right)^m J_{\nu+m}(z). \tag{5.12}$$

To show this, we consider the right side of this expression. Replacing the Bessel function $J_{\nu+m}(z)$ with the incomplete cylindrical function $\varepsilon_{\nu+m}(w,z)$, we have

$$\sum_{m=0}^{\infty}\frac{(-1)^m\left(\frac{kz}{2}\right)^m}{m!}\varepsilon_{\nu+m}(w,z)=\frac{1}{\pi i}\sum_{m=0}^{\infty}\frac{(-1)^m\left(\frac{kz}{2}\right)^m}{m!}\int_0^w e^{z\sinh u-(\nu+m)u}\,du$$

$$=\frac{1}{\pi i}\int_0^w e^{z\sinh u-\nu u}\left(\sum_{m=0}^{\infty}\frac{(-1)^m\left(\frac{kz}{2}\right)^m e^{-mu}}{m!}\right)du.$$

Here the interchange of the order of integration and summation is clearly legitimate. Carrying out the summation under the integral sign, we obtain

$$\sum_{m=0}^{\infty}\frac{(-1)^m\left(\frac{kz}{2}\right)^m}{m!}\varepsilon_{\nu+m}(w,z)=\frac{1}{\pi i}\int_0^w e^{z\sinh u-\nu u-1/2\,kz(\cosh u-\sinh u)}\,du.$$

The right side of this expression may be simplified if we introduce an auxilliary quantity ψ, defined by the relation

$$\exp(2\psi)=1+k, \tag{5.13}$$

so that

$$\left.\begin{aligned} 1+\frac{1}{2}k &= \sqrt{1+k}\cosh\psi; \\ \frac{1}{2}k &= \sqrt{1+k}\sinh\psi. \end{aligned}\right\} \tag{5.14}$$

Then after some simplifications we have for the right side

$$\frac{1}{\pi i}\int_0^w e^{z\sqrt{1+k}\sinh(u-\psi)-\nu u}\,du.$$

This latter expression, after the change of integration variable $u-\psi=t$ can be represented as a difference between two incomplete cylindrical functions. We have finally

$$\begin{aligned} &\varepsilon_\nu\left(w-\psi, z\sqrt{1+k}\right)-\varepsilon_\nu\left(-\psi, z\sqrt{1+k}\right) \\ &= (1+k)^{\nu/2}\sum_{m=0}^{\infty}\frac{(-1)^m\left(\frac{kz}{2}\right)^m}{m!}\varepsilon_{\nu+m}(w, z). \end{aligned} \tag{5.15}$$

The formula just obtained constitutes an addition theorem for incomplete cylindrical functions, analogous to the Lommel addition theorem for Bessel functions. For integer $\nu=n$ and $w=2\pi i$ it immediately reduces to formula (5.12), if use is made of the relations

$$\left.\begin{aligned} \varepsilon_n\left(2\pi i-\psi, z\sqrt{1+k}\right)-\varepsilon_n\left(-\psi, z\sqrt{1+k}\right) &= 2J_n\left(z\sqrt{1+k}\right); \\ \varepsilon_n(2\pi i, z) &= 2J_n(z). \end{aligned}\right\} \tag{5.16}$$

These results may be used for finding similar addition theorems for incomplete cylindrical functions of the Poisson form with integer indices, since for integer indices there is a simple one-to-one connection between these sets of functions (c.f. § 3 of this Chapter).

Thus, the incomplete cylindrical functions posess a series of relations which are analogous to the relations for cylindrical functions. These results emphasize the intimate relationship between these functions.

We turn now to the construction of series for $\varepsilon_\nu(w, z)$. For integer indices $\nu=n$, such series may be obtained from the relation between these functions and the functions $E_n^\pm(w, z)$, expansions for which have already been obtained. For half-odd indices $\nu=n+1/2$ representations may be obtained from the relation between $\varepsilon_{n+1/2}(w, z)$ and the probability integral. Some of these expansions, however, can be obtained directly from the definition

$$\varepsilon_\nu(w, z)=\frac{1}{\pi i}\int_0^w e^{z\sinh u-\nu u}\,du. \tag{5.17}$$

Expanding exp $(z \sinh u)$ in powers of $z \sinh u$, we find

$$\varepsilon_\nu(w, z) = \sum_{m=0}^{\infty} \frac{a^m}{m!} z_m, \tag{5.18}$$

where the coefficients a_m are determined by

$$\begin{aligned} a_m &= \frac{1}{\pi i} \int_0^w \sinh^m u e^{-\nu u} \, du \\ &= \frac{2^{-m}}{\pi i} \sum_{k=1}^{m} \frac{\Gamma(m+1)}{\Gamma(k+1)\,\Gamma(m-k+1)} \, \frac{e^{(m-\nu-2k)w} - 1}{(m-\nu-2k)}. \end{aligned} \tag{5.19}$$

The first two coefficients have the form

$$\begin{aligned} a_0 &= \frac{1}{\pi i \nu} (1 - e^{-\nu w}); \\ a_1 &= \frac{1}{\pi i (1-\nu^2)} [e^{-\nu w}(\cosh w + \nu \sinh w) - 1]. \end{aligned} \tag{5.20}$$

The functions $\varepsilon_\nu(w, z)$, can also be expanded in series of Bessel functions $J_m(z)$. For this purpose one may use the relation

$$\begin{aligned} \exp(z \sinh u) = J_0(z) &+ 2 \sum_{k=1}^{\infty} J_{2k}(z) \cosh(2ku) \\ &+ 2 \sum_{k=0}^{\infty} J_{2k+1}(z) \sinh(2k+1) u. \end{aligned} \tag{5.21}$$

Substituting it in formula (5.17) and changing the order of integration and summation, we find after some transformations

$$\begin{aligned} \varepsilon_\nu(w, z) &= \frac{1}{\pi i \nu} (1 - e^{-\nu w}) J_0(z) \\ &+ \frac{2}{\pi i} \sum_{k=1}^{\infty} \frac{e^{-\nu w}[2k \sinh 2kw + \nu \cosh 2kw] - \nu}{4k^2 - \nu^2} J_{2k}(z) \\ &+ \frac{2}{\pi i} \sum_{k=0}^{\infty} \frac{e^{-\nu w}[(2k+1) \cosh(2k+1) w + \nu \sinh(2k+1) w] - (2k+1)}{(2k+1)^2 - \nu^2} J_{2k+1}(z). \end{aligned} \tag{5.22}$$

Finally, a series in powers of w has the form

$$\varepsilon_\nu(w, z) = \sum_{n=1}^{\infty} \left\{ \left(\frac{d^{n-1}}{dw^{n-1}} \right) [e^{z \sinh w - \nu w}]_{w=0} \right\} \frac{w^n}{n!}. \tag{5.23}$$

Addition formulae and series representations in terms of $\varepsilon_\nu(w, z)$ for other incomplete cylindrical functions of Bessel form are not difficult to obtain from their definitions, (c.f. § 1 of the present Chapter).

6. Asymptotic Expansions for Incomplete Cylindrical Functions of the Bessel Form

As is clear from their definition, the incomplete cylindrical functions of Bessel form

$$\varepsilon_\nu(w, z) = \frac{1}{\pi i} \int_0^w e^{z \sinh t - \nu t}\, dt \tag{6.1}$$

belong to the class of functions $F(z)$ considered in § 9 of Chapter II, i.e.

$$F(z) = \int_a^b e^{z h(t)} \varphi(t)\, dt \tag{6.2}$$

where here $a = 0$, $b = w$, $h(t) = \sinh t$ and $\varphi(t) = e^{-\nu t}$. Therefore, the procedure of finding asymptotic expansions for them leads to determination of contributions from the vicinity of the ends of the interval of integration, and from extrema determined by

$$h'(t) = \cosh t = 0. \tag{6.3}$$

The extrema are obviously distributed along the imaginary axis at the points $\pm i\pi/2$, $\pm 3/2 i\pi$, etc. If we limit ourselves to the strip $-3/2\pi < \operatorname{Im} w < 3/2\pi$, it is necessary to consider only two extremal points,

$$\left.\begin{aligned} c_1 &= i\frac{\pi}{2}; \\ c_2 &= -i\frac{\pi}{2}. \end{aligned}\right\} \tag{6.4}$$

For convenience in the following investigation we introduce the following auxilliary functions:

$$\Phi_\nu(b, z) = \frac{1}{\pi i} \int_{-\infty + i\lambda}^{b} e^{z \sinh t - \nu t}\, dt, \tag{6.5}$$

where, to guarantee convergence, it is sufficient that

$$|\arg z - \lambda| < \frac{\pi}{2}. \tag{6.6}$$

Formula (6.1) then takes the form

$$\varepsilon_\nu(w, z) = \Phi_\nu(w, z) - \Phi_\nu(0, z). \tag{6.7}$$

The auxilliary functions $\Phi_\nu(b, z)$ are quite convenient for finding asymptotic expansions, because for $w \to \infty + i(\pi - \lambda)$ or $w \to \infty - i(\pi + \lambda)$ they reduce respectively to the Hankel functions of the first or second kind, (c.f. § 4 of Chapter I) i.e.

$$\lim_{b \to \infty + i(\pi - \lambda)} \Phi_\nu(b, z) = \frac{1}{\pi i} \int_{-\infty + i\lambda}^{\infty + i(\pi - \lambda)} e^{z \sinh t - \nu t}\, dt = \mathscr{H}_\nu^{(1)}(z);$$

$$\lim_{b \to \infty - i(\pi + \lambda)} \Phi_\nu(b, z) = \frac{1}{\pi i} \int_{-\infty + i\lambda}^{\infty - i(\pi + \lambda)} e^{z \sinh t - \nu t}\, dt = -\mathscr{H}_\nu^{(2)}(z),$$

where the contour of integration is shown in Fig. 7. Moreover, these functions obey the following interesting and easily proved relations:

$$\left.\begin{aligned} \Phi_\nu(b, z) &= \mathscr{H}_\nu^{(1)}(z) - e^{-i\nu\pi}\,\Phi_{-\nu}(i\pi - b, z); \\ \Phi_\nu(b, z) &= -\mathscr{H}_\nu^{(2)} - (z) - e^{i\nu\pi}\Phi_{-\nu}(-i\pi - b, z). \end{aligned}\right\} \quad (6.8)$$

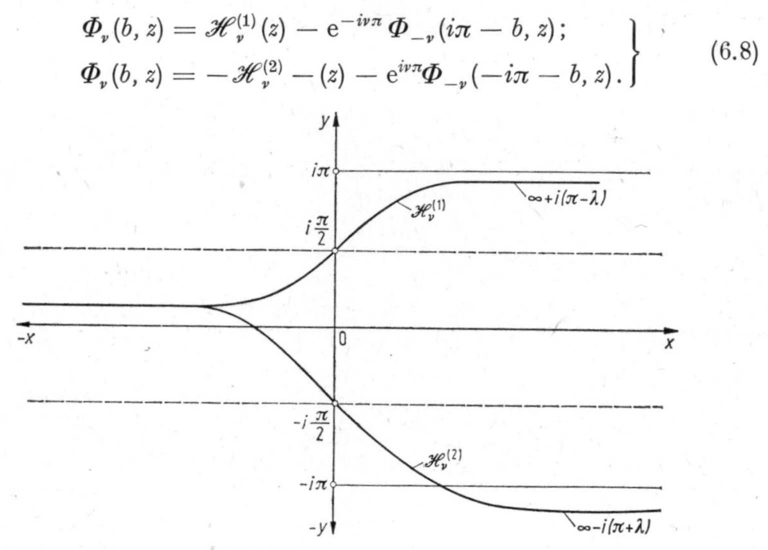

Fig. 7. Integration contours passing through the points $\pm i\pi/2$.

The functions $\Phi_\nu(b, z)$ are also important because, as we shall see below, they appear in the solutions of a series of theoretical and applied problems. We shall therefore study in detail the behavior of these functions for large $|z|$. Because of Eq. (6.8) we need only investigate them only in the region $|\operatorname{Im} b| \leq \pi/2$. However, for the sake of generality we shall consider them in the larger region

$$\left.\begin{aligned} |\operatorname{Im} b| &< \frac{3}{2}\pi; \\ |\arg z| &< \pi, \end{aligned}\right\} \quad (6.9)$$

so that the parameter λ, by expression (6.6), may be limited to the interval $|\lambda| < \pi/2$.

From the given formula it is evident that the points $c_1 = i\pi/2$ and $c_2 = -i\pi/2$ are the extrema to be considered in finding the asymptotic expansion of $\Phi_\nu(b, z)$, as well as for finding the asymptotic expansion for the Hankel functions of first and second kind. With this in mind, the contribution of the extrema to the asymptotic behavior of $\Phi_\nu(b, z)$ for $|z| \gg 1$ and $|\arg z| < \pi$ can be completely characterized by the asymptotic behavior of the corresponding Hankel function, as in § 10 of Chapter II. For finding asymptotic expansions for incomplete cylindrical functions of the Poisson form, it was necessary to determine the contribution from the extremal point separately since this point coincided with one of the limits of interation. In the present case the contribution from the extrema depends on value of the imaginary part of the variable w. Therefore, we consider a separately the cases $|\mathrm{Im}\, b| < \pi/2$ and $\pi/2 < |\mathrm{Im}\, b| < 3/2\pi$.

We assume first that $|\mathrm{Im}\, b| < \pi/2$, i.e. that the upper limit of integration in the integral (6.5) lies within the strip whose boundaries are shown by the dashed line of Fig. 7. In this case the extremal point $c_1 = i\pi/2$ (or $c_2 = -i\pi/2$) contributes nothing to the asymptotic behavior, since all contours passing through this point and the ends of the interval of integration will cross the line $t = i\pi/2$ (or $t = -i\pi/2$) twice travelling in opposite directions, so that by the Cauchy theorem they are equivalent to a single contour entering the extremal point and leaving it in the opposite direction. For such values of the upper integration limit b under the condition (6.6), we need only take account of the contribution to the asymptotic behavior from the upper end of the interval of integration, itself i.e.

$$\Phi_\nu(b, z) \approx \text{(contribution from upper limit)}$$

Now let b lie in the region $\pi/2 < \mathrm{Im}\, b < 3/2\pi$. In this case the contour of integration in Eq. (6.5) will always cross the line $t = i\pi/2$ in one direction only, which is equivalent to entering the stationary point $c_1 = i\pi/2$ and leaving it in the same direction. According to the remarks above, the contribution at this point is completely determined by the asymptotic behavior of the Hankel function of the first kind; consequently, for $\pi/2 < \mathrm{Im}\, b < 3/2\pi$

$$\Phi_\nu(b, z) \approx \mathscr{H}_\nu^{(1)} + \text{(contribution from upper limit)}.$$

If in fact b lies the region $-3/2\pi < \mathrm{Im}\, b < -\pi/2$, then the contribution from the extremal point $c_2 = -i\pi/2$ is completely characterized by the asymptotic behavior of the Hankel function of the second kind, and have in this case

$$\Phi_\nu(b, z) \sim -\mathscr{H}_\nu^{(2)} + \text{(contribution from upper limit)}.$$

Using the fundamental expression (9.23) of Chapter II we may write the following summarizing equation for the asymptotic expansion of $\Phi_\nu(b, z)$ for $|z| \gg 1$ and $|\arg z| < \pi$:

$$\Phi_\nu(b, z) \approx \frac{e^{z \sinh b}}{\pi i} \sum_{k=0}^{\infty} \frac{A_k(b)\, \Gamma(k+1)}{z^{k+1}}$$

$$+ \left\{ \begin{array}{ll} 0; & -\pi/2 < \mathrm{Im}\, b < \pi/2; \\ \mathscr{H}_\nu^{(1)}; & \pi/2 < \mathrm{Im}\, b < 3/2\pi; \\ -\mathscr{H}_\nu^{(2)}; & -3/2\pi < \mathrm{Im}\, b < -\pi/2, \end{array} \right\} \tag{6.10}$$

where the expansion coefficients $A_k(b)$ are determined from the system of Eqs. (9.14) of Chapter II

$$h[b + \Psi(u)] = h(b) - u;$$

$$\varphi[b + \Psi(u)]\,\Psi'(u) = \sum_{k=0}^{\infty} A_k(b) \cdot u^k.$$

In our case $h(t) = \sinh(t)$, $\varphi(t) = e^{-\nu t}$, and this system takes the form

$$\left.\begin{aligned} \sinh[b + \Psi(u)] &= \sinh b - u; \\ e^{-\nu[b+\Psi(u)]}\,\Psi'(u) &= \sum_{k=0}^{\infty} A_k(b)\, u^k. \end{aligned}\right\} \tag{6.11}$$

The first of these equations yields

$$b + \Psi(u) = \ln\{\sinh b - u + \sqrt{(\sinh b - u)^2 + 1}\}.$$

so that, by the second equation of system (6.11), the problem reduces to finding the coefficients of the power series in u for the following function:

$$e^{-\nu[b+\Psi(u)]}\,\Psi'(u) = \frac{-1}{\sqrt{(\sinh b - u)^2 + 1}}\;\frac{1}{[\sinh b - u + \sqrt{(\sinh b - u)^2 + 1}]^\nu}\,. \tag{6.12}$$

We now find explicit expressions for these coefficients. For this purpose we introduce for the moment the following notation:

$$\left.\begin{aligned} x &= \tanh b; \\ t &= \frac{u}{\cosh b}\,; \\ R &= \sqrt{1 + t^2 - 2tx}\,. \end{aligned}\right\} \tag{6.13}$$

In such a case (6.12) can be written in the form

$$\begin{aligned} e^{-\nu[b+\Psi(u)]}\,\Psi'(u) &= -(\cosh b)^{-\nu-1} R^{-1} (x - t + R)^{-\nu} \\ &\equiv \frac{e^{-\nu b}}{\cosh b}\left\{\left[\frac{e^b}{(x - t + R)\cosh b}\right]^\nu \frac{1}{R}\right\}. \end{aligned} \tag{6.14}$$

It is not difficult to verify that the expression in curved brackets is a generating function for the Jacobi polynomials $P_k^{(\alpha,\beta)}(x)$, determined by the following hypergeometric series (c.f. [11, p. 1050]):

$$P_k^{(\alpha,b)}(x) = \frac{(-1)^k \Gamma(k + 1 + \beta)}{\Gamma(k + 1)\,\Gamma(1 + \beta)}\, F\left(k + \alpha + \beta + 1, -k; 1 + \beta; \frac{1 + x}{2}\right) \tag{6.15}$$

$$= \frac{\Gamma(k + 1 + \alpha)}{\Gamma(k + 1) \cdot \Gamma(1 + \alpha)}\, F\left(k + \alpha + \beta + 1, -k; 1 + \alpha; \frac{1 - x}{2}\right).$$

In fact, the generating function for these polynomials has the form

$$\sum_{k=0}^{\infty} P_k^{(\alpha,\beta)}(x)\, t^k = 2^{\alpha+\beta} R^{-1} (1 - t + R)^{-\alpha} (1 + t + R)^{-\beta}, \quad (6.16)$$

where $|t| < 1$ and $R = \sqrt{1 - 2tx + t^2}$. Substituting here $\alpha = \nu$, $\beta = -\nu$, we have

$$\sum_{k=0}^{\infty} P_k^{(\nu,-\nu)}(x) \cdot t^k = \frac{1}{R} \left(\frac{1 + t + R}{1 - t + R}\right)^{\nu}, \quad (6.17)$$

but

$$\frac{1 + t + R}{1 - t + R} = \frac{1 + t + R}{1 - t + R} \cdot \frac{1 + t - R}{1 + t - R} = \frac{(1 + t)^2 - R^2}{1 - (t - R)^2} = \frac{1 + x}{x - t + R}.$$

Therefore, in our case for $x = \tanh(b)$ formula (6.17) can be written as

$$\sum_{k=0}^{\infty} P_k^{(\nu,-\nu)}(\tanh b)\, t^k = \frac{1}{R} \left[\frac{e^b}{(x - t + R)\cosh b}\right]^{\nu}. \quad (6.18)$$

The comparison of the right side of this formula with expression (6.14) immediately shows the validites of our assertion.

Returning now to (6.12), we find from (6.13), (6.14), and (6.18) that

$$-\frac{e^{-\nu b}}{\cosh b} \sum_{k=0}^{\infty} P_k^{(\nu,-\nu)}(\tanh b) \left(\frac{u}{\cosh b}\right)^k = \sum_{k=0}^{\infty} A_k(b)\, u^k. \quad (6.19)$$

From this we readily obtain the desired coefficients $A_k(b)$:

$$A_k(b) = -\frac{e^{-\nu b}}{\cosh^{k+1} b} P_k^{(\nu,-\nu)}(\tanh b). \quad (6.20)$$

Or, using formula (6.15)

$$A_k(b) = -\frac{e^{-\nu b}}{\cosh^{k+1} b} \frac{\Gamma(k + 1 + \nu)}{\Gamma(k + 1)\,\Gamma(1 + \nu)} F\left(k + 1, -k; 1 + \nu; \frac{1 - \tanh b}{2}\right). \quad (6.21)$$

Substituting these coefficients in formula (6.10), we obtain finally an asymptotic expansion for the incomplete cylindrical function $\Phi_\nu(b, z)$ for $|z| \gg 1$ and $|\arg z| < \pi$:

$$\Phi_\nu(b, z) \approx \frac{1}{\pi i} \frac{e^{z\sinh b - \nu b}}{z \cdot \cosh b} \sum_{k=0}^{\infty} \frac{\Gamma(k + 1 + \nu)}{\Gamma(1 + \nu)} \frac{1}{(z \cosh b)^k}$$
$$\times F\left(k + 1, -k; 1 + \nu; \frac{1 - \tanh b}{2}\right)$$
$$+ \begin{cases} 0; & -\pi/2 < \operatorname{Im} b < \pi/2; \\ \mathscr{H}_\nu^{(1)}; & \pi/2 < \operatorname{Im} b < 3/2\pi; \\ -\mathscr{H}_\nu^{(2)}; & -3/2\pi < \operatorname{Im} b < -\pi/2. \end{cases} \quad (6.22)$$

Some particular cases of this development are of interest: taking $b = 0$, we have

$$\Phi_\nu(0, z) \approx \frac{1}{\pi i z} \sum_{k=0}^{\infty} \frac{\Gamma(k+1+\nu)}{z^k \Gamma(1+\nu)} F\left(k+1, -k; 1+\nu; \frac{1}{2}\right). \tag{6.23}$$

This expression can be simplified somewhat by using the following relations for the Legendre Polynomials (c.f. [11, pp. 1013 and 1023]):

$$\left.\begin{aligned} P_k^{-\nu}(x) &= \frac{1}{\Gamma(1+\nu)} \left(\frac{1-x}{1+x}\right)^{\nu/2} F\left(k+1; -k; 1+\nu; \frac{1-x}{2}\right); \\ P_k^{-\nu}(0) &= \frac{2^{-\nu}\sqrt{\pi}}{\Gamma\left(\frac{\nu+k}{2}+1\right)\Gamma\left(\frac{\nu-k+1}{2}\right)}. \end{aligned}\right\} \tag{6.24}$$

We find

$$F\left(k+1, -k; 1+\nu; \frac{1}{2}\right) = \frac{2^{-\nu}\sqrt{\pi}\,\Gamma(1+\nu)}{\Gamma\left(\frac{k+\nu}{2}+1\right)\Gamma\left(\frac{1+\nu-k}{2}\right)}. \tag{6.25}$$

Applying now the duplication formula for the gamma-function ([11, p. 932])

$$\frac{\Gamma(1+\nu+k)}{\Gamma\left(\frac{k+\nu}{2}+1\right)} = \frac{2^{\nu+k}}{\sqrt{\pi}} \Gamma\left(\frac{\nu+k+1}{2}\right), \tag{6.26}$$

and substituting (6.25) into (6.23) we find

$$\Phi_\nu(0, z) \approx \frac{1}{\pi i z} \sum_{k=0}^{\infty} \frac{2^k}{z^k} \frac{\Gamma\left(\frac{\nu+k+1}{2}\right)}{\Gamma\left(\frac{\nu-k+1}{2}\right)}. \tag{6.27}$$

For immediate numerical evaluation it is convenient also to use the following formula ([11], loc. cit.) for even k $(=2m)$

$$\frac{2^k \Gamma\left(\frac{\nu+k+1}{2}\right)}{\Gamma\left(\frac{\nu-k+1}{2}\right)} = (\nu^2 - 1^2)(\nu^2 - 3^2) \cdots [\nu^2 - (k-1)^2] \tag{6.28}$$

and for odd k $(=2m+1)$, $m = 0, 1, 2, \ldots$

$$2^k \frac{\Gamma\left(\frac{\nu+k+1}{2}\right)}{\Gamma\left(\frac{\nu-k+1}{2}\right)} = (\nu+k-1)\,[(\nu-1)^2 - 1^2] \cdots [(\nu-1)^2 - (k-2)^2]. \tag{6.29}$$

Further setting $b = \pm i\pi$ in (6.22), and taking (6.25) and (6.26) into account we obtain

$$\begin{aligned} \Phi_\nu(\pm i\pi, z) \approx &-\frac{e^{\mp i\nu\pi}}{\pi i z} \sum_{k=0}^{\infty} \left(-\frac{1}{z}\right)^k \frac{2^k \Gamma\left(\frac{\nu+k+1}{2}\right)}{\Gamma\left(\frac{\nu-k+1}{2}\right)} \\ &+ \begin{cases} \mathscr{H}_\nu^{(1)}(z) & \text{for} \quad b = i\pi; \\ -\mathscr{H}_\nu^{(2)}(z) & \text{for} \quad b = -i\pi. \end{cases} \end{aligned} \tag{6.30}$$

We note in passing that the results obtained above permit one to find asymptotic expansions for the integrals $\int_0^\infty e^{\pm \nu t - z \sinh t}\, dt$, which appear in the definition of the cylindrical functions of the Bessel-Schlaefli form and which are usually used for finding asymptotic expansions of Anger and Weber functions. It is evident that these integrals are particular cases of the auxiliary function $\Phi_\nu(b, z)$ for $b = 0$ and $|\arg z| < \pi/2$. Therefore, on the basis of (6.23) and (6.24) we have

$$\int_0^\infty e^{\pm \nu t - z \sinh t}\, dt = i\pi \Phi_{\pm\nu}(0, z) \approx \frac{1}{z} \sum_{k=0}^{\infty} \left(\frac{2}{z}\right)^k \frac{\Gamma\left(\frac{\pm\nu + k + 1}{2}\right)}{\Gamma\left(\frac{\pm\nu - k + 1}{2}\right)}.$$

This method of finding asymptotic expansions for these integrals is different from that used in the [10, p. 342] and it appears to be more direct and simpler.

Formula (6.22) was obtained under the hypotheses that $\operatorname{Im} b \neq \pm i\pi/2$. We consider now the cases $b = \alpha \pm i\pi/2$ separately. If $\operatorname{Re} b = \alpha > 0$, then the integration contour in (6.5) may always be made to pass through the extrema $c_1 = i\pi/2$ and $c_2 = -i\pi/2$, and their contributions as before are determined by the asymptotic behavior of the corresponding Hankel functions. In this case the asymptotic expansion for $\Phi_\nu(b, z)$ may be obtained as a limiting form of (6.22) taking in it $b = \alpha \pm i(\pi/2 + 0)$. If in fact $b = -\alpha \pm i\pi/2$ and $\operatorname{Re} b = -\alpha < 0$, then it follows from the definition (6.5) that the contour of integration can not pass through the stationary points in just one direction and according to (6.8), the asymptotic expansion of $\Phi_\nu(b, z)$ is wholly determined by the contribution from the upper limit of integration. This expansion may also be obtained from (6.22) as a limiting case, taking there $b = -\alpha \pm i(\pi/2 - 0)$. In other words, for $b = \alpha \pm i\pi/2$

$$\Phi_\nu(\alpha \pm i\pi/2, z) \approx \frac{e^{\pm i(z\cosh\alpha - \nu\pi/2) - \alpha\nu}}{\mp \pi z \sinh\alpha}$$

$$\times \sum_{k=0}^{\infty} \frac{\Gamma(k+1+\nu)}{\Gamma(1+\nu)} \left(\frac{1}{\pm iz\sinh\alpha}\right)^k F\left(k+1, -k; 1+\nu; \frac{1-\tanh\alpha}{2}\right)$$

$$+ \begin{Bmatrix} \mathscr{H}_\nu^{(1)}(z) & \text{for} & b = \alpha + i\pi/2; & \alpha > 0, \\ -\mathscr{H}_\nu^{(2)}(z) & \text{for} & b = \alpha - i\pi/2; & \alpha > 0, \\ 0 & \text{for} & b = \alpha \pm i\pi/2; & \alpha < 0. \end{Bmatrix} \tag{6.31}$$

The cases $b = \pm i\pi/2$ require different considerations. The essential difference is that one or the other of the extrema $c_2 = i\pi/2$, and $c_2 = -i\pi/2$ coincides with the upper limit of integration, and the corresponding contributions must be evaluated by formula (9.25) of Chapter II, which here takes the form for $c_2 = i\pi/2$

$$\begin{aligned}\Phi_\nu(i\pi/2, z) \approx -\frac{1}{2\pi i} e^{iz} \sum_{k=0}^{\infty} B_{2k+1}(i\pi/2) \frac{\Gamma(k+1)}{z^{k+1}} \\ + \frac{e^{iz}}{2\pi i \sqrt{z}} \sum_{k=0}^{\infty} B_{2k}(i\pi/2) \frac{\Gamma(k+1/2)}{z^k},\end{aligned} \tag{6.32}$$

The coefficients B_m are determined from system (9.16) of Chapter II, which in this case can be written in the form

$$\left.\begin{aligned}\sinh[i\pi/2 + \Psi(u)] = \sinh i\pi/2 - u^2; \\ e^{-\nu[i\pi/2+\Psi(u)]}\Psi'(u) = \sum_{m=0}^{\infty} B_m(i\pi/2)\, u^m.\end{aligned}\right\} \tag{6.33}$$

From this we find

$$\begin{aligned}e^{-\nu[i\pi/2+\Psi(u)]}\,\Psi'(u) = e^{-i\nu\pi/2} \frac{2iu}{[1 + iu^2 + \sqrt{(1+iu^2)^2 - 1}]^\nu} \\ \times \frac{1}{\sqrt{(1+iu^2)^2 - 1}} \equiv \frac{e^{-i\nu\pi/2}\sqrt{2i}\,(\sqrt{iu/2} + \sqrt{1 + iu^2/2})^{-2\nu}}{\sqrt{1 + iu^2/2}}.\end{aligned} \tag{6.34}$$

On the other hand, substituting $\alpha = 2\nu$, $\beta = -2\nu$, $x = 0$ and $t = -\sqrt{iu/2}$, in (6.16) we have

$$\sum_{m=0}^{\infty} P_m^{(2\nu,-2\nu)}(0)\,(\sqrt{iu/2})^m = \frac{(\sqrt{iu/2} + \sqrt{1 + iu^2/2})^{-2\nu}}{\sqrt{1 + iu^2/2}}. \tag{6.35}$$

In other words, the function (6.34) is a generating function for the Jacobi polynomials $P_k^{(2\nu,-2\nu)}(0)$. Thus the desired coefficients for the expansion in formulae (6.32), (6.33) will be

$$B_m(i\pi/2) = e^{i\pi/2\left(-\nu + \frac{m+1}{2}\right)} \frac{(-1)^m}{(\sqrt{2})^{m-1}} P_m^{(2\nu,-2\nu)}(0)$$

or, on the basis of formulae (6.15), and (6.25), (6.26), we get finally

$$\begin{aligned}B_m(i\pi/2) &= e^{i\pi/2\left(-\nu + \frac{m+1}{2}\right)} \frac{(-1)^m}{(\sqrt{2})^{m-1}} \frac{\Gamma(m+1+2\nu)}{\Gamma(m+1)\,\Gamma(1+2\nu)} \\ &\quad \times F\left(m+1, -m, 1+2\nu; \frac{1}{2}\right) \\ &= e^{i\pi\left(-\nu/2 + \frac{m+1}{4}\right)} \frac{(-1)^m (\sqrt{2})^{m+1}}{\Gamma(m+1)} \frac{\Gamma\left(\nu + \frac{m+1}{2}\right)}{\Gamma\left(\nu + \frac{1-m}{2}\right)}.\end{aligned} \tag{6.36}$$

Substituting the values of these coefficients in (6.32), we see that the second term, as would be expected, represents the asymptotic expansion for $1/2 \cdot \mathscr{H}_\nu^{(1)}(z)$ (c.f. expression (10.6) of Chapter II) and so the asymptotic expansion for $\Phi_\nu(i\pi/2, z)$ takes the form

$$\Phi_\nu(i\pi/2, z) \approx \frac{1}{2\pi i} e^{iz - i\nu\pi/2} \times \sum_{k=0}^{\infty} \left(\frac{2i}{z}\right)^{k+1} \frac{\Gamma(k+1)\,\Gamma(\nu+k+1)}{\Gamma(2k+2)\,\Gamma(\nu-k)} + \frac{1}{2}\mathscr{H}_\nu^{(1)}(z). \tag{6.37}$$

By similar considerations we obtain

$$\Phi_\nu(-i\pi/2, z) \approx \frac{1}{2i\pi} e^{-iz + i\nu\pi/2} \times \sum_{k=0}^{\infty} \left(\frac{2}{iz}\right)^{k+1} \frac{\Gamma(k+1)\,\Gamma(\nu+k+1)}{\Gamma(2k+2)\,\Gamma(\nu-k)} - \frac{1}{2}\mathscr{H}_\nu^{(2)}(z). \tag{6.38}$$

For integer values of the index $\nu = n$ the series on the right sides of the last two formulae terminate, and reduce to finite sums of n terms. This arises from the fact that for integral values of ν, the incomplete cylindrical functions of Bessel and Poisson forms can be expressed in terms of each other and in finite form, (c.f., e.g. formulae (3.11) and (3.13) of this Chapter).

We have thus obtained asymptotic expansions for the auxiliary functions $\Phi_\nu(b, z)$ for all values of b and z in the region

$$-3\pi/2 < \operatorname{Im} b < 3\pi/2; \qquad -\pi < \arg z < \pi.$$

This makes it possible to construct asymptotic expansions for the $\varepsilon_\nu(w, z)$, as well as for other incomplete cylindrical functions of Bessel form.

7. Asymptotic Expansions for Incomplete Cylindrical Functions, the General Case

In the preceeding section the asymptotic expansion of the functions $\Phi_\nu(w, z)$ was given for the cases when the variables w and z were restricted to the regions

$$\left.\begin{aligned} -3\pi/2 < \operatorname{Im} w < 3\pi/2; \\ -\pi < \arg z < \pi. \end{aligned}\right\} \tag{7.1}$$

In these regions, according to formulae (6.7), (6.22), (6.27), (6.35), (6.37) and (6.38), we obtain the following asymptotic expansion for the incomplete cylindrical function $\varepsilon_\nu(w, z)$:

$$\varepsilon_\nu(w, z) \approx \frac{1}{\pi i} \frac{e^{z\sinh w - \nu w}}{z \cosh w} \sum_{k=0}^{\infty} \frac{\Gamma(k+1+\nu)}{\Gamma(1+\nu)} \frac{1}{(z \cosh w)^k}$$

$$\times F\left(k+1, -k, 1+\nu; \frac{1-\tanh w}{2}\right) - \frac{1}{\pi i z} \sum_{k=0}^{\infty} \left(\frac{2}{z}\right)^k \frac{\Gamma\left(\frac{\nu+k+1}{2}\right)}{\Gamma\left(\frac{\nu-k+1}{2}\right)}$$

$$+ \begin{Bmatrix} \mathscr{H}_\nu^{(1)}(z); & \pi/2 < \operatorname{Im} w < 3\pi/2; \\ \mathscr{H}_\nu^{(1)}(z); & \operatorname{Im} w = \frac{\pi}{2}, \ \operatorname{Re} w > 0; \\ 0; & \operatorname{Im} w = \frac{\pi}{2}, \ \operatorname{Re} w < 0; \\ 0; & -\pi/2 < \operatorname{Im} w < \pi/2; \\ 0; & \operatorname{Im} w = -\pi/2, \ \operatorname{Re} w < 0; \\ -\mathscr{H}_\nu^{(2)}(z); & \operatorname{Im} w = -\pi/2, \ \operatorname{Re} w > 0; \\ -\mathscr{H}_\nu^{(2)}(z); & -3\pi/2 < \operatorname{Im} w < -\pi/2; \end{Bmatrix} \tag{7.2}$$

while for the special case $w = \pm i\pi/2$

$$\varepsilon_\nu(\pm i\pi/2, z) \approx \frac{1}{2\pi i} e^{\pm i(z - \nu\pi/2)} \sum_{k=0}^{\infty} \left(\frac{\pm 2i}{z}\right)^{k+1} \frac{\Gamma(k+1)\,\Gamma(1+k+\nu)}{\Gamma(2k+2)\,\Gamma(\nu-k)}$$

$$- \frac{1}{\pi i z} \sum_{k=0}^{\infty} \left(\frac{2}{z}\right)^k \frac{\Gamma\left(\frac{\nu+k+1}{2}\right)}{\Gamma\left(\frac{\nu-k+1}{2}\right)} + \frac{1}{2} \begin{Bmatrix} \mathscr{H}_\nu^{(1)}(z); w = i\pi/2; \\ -\mathscr{H}_\nu^{(2)}(z); w = -i\pi/2. \end{Bmatrix} \tag{7.3}$$

Explicit expressions for the dominant terms of these developments have the forms

$$\varepsilon_\nu(w, z) \approx \frac{1}{\pi i z}\left(\frac{e^{z\sinh w - \nu w}}{\cosh w} - 1\right) + \frac{1}{\pi i z^2}\left[\frac{e^{z\sinh w - \nu w}}{\cosh^2 w}(\nu + \tanh w) - \nu\right]$$

$$+ \frac{1}{\pi i z^3}\left[\frac{e^{z\sinh w - \nu w}}{\cosh^3 w}(\nu^2 - 1 + 3\nu \tanh w - 3\tanh^2 w) - (\nu^2 - 1)\right]$$

+ term in curved brackets of formula (7.2); (7.4)

$$\begin{aligned} \varepsilon_\nu(\pm i\pi/2, z) \approx {} & \frac{1}{i\pi z}(\pm i\nu\, e^{\pm i(\nu - \nu\pi/2)} - 1) \\ & - \frac{\nu}{i\pi z^2}\left(\frac{\nu^2 - 1}{3} e^{\pm(-\nu\pi/2)} + 1\right) \\ & - \frac{\nu^2 - 1}{i\pi z^3}\left(\pm \frac{i\nu(\nu+3)(\nu^2-4)}{15} e^{\pm i(z - \nu\pi/2)} + 1\right) \\ & + \frac{1}{2}\begin{Bmatrix} \mathscr{H}_\nu^{(1)}(z), w = i\pi/2; \\ -\mathscr{H}_\nu^{(2)}(z), w = -i\pi/2. \end{Bmatrix} \end{aligned} \tag{7.5}$$

Asymptotic expansions for other incomplete cylindrical functions of Bessel form can be obtained from their representations in terms of $\varepsilon_\nu(w, z)$, or in terms of $\Phi_\nu(w, z)$ (c.f. Sections 1 and 6 of this Chapter):

$$j_\nu(w, z) = \frac{1}{2}[\varepsilon_\nu(w, z) - \varepsilon_\nu(\bar{w}, z)] = \frac{1}{2}[\Phi_\nu(w, z) - \Phi_\nu(\bar{w}, z)]; \tag{7.6}$$

$$\begin{aligned} i_\nu(w, z) &= \frac{e^{-i\nu\pi/2}}{2}[\varepsilon_\nu(w - i\pi/2, iz) - \varepsilon_\nu(\bar{w} - i\pi/2, iz)] \\ &= \frac{e^{-i\nu\pi/2}}{2}[\Phi_\nu(w - i\pi/2, iz) - \Phi_\nu(\bar{w} - i\pi/2, iz)]; \end{aligned} \tag{7.7}$$

$$\begin{aligned} h_\nu(w, z) &= \varepsilon_\nu(w + i\pi/2, z) - \varepsilon_\nu(-w + i\pi/2, z) \\ &= \Phi_\nu(w + i\pi/2, z) - \Phi_\nu(-w + i\pi/2, z); \end{aligned} \tag{7.8}$$

$$\begin{aligned} k_\nu(w, z) &= i\pi/2 e^{i\nu\pi/2}[\varepsilon_\nu(w + i\pi/2, iz) - \varepsilon_\nu(-w + i\pi/2, iz)] \\ &= i\pi/2 e^{i\nu\pi/2}[\Phi_\nu(w + i\pi/2, iz) - \Phi_\nu(-w + i\pi/2, iz)]. \end{aligned} \tag{7.9}$$

In order to find asymptotic expansions for the incomplete Anger function $A_\nu(\beta, z)$ and for the Weber function $B_\nu(\beta, z)$, we set $w = i\beta$ in formulae (7.2)—(7.5), writing $\varepsilon_\nu(i\beta, z)$ in the form

$$\varepsilon_\nu(i\beta, z) = A_\nu(\beta, z) - iB_\nu(\beta, z),$$

and then separating real and imaginary parts, proceeding formally as if ν, β and z were real quantities. The principal terms of the expansions of these functions for real β then have the form

$$A_\nu(\beta, z) \approx \frac{1}{\pi z}\frac{\sin(z\sin\beta - \nu\beta)}{\cos\beta} + \begin{Bmatrix} J_\nu(z); & \pi/2 < \beta < 3/2\pi; \\ 0; & -\pi/2 < \beta < \pi/2; \\ -J_\nu(z); & -3/2\pi < \beta < -\pi/2; \end{Bmatrix} \tag{7.10}$$

$$A_\nu(\pm\pi/2, z) \approx \pm\left[\frac{1}{\pi z}\cos\left(z - \frac{\nu\pi}{2}\right) + \frac{1}{2}J_\nu(z)\right]; \tag{7.11}$$

$$B_\nu(\beta, z) \approx \frac{1}{\pi z}\left[\frac{\cos(z\sin\beta - \nu\beta)}{\cos\beta} - 1\right] + \begin{Bmatrix} 0, & |\beta| < \frac{\pi}{2}; \\ N_\nu(z), & \frac{\pi}{2} < |\beta| < \frac{3}{2}\pi; \end{Bmatrix} \tag{7.12}$$

$$B_\nu(\pm\pi/2, z) = -\frac{1}{\pi z}\left[1 \pm \nu\sin\left(z - \nu\frac{\pi}{2}\right)\right] - \frac{1}{2}N_\nu(z). \tag{7.13}$$

In all of these formulae, the dominant terms of the relevant Hankel functions $\mathscr{H}_\nu^{(1)}(z)$, $\mathscr{H}_\nu^{(2)}(z)$, Bessel function $J_\nu(z)$, and Neumann function $N_\nu(z)$ are denoted by the corresponding symbol.

As we have already pointed out, these expansions are valid for w and z lying in regions (7.1).

If w and z lie outside these regions, then it is also necessary to take into account contributions from the extrema $c_m = \pm i \times (2m+1)\,\pi/2$, $m = 1, 2, 3, \ldots$

To handle these additional contributions, we proceed as follows. Let $|\arg z| < \pi$, and assume w lies in the region $(2k-1)\,\pi < \operatorname{Im} w < (2k+1)\,\pi$, where k is an arbitrary integer. We have

$$\begin{aligned} \varepsilon_\nu(w, z) = \frac{1}{\pi i}\int_0^w \mathrm{e}^{z\sinh u - \nu u}\,du &= \frac{1}{\pi i}\int_0^{2k\pi i} \mathrm{e}^{z\sinh u - \nu u}\,du \\ &+ \frac{1}{\pi i}\int_{2k\pi i}^{w} \mathrm{e}^{z\sinh u - \nu u}\,du. \end{aligned} \tag{7.14}$$

After the introduction of the new integration variable $u - 2k\pi i$, the second term in this formula takes the form

$$\frac{1}{\pi i}\int_{2k\pi i}^{w} \mathrm{e}^{z\sinh u - \nu u}\,du = \mathrm{e}^{-i^2 k\pi\nu}\,\varepsilon_\nu(w - 2ik\pi, z), \tag{7.15}$$

where $-\pi \leq \operatorname{Im}(w - 2k\pi i) \leq \pi$. We note that a previously obtained expansion applies to the function $\varepsilon_\nu(w - 2k\pi i, z)$.

We turn now to the first term in (7.14), writing it in the form

$$\begin{aligned} \frac{1}{\pi i}\int_0^{2ik\pi} \mathrm{e}^{z\sinh u - \nu u}\,du &= \sum_{l=0}^{k-1} \frac{1}{\pi i}\int_{2il\pi}^{2i(l+1)\pi} \mathrm{e}^{z\sinh u - \nu u}\,du \\ &= \frac{1}{\pi i}\int_0^{2i\pi} \mathrm{e}^{z\sin th - \nu t}\,dt \sum_{l=0}^{k-1} \mathrm{e}^{-2il\nu\pi}. \end{aligned}$$

But

$$\sum_{l=0}^{k-1} \mathrm{e}^{-2il\nu\pi} = \frac{1 - \mathrm{e}^{-2i\nu k\pi}}{1 - \mathrm{e}^{-2i\nu\pi}} = \mathrm{e}^{-i\nu(k-1)\pi}\,\frac{\sin \nu k\pi}{\sin \nu\pi}$$

and, by the definition of the function $\Phi_\nu(w, z)$, we have

$$\frac{1}{\pi i}\int_0^{2i\pi} \mathrm{e}^{z\sinh t - \nu t}\,dt = \frac{\mathrm{e}^{-i\nu\pi}}{\pi i}\int_{-i\pi}^{i\pi} \mathrm{e}^{z\sinh u + \nu u}\,du = \frac{\mathrm{e}^{-i\nu\pi}}{\pi i}\left[\Phi_{-\nu}(i\pi, z) - \Phi_{-\nu}(-i\pi, z)\right].$$

By virtue of (6.8) we find

$$\frac{1}{i\pi}\int_0^{2i\pi} \mathrm{e}^{z\sinh t - \nu t}\,dt = \mathrm{e}^{-i\nu\pi}\left[\mathscr{H}_{-\nu}^{(1)}(z) + \mathscr{H}_{-\nu}^{(2)}(z) - 2i\sin\nu\pi\Phi_\nu(0, z)\right].$$

Consequently, the first term in (7.14) can be written in the form

$$\frac{1}{\pi i}\int_0^{2ik\pi} e^{z\sinh t-\nu t}\,dt = 2e^{-i\nu k\pi}\sin\nu k\pi\left[\frac{J_{-\nu}(z)}{\sin\nu\pi} - i\Phi_\nu(0,z)\right].$$

Substituting this, together with expression (7.15), into (7.14), we obtain finally

$$\varepsilon_\nu(w,z) = e^{-2i\nu k\pi}\varepsilon_\nu(w-2ik\pi,z) + 2e^{-i\nu k\pi}\sin\nu k\pi\left[\frac{J_{-\nu}(z)}{\sin\nu\pi} - i\Phi_\nu(0,z)\right], \tag{7.16}$$

where $-\pi \leq \operatorname{Im}(w-2ik\pi) \leq \pi$. This formula expresses $\varepsilon_\nu(w,z)$ for arbitrary values of w in terms of $\varepsilon_\nu(w-2ik\pi,z)$, for which the asymptotic expansions (7.2)—(7.5) hold.

Formula (7.16) thus extends the region of validity of the expansions of $\varepsilon_\nu(w,z)$ and $\Phi_\nu(w,z)$ to include all values of w for z in the region $|\arg z| < \pi$.

Now let $z = e^{\pm i\pi}z_0$ and $|\arg z_0| < \pi$. Then by definition we have

$$\varepsilon_\nu(w, e^{\pm i\pi}z_0) = \frac{1}{\pi i}\int_0^w \exp(e^{\pm i\pi}z_0\sinh t - \nu t)\,dt.$$

Making use of (3.7) of Chapter II, which in this case takes the form

$$\sinh(u \pm i\pi) = e^{\mp i\pi}\sinh u,$$

we find by means of an obvious transformation

$$\begin{aligned}\varepsilon_\nu(w, e^{\pm i\pi}z_0) &= \frac{e^{\mp i\pi\nu}}{i\pi}\int_{\pm i\pi}^{w\pm i\pi}\exp(z_0\sinh\xi - \nu\xi)\,d\xi \\ &= e^{\mp i\nu\pi}[\varepsilon_\nu(w\mp i\pi, z_0) - \varepsilon_\nu(\mp i\pi, z_0)].\end{aligned} \tag{7.17}$$

This formula expresses $\varepsilon_\nu(w, e^{\pm i\pi}z_0)$ in terms of $\varepsilon_\nu(w\mp i\pi, z_0)$ and $\varepsilon_\nu(\mp i\pi, z_0)$, for which the developments given above are valid.

Thus, by means of (7.16) and (7.17) together with (6.27) and (7.2)—(7.5), one can write down asymptotic expansions for all incomplete cylindrical functions of Bessel form, for all values of the variables w and z. It must be emphasized that when constructing asymptotic expansions for incomplete Bessel functions $j_\nu(w,z)$ and $i_\nu(w,z)$, Hankel functions $h_\nu(w,z)$ or MacDonald functions $k_\nu(w,z)$ it is necessary to observe caution, since all these functions are expressed as differences between $\varepsilon_\nu(w,z)$ or $\Phi_\nu(w,z)$ valid for different values of w and z.

For example, assume we want to find an asymptotic expansion for the incomplete MacDonald function valid for $|\arg z| < \pi$ and $|\operatorname{Im} w| < \pi$;

we have

$$k_\nu(w, z) = \frac{1}{2} \int_{-w}^{w} \exp(-z \cosh t - \nu t)\, dt$$

$$= \frac{i\pi}{2} e^{i\nu\pi/2} [\varepsilon_\nu(w + i\pi/2, iz) - \varepsilon_\nu(-w + i\pi/2; iz)] \qquad (7.18)$$

$$= \frac{i\pi}{2} e^{i\nu\pi/2} [\Phi_\nu(w + i\pi/2, iz) - \Phi_\nu(-w + i\pi/2, iz)].$$

Now if $|\arg iz| < \pi$, i.e. $-\pi < \arg z < \pi/2$, then asymptotic expansions (7.2)—(7.5) may be applied immediately to the functions on the right side of this formula, since $|\pm \mathrm{Im}\, w + \pi/2| < 3/2\pi$. If on the other hand $\pi/2 < \arg z < \pi$ $(\pi < \arg iz < 3/2\pi)$, it is necessary to first represent iz as:

$$iz = e^{i\pi} z_0;\ z_0 = |z|\, e^{i(\arg z - \pi/2)} \qquad (7.19)$$

so that formula (7.18) takes the form

$$k_\nu(w, z) = (i\pi/2)\, e^{-i\nu\pi/2} [\varepsilon_\nu(w + i\pi/2, e^{i\pi} z_0) - \varepsilon_\nu(-w + i\pi/2, e^{i\pi} z_0)],$$

or, by (7.17)

$$k_\nu(w, z) = (i\pi/2)\, e^{-i\nu\pi/2} [\varepsilon_\nu(w - i\pi/2, z_0) - \varepsilon_\nu(-w - i\pi/2, z_0)]. \qquad (7.20)$$

This is valid for $|\arg z_0| < \pi$, $|\pm \mathrm{Im}\, w - \pi/2| < 3/2\pi$ and therefore the asymptotic expansions (7.2)—(7.5) may now be applied to the functions on the right side.

8. Incomplete Airy Integrals and their Generalizations

In many theoretical and applied problems there appear the integrals

$$\int_0^\infty \cos(t^3 \pm xt)\, dt, \qquad (8.1)$$

which are known as Airy integrals. These integrals satisfy the differential equation

$$\frac{d^2 v}{dx^2} \pm \frac{1}{3} x v = 0,$$

which may be transformed into a form of Bessel's equation. The integrals (8.1) therefore belong to the class of cylindrical functions. They are related to the Bessel functions by (c.f. [10, p. 211]):

$$\int_0^\infty \cos(t^3 - xt)\, dt = \frac{1}{2} \pi \sqrt{\frac{x}{3}} \left[J_{-1/3}\left(\frac{2x\sqrt{x}}{3\sqrt{3}}\right) + J_{1/3}\left(\frac{2x\sqrt{x}}{3\sqrt{3}}\right)\right]; \qquad (8.2)$$

$$\int_0^\infty \cos(t^3 + xt)\, dt = \frac{\sqrt{x}}{3} K_{1/3}\left(\frac{2x\sqrt{x}}{3\sqrt{3}}\right). \qquad (8.3)$$

We now show that the Airy integrals (8.1) can be expressed in terms of incomplete cylindrical functions of Bessel form. We consider the integral

$$Ei(\alpha, z) = \int_0^\alpha e^{-i(t^3 - zt)}\, dt. \tag{8.4}$$

Making the change of variable $t = 2\sqrt{z/3}\sin u/3$ we have

$$Ei(\alpha, z) = \frac{2}{3}\sqrt{\frac{z}{3}} \int_0^w e^{-i\left(\frac{z}{3}\right)^{3/2}[8\sin^3(u/3) - 6\sin(u/3)]} \cos(u/3)\, du,$$

where for brevity we have put

$$w = 3\arcsin\left(\frac{\alpha}{2}\sqrt{3/z}\right).$$

Applying the identity

$$3\sin(u/3) - 4\sin^3(u/3) = \sin u,$$

we have

$$Ei(\alpha, z) = \frac{1}{3}\sqrt{\frac{z}{3}} \int_0^w e^{i2(z/3)^{3/2}\sin u} \left(e^{iu/3} + e^{-iu/3}\right) du. \tag{8.5}$$

Recalling now the definition of the incomplete cylindrical functions in Bessel form, we obtain finally

$$Ei(\alpha, z) = \int_0^\alpha e^{-i(t^3 - zt)}\, dt = \frac{\pi}{3}\sqrt{\frac{z}{3}}\left[\varepsilon_{1/3}(iw, \zeta) + \varepsilon_{-1/3}(iw, \zeta)\right], \tag{8.6}$$

where $\zeta = \dfrac{2z\sqrt{z}}{3\sqrt{3}}$.

If we now make use of the relation $\varepsilon_\nu(iw, z) = A_\nu(w, z) - iB_\nu(w, z)$ (c.f. formula (1.11)) then from (8.6) we find

$$Ec(\alpha, z) = \int_0^\alpha \cos(t^3 - zt)\, dt = \frac{\pi}{3}\sqrt{\frac{z}{3}}\left[A_{1/3}(w, \zeta) + A_{-1/3}(w, \zeta)\right]; \tag{8.7}$$

$$Es(\alpha, z) = \int_0^\alpha \sin(t^3 - zt)\, dt = \frac{\pi}{3}\sqrt{\frac{z}{3}}\left[B_{1/3}(w, \zeta) + B_{-1/3}(w, \zeta)\right], \tag{8.8}$$

where $A_{\pm 1/3}(w, \zeta)$ and $B_{\pm 1/3}(w, \zeta)$ are incomplete Anger and Weber functions. We will call (8.7) an incomplete Airy function.

Formulae (8.6)—(8.8) show that the incomplete Airy function $Ec(\alpha, z)$, as well as the function $Es(\alpha, z)$, are expressible in terms of incomplete Anger and Weber functions and belong to the class of incomplete cylindrical functions of the Bessel form.

For real, negative $z = x \exp i\pi = -x$, formula (8.7) reduces to formula (8.3) as $\alpha \to \infty$. This is clear from the asymptotic expansions for incomplete cylindrical functions of Bessel form which were obtained in the preceeding section. An alternative demonstration runs as follows: Since

$$A_{\pm 1/3}(w, z) = \frac{1}{\pi} \int_0^w \cos(z \sin t \pm 1/3t)\, dt,$$

formula (8.7) may be written as

$$Ec(\alpha, -x) = i \frac{2}{3} \sqrt{\frac{x}{3}} \int_0^{-i\beta} \cosh(\varrho \sinh t) \cos\left(\frac{t}{3}\right) dt$$

$$= \frac{2}{3} \sqrt{\frac{x}{3}} \int_0^{\beta} \cos(\varrho \sin u) \cosh\left(\frac{u}{3}\right) du,$$

where $\varrho = \frac{2}{3} \frac{x\sqrt{x}}{\sqrt{3}}$, $\beta = 3 \sinh^{-1} \frac{\alpha}{2} \sqrt{\frac{3}{x}}$. But for $\alpha \to \infty$, $\beta \to \infty$, and so

$$\lim_{\alpha \to \infty} Ec(\alpha, -x) = \frac{2}{3} \sqrt{\frac{x}{3}} \int_0^{\infty} \cos(\varrho \sinh u) \cosh\left(\frac{u}{3}\right) du = \frac{1}{3} \sqrt{x} K_{1/3}(\varrho). \tag{8.9}$$

Here we have used the fact that the MacDonald function has the integral representation

$$K_\nu(\varrho) = \frac{1}{\cos(\nu\pi/2)} \int_0^{\infty} \cos(\varrho \sinh t) \cosh \nu t\, dt, \tag{8.10}$$

valid for $\varrho > 0$ and $|\mathrm{Re}\, \nu| < 1$.

Similarly one can verify, that for real and positive values of $z = x$ and for $\alpha \to \infty$, formula (8.7) reduces to (8.2).

We turn now to the consideration of more general integrals of the Airy type. If for the moment we write $\sigma = \sinh \varphi$, then the following relations hold:

$$2 \cosh 2\varphi = 4\sigma^2 + 2;$$

$$2 \sinh 3\varphi = 8\sigma^2 + 6\sigma;$$

$$2 \cosh 4\varphi = 16\sigma^4 + 16\sigma^2 + 2$$

etc. which can be written in general as

$$2 {}^{\cosh}_{\sinh} n\varphi = (2\sigma)^n F\left(-\frac{n}{2};\ \frac{1-n}{2};\ 1-n; -\frac{1}{\sigma^2}\right). \tag{8.11}$$

Here either cosh or sinh is taken, according as n is even or odd, and $F(\alpha; \beta; \gamma; z)$ is the hypergeometric function. We also introduce the notation

$$T_n(t, s) = t^u F\left(-\frac{n}{2};\ \frac{1-n}{2};\ 1-n; -\frac{4s}{t^2}\right) \tag{8.12}$$

and consider the following integrals:

$$Ei_n(\alpha, s) = \int_0^\alpha \exp\left[-T_n(t, s)\right] dt; \tag{8.13}$$

$$Ci_n(\alpha, s) = \int_0^\alpha \cos\left[T_n(t, s)\right] dt; \tag{8.14}$$

$$Si_n(\alpha, s) = \int_0^\alpha \sin\left[T_n(t, s)\right] dt; \tag{8.15}$$

$$Ci_n(\alpha, s) + iSi_n(\alpha, s) = \int_0^\alpha \exp\left[iT_n(t, s)\right] dt. \tag{8.16}$$

For $\alpha \to \infty$ these integrals reduce to the generalized Airy integrals which were studied by Hardy (c.f. [10, p. 349]). In this case they can be expressed in terms of cylindrical functions with indices $\nu = 1/n$. For example, for even $n = 2k$ one can find the following relations:

$$\begin{aligned} Ci_n(\infty, s) + iSi_n(\infty, s) &= 2\frac{s^{1/2}}{n} \int_0^\infty \exp\left(2s^{n/2} i \cosh u\right) \cosh\left(\frac{u}{n}\right) du \\ &= \frac{\pi i}{n} s^{1/2} e^{\pi i/2n} \mathscr{H}^{(1)}_{1/n}\left(2s^{n/2}\right); \end{aligned} \tag{8.17}$$

$$\begin{aligned} Ei_n(\infty, s) &= \frac{2}{n} s^{1/2} \int_0^\infty \exp\left(-2s^{n/2} \cosh u\right) \cosh\left(\frac{u}{n}\right) du \\ &= \frac{2}{n} s^{1/2} K_{1/n}\left(2s^{n/2}\right). \end{aligned} \tag{8.18}$$

We now show that generalized incomplete integrals of the Airy-Hardy type (8.13)–(8.15) can be expressed in terms of incomplete cylindrical functions with the same indices, ($\nu = 1/n$). We consider first the case of even n. Introducing in the integrals (8.13)–(8.16) new variables of integration $t = 2s^{1/2} \sinh(u/n)$, and using relations (8.11) and (8.12) for polynomials $T_n(t, s)$, we find

$$\begin{aligned} Ei_n(\alpha, s) &= \frac{2s^{1/2}}{n} \int_0^w \exp\left(-2s^{n/2} \cosh u\right) \cosh\left(\frac{u}{n}\right) du \\ &= \frac{2s^{1/2}}{n} k_{1/n}\left(w, 2s^{n/2}\right); \end{aligned} \tag{8.19}$$

$$\begin{aligned} Ci_n(\alpha, s) + iSi_n(\alpha, s) &= \frac{2s^{1/2}}{n} \int_0^w \exp\left(2s^{n/2} i \cosh u\right) \cosh\left(\frac{u}{n}\right) du \\ &= \pi i \frac{s^{1/2}}{n} e^{i\pi/2n} h_{1/n}\left(w, 2s^{n/2}\right), \end{aligned} \tag{8.20}$$

where $n = 2k$ is even, $w = \sinh^{-1}(\alpha/2s^{1/2})$, and $k_\nu(w, z)$ and $h_\nu(w, z)$ — are, respectively, incomplete MacDonald and Hankel functions of the Bessel form.

These formulae show that the relation between incomplete integrals of the Airy-Hardy type and the corresponding incomplete cylindrical functions is identical with that between the Airy-Hardy integrals and the corresponding cylindrical functions. Obviously, as $\alpha \to \infty$ formulae (9.19) and (8.20) reduce to (8.17) and (8.18).

We turn now to the consideration of these case of odd $n = 2k + 1$. Repeating the same computations, but using relations (8.11), (8.12) for odd n, we find

$$\left.\begin{aligned} &Ci_n(\alpha, s) + iSi_n(\alpha, s) \\ &\quad = \frac{2}{n} s^{1/2} \int_0^w \exp(i2s^{n/2} \sinh u) \cosh\left(\frac{u}{n}\right) du \\ &\quad = \frac{\pi i}{n} s^{1/2} [\varepsilon_{1/n}(w, 2is^{n/2}) + \varepsilon_{-1/n}(w, 2is^{n/2})]; \\ Ei_n(\alpha, s) &= \frac{2s^{1/2}}{n} \int_0^w \exp(-2s^{n/2} \sinh u) \cosh\left(\frac{u}{n}\right) du \\ &= \frac{\pi i s^{1/2}}{n} [\varepsilon_{1/n}(w, -2s^{n/2}) + \varepsilon_{-1/n}(w, -2s^{n/2})], \end{aligned}\right\} \tag{8.21}$$

where $n = 2k + 1$ is an odd number, $\varepsilon_\nu(w, z)$ is an incomplete cylindrical function of Bessel form, and w as before is given by $w = \sinh^{-1}(\alpha/2s^{1/2})$.

Thus the incomplete Airy-Hardy integrals considered here can be expressed in terms of incomplete cylindrical functions of Bessel form and analysed by methods applicable to them.

9. Asymptotic Expansions of Incomplete Cylindrical Functions of the Bessel Form for Large Indices

We consider the behavior of the incomplete cylindrical functions of Bessel form

$$\Phi_\nu(w, z) = \frac{1}{\pi i} \int_{-\infty + i\lambda}^{w} e^{z \sinh t - \nu t} dt, \tag{9.1}$$

where $|\arg z - \lambda| < \pi/2$, and

$$\varepsilon_\nu(w, z) = \frac{1}{\pi i} \int_0^w e^{z \sinh t - \nu t} dt \tag{9.2}$$

for large values of the index ν and fixed values of the variables w and z. For this purpose we rewrite these in the simpler form set forth in § 9 of Chapter II. We set

$$\left.\begin{aligned} \varphi(t) &= e^{z \sinh t}; \\ h(t) &= -t. \end{aligned}\right\} \tag{9.3}$$

Since $h'(t) \neq 0$, there are no extremal points, and so in order to find asymptotic expansions it is only necessary to take account of the contributions from the neighborhoods of the limits of integration, i.e.

$$\Phi_\nu(w, z) \approx -\frac{1}{\pi i} e^{-\nu w} \sum_{k=0}^{\infty} \frac{A_k(w)\, \Gamma(k+1)}{\nu^{k+1}}; \tag{9.4}$$

$$\varepsilon_\nu(w, z) \approx -\frac{1}{\pi i} e^{-\nu w} \sum_{k=0}^{\infty} \frac{A_k(w)\, \Gamma(k+1)}{\nu^{k+1}} + \frac{1}{\pi i} \sum_{k=0}^{\infty} \frac{A^k(0)\, \Gamma(k+1)}{\nu^{k+1}}, \tag{9.5}$$

where the expansion coefficients $A_k(b)$ are found from the system of equations (9.4) of Chapter II, and in the present case take the form

$$A_k(b) = \frac{1}{\Gamma(k+1)} \left(\frac{d}{db}\right)^k \exp\,[z \sinh b]. \tag{9.6}$$

Finding explicit expressions for the coefficients $A_k(b)$ is straightforward. Limiting ourselves to the first three terms in (9.4) and (9.5) we have

$$\Phi_\nu(w, z) \approx -\frac{1}{\pi i \nu} e^{z \sinh w - \nu w} \left(1 + \frac{z \cosh w}{\nu} + \frac{z^2 \cosh^2 w + z \sinh w}{\nu^2}\right); \tag{9.7}$$

$$\begin{aligned} \varepsilon_\nu(w, z) \approx \frac{1}{\pi i \nu} (1 - e^{z \sinh w - \nu w}) + \frac{z}{\pi i \nu^2} (1 - e^{z \sinh w - \nu w} \cosh w) \\ + \frac{z^2}{\pi i \nu^3} \left[1 - e^{z \sinh w - \nu w} \left(\cosh^2 w + \frac{\sinh w}{z}\right)\right]. \end{aligned} \tag{9.8}$$

It is evident from the last formulae that (9.4) and (9.5) are inconvenient when $z \cosh w$ and ν are both large and nearly equal. For this case, the saddle point method in the form used above is inadequate. In order to obtain asymptotic expansions which reflect the behavior of these functions for large $z \cosh w$ and ν such that $\nu/z \cosh w$ is of order unity, it is first necessary to represent the functions in the form

$$\left.\begin{aligned} \Phi_\nu(w, z) &= \frac{1}{\pi i} \int_{-\infty + i\lambda}^{w} e^{\nu h(t)}\, dt; \\ \varepsilon_\nu(w, z) &= \frac{1}{\pi i} \int_{0}^{w} e^{\nu h(t)}\, dt, \end{aligned}\right\} \tag{9.9}$$

where $h(t) = z/\nu \sinh t - t$, and to estimate the additional contributions from the extrema which are determined by

$$h'(t) = (z/\nu) \cosh t - 1 = 0. \tag{9.10}$$

The fundamental difficulty here is the correct choice of the contour in the neighborhoods of the extrema for arbitrary w, z and ν. We note, however, that in finding asymptotic expansions of the cylindrical functions with large indices it is usual to use their integral representations in the Bessel-Schlaefli form, whose contours have been thoroughly studied (c.f. [10, p. 289] and also [18]). Therefore, as in the preceeding sections, the contribution of the extrema of (9.10) to the expansions for $\Phi_\nu(w, z)$ and $\varepsilon_\nu(w, z)$ may be characterized by the asymptotic behavior of the Hankel functions for large ν. For real $z = x > 0$ and $\nu > 0$ we have for example (c.f. [11, p. 978])

$$\left.\begin{aligned} \mathscr{H}_\nu^{(1)}(x) &\approx \frac{\alpha}{\sqrt{3}} \mathscr{H}_{1/3}^{(1)}\left(\frac{\nu\alpha^3}{3}\right) \\ &\quad \times \exp\left\{i\left[\pi/6 + \nu\left(\alpha - \frac{\alpha^3}{3} - \cot^{-1}\alpha\right)\right]\right\}; \\ \mathscr{H}_\nu^{(2)}(x) &\approx \frac{\alpha}{\sqrt{3}} \mathscr{H}_{1/3}^{(2)}\left(\frac{\nu\alpha^3}{3}\right) \\ &\quad \times \exp\left\{i\left[-\frac{\pi}{6} - \nu\left(\alpha - \frac{\alpha^3}{3} - \cot^{-1}\alpha\right)\right]\right\}, \end{aligned}\right\} \tag{9.11}$$

where $\alpha = \sqrt{(x/\nu)^2 - 1}$. The contributions to the asymptotic expansions of (9.1) and (9.2) from the neighborhoods of the limits of integration are characterized as before by formulae (9.4) and (9.5).

Asymptotic developments for $\Phi_\nu(w, z)$ and $\varepsilon_\nu(w, z)$ for large ν may also be obtained by methods similar to that of Fok for finding asymptotic expansions of the complete cylindrical functions (c.f., e.g. [10, p. 717] and ref 18, p. 69). To illustrate this method we limit ourselves to the simple case of real $z = x$ and purely imaginary $w = i\beta$, ($|\beta| < \pi$). In this case the most important contribution to the integral

$$\varepsilon_\nu(i\beta, x) = \frac{1}{\pi}\int_0^\beta e^{ix\sin\theta - i\nu\theta}\, d\theta$$

for $|\nu| \sim |x| \gg 1$ is made by small values of θ, for which $\sin\theta \approx \theta - 1/6\,\theta^3$, so that

$$\varepsilon_\nu(i\beta, x) \approx \frac{1}{\pi}\int_0^\beta e^{i[(x-\nu)\theta - 1/6 x\theta^3}\, d\theta. \tag{9.12}$$

The integral on the right side of this formula is an incomplete integral of the Airy-Hardy type, considered in the preceeding section. Therefore, we find after some transformations

$$\varepsilon_\nu(i\beta, x) \approx \left(\frac{6}{x}\right)^{1/3} \frac{1}{3} \gamma^{1/2} [\varepsilon_{1/3}(i\lambda, 2\gamma^{3/2}) + \varepsilon_{-1/3}(i\lambda, 2\gamma^{3/2})], \tag{9.13}$$

where for brevity we have put

$$\gamma = \left(\frac{6}{x}\right)^{1/3} \frac{x-\nu}{3};\; \lambda = 3\sin^{-1}\left[\beta\left(\frac{x}{6}\right)^{1/3} \frac{1}{2\gamma^{1/2}}\right]. \tag{9.14}$$

Similar computations lead us to an estimation for $\Phi_\nu(w, z)$ of the form

$$\Phi_\nu(i\beta, z) = \frac{1}{3}\left(\frac{6}{x}\right)^{1/3} \gamma^{1/2} [\Phi_{1/3}(i\lambda, 2\gamma^{3/2}) + \Phi_{-1/3}(i\lambda, 2\gamma^{3/2})]. \tag{9.15}$$

Thus approximate values of $\varepsilon_\nu(i\beta, x)$ and $\Phi_\nu(i\beta, x)$ for large ν and x may be found with the aid of the functions $\varepsilon_{\pm 1/3}(w, z)$ and $\Phi_{\pm 1/3}(w, z)$, series and asymptotic expansions for which have already been derived. We remark that these results also hold for complex values of $z = x + iy$, provided $y = \operatorname{Im} z > 0$. If this latter condition is fulfilled, the dominant contribution to the integrals (9.1) and (9.2) is determined as before by small values of θ.

Chapter IV

Incomplete Cylindrical Functions of Sonine-Schlaefli Form and the Incomplete Weber Integrals

1. The Concept of Incomplete Cylindrical Functions of Whittaker Form

Up to this point we have considered the incomplete cylindrical functions starting from their most general integral representations in either Poisson form

$$z^{\nu} \int_a^b e^{izt} (t^2 - 1)^{\nu - 1/2} \, dt \tag{1.1}$$

or in Bessel form

$$\int_a^b e^{z\sinh t - \nu t} \, dt. \tag{1.2}$$

These integrals have been shown to satisfy a non-homogeneous Bessel differential equation with a uniquely defined right side, and have allowed the introduction of the incomplete cylindrical functions of Poisson and Bessel form, as well as the incomplete Lipschitz-Hankel integrals. For integral values of the index ν, these functions are connected by simple relations, but for arbitrary values of ν they represent functions of entirely different classes, as was shown in § 4, Chapter III.

One can also introduce other types of incomplete cylindrical functions by means of different integral representations.

We can examine, for example, integrals of the form

$$z^{1/2} \int_a^b e^{izt} T(t) \, dt. \tag{1.3}$$

where $T(t)$ is an as yet unspecified function. Applying to (1.3) the Bessel operator

$$\nabla_{\nu} = z^2 \frac{d^2}{dz^2} + z \frac{d}{dz} + (z^2 - \nu^2), \tag{1.4}$$

we obtain, after some manipulation,

$$\begin{aligned}\nabla_\nu\left[z^{1/2}\int_a^b e^{izt}T(t)\,dt\right] &= z^{1/2}e^{izt}(1-t^2)\left(\frac{dT}{dt}-izT\right)\Big|_{t=a}^{t=b}\\ &- z^{1/2}\int_a^b e^{izt}\left[(1-t^2)\frac{d^2T}{dt^2}-2t\frac{dT}{dt}+\left(\nu^2-\frac{1}{4}\right)T\right]dt.\end{aligned} \tag{1.5}$$

Hence, if $T(t)$ satisfies the differential equation

$$(1-t^2)\frac{d^2T}{dt^2}-2t\frac{dT}{dt}+\left(\nu^2-\frac{1}{4}\right)T=0 \tag{1.6}$$

and if the contour in (1.3) is so chosen that the first term on the right in (Eq. 1.5) vanishes,

$$\left[z^{1/2}e^{izt}(1-t^2)\left(\frac{dT}{dt}-izT\right)\right]_{t=a}^{t=b}=0, \tag{1.7}$$

then (1.3) is a solution of Bessel's equation and can be used as an integral representation for the cylindrical functions.

As is well known, the solutions of Eq. (1.6) are the Legendre functions of first and second kind $P_{\nu-1/2}(t)$ and $Q_{\nu-1/2}(t)$. Thus, with an appropriate choice of integration contour we can find the following integral representations for complete cylindrical functions in what is known as the Whittaker form (c.f. [10, p. 194])

$$\mathscr{H}_\nu^{(1)}(z)=\left(\frac{2z}{\pi}\right)^{1/2}\frac{e^{-i\pi(\nu/2+1/4)}}{\pi}\int_{\infty i\exp(-i\omega)}^{(1+)} e^{izt}Q_{\nu-1/2}(t)\,dt; \tag{1.8}$$

$$\mathscr{H}_\nu^{(2)}(z)=\left(\frac{z}{2\pi}\right)^{1/2}\frac{e^{i\pi(\nu/2+1/4)}}{\pi\cos\nu\pi}\int_{\infty i\exp(-i\omega)}^{(-1+)} e^{izt}P_{\nu-1/2}(t)\,dt. \tag{1.9}$$

The contour comes in from infinity a long $i\exp(-iw)$ and encircles the branch points $+1$ and -1 in the counterclockwise direction. If $T(t)$ satisfies Eq. (1.6) but a contour is chosen such that (1.7) is not satisfierd, then (1.3) satisfies a non-homogeneous Bessel differential equation with a uniquely determined right side. Such functions are accordingly incomplete cylindrical functions and shall be called incomplete cylindrical functions of Whittaker form. By analogy, one can construct a general theory for this class of functions. However, since they are met relatively infrequently we shall not pursue themfurther here.

2. Incomplete Cylindrical Functions of Sonine-Schlaefli Form and their Basic Properties

In the first chapter we noted that in addition to the Poisson and Bessel forms, there is wide application of integral representation in the form of Sonine-Schlaefli.

For the Bessel functions, this representation takes the form

$$J_\nu(z) = \frac{1}{2\pi i}\left(\frac{z}{2}\right)^\nu \int\limits_{c-i\infty}^{c+i\infty} t^{-\nu-1} \exp\left(t - \frac{z^2}{4t}\right) dt, \tag{2.1}$$

where c is an arbitrary positive constant, and $\operatorname{Re}\nu > -1$. Let us consider a different contour of integration, and write

$$S_\nu(p, q; z) = \frac{1}{2\pi i}\left(\frac{z}{2}\right)^\nu \int\limits_p^q t^{-\nu-1} \exp\left(t - \frac{z^2}{4t}\right) dt, \tag{2.2}$$

where p and q are complex numbers, one of which may be regarded as fixed, say $p = 1$.

In accordance with Eq. (2.1) these functions reduce for $q = \bar{p} \to c + i\infty$ to the Bessel function. In what follows, we shall call them incomplete cylindrical functions of Sonine-Schlaefli form. It is also easily seen that for $p = z/2$; $q = (z/2)\exp w$ they reduce to incomplete cylindrical functions of Bessel form. In fact, from the definition we have

$$S_\nu\left(\frac{z}{2}, \frac{z}{2}\mathrm{e}^w; z\right) = \frac{1}{2\pi i}\left(\frac{z}{2}\right)^\nu \int\limits_{z/2}^{(z/2)\exp w} t^{-\nu-t} \exp\left(t - \frac{z^2}{4t}\right) dt,$$

which, under the change of variable $t = (z/2)\exp u$ becomes

$$S_\nu\left(\frac{z}{2}, \frac{z}{2}\mathrm{e}^w; z\right) = \frac{1}{2\pi i}\int\limits_0^w \mathrm{e}^{z\sinh u - \nu u}\, du = \frac{1}{2}\,\varepsilon_\nu(w, z). \tag{2.3}$$

In this sense, incomplete cylindrical functions of Sonine-Schlaefli form constitute a more general class of functions than do the functions $\varepsilon_\nu(w, z)$. Thus by studying the properties of $S_\nu(p, q; z)$ one can obtain those of $\varepsilon_\nu(w, z)$ as a special case.

We shall first derive a differential equation and recurrence relations for the $S_\nu(p, q; z)$. To this end we apply the Bessel operator (1.4) and obtain

$$\frac{dS_\nu}{dz} = \frac{1}{2\pi i 2^\nu}\left[\nu z^{\nu-1}\int\limits_p^q t^{-\nu-1}\exp\left(t - \frac{z^2}{4t}\right) dt - \frac{z^{\nu+1}}{2}\int\limits_p^q t^{-\nu-2}\exp\left(t - \frac{z^2}{4t}\right) dt\right];$$

$$\frac{d^2S_\nu}{dz^2} = \frac{1}{2\pi i 2^\nu}\left[\nu(\nu - 1)\, z^{\nu-2}\int\limits_p^q t^{-\nu-1}\exp\left(t - \frac{z^2}{4t}\right) dt\right.$$

$$\left. - \frac{2\nu + 1}{2} z^\nu \int\limits_p^q t^{-\nu-2}\exp\left(t - \frac{z^2}{4t}\right) dt + \frac{z^{\nu+2}}{4}\int\limits_p^q t^{-\nu-3}\exp\left(t - \frac{z^2}{4t}\right) dt\right].$$

The last integral of this formula can be written as

$$\frac{z^{\nu+2}}{4}\int_p^q t^{-\nu-3}\exp\left(t-\frac{z^2}{4t}\right)dt = z^\nu\int_p^q e^t t^{-\nu-1}\frac{d}{dt}\left[\exp\left(-\frac{z^2}{4t}\right)\right]dt$$

$$= z^\nu t^{-\nu-1}\exp\left(t-\frac{z^2}{4t}\right)\Big|_{t-p}^{t=q}$$

$$-z^\nu\left[\int_p^q t^{-\nu-1}\exp\left(t-\frac{z^2}{4t}\right)dt-(\nu+1)\int_p^q t^{-\nu-2}\exp\left(t-\frac{z^2}{4t}\right)dt\right].$$

Forming the Bessel operator, we find after some transformation:

$$\nabla_\nu S_\nu(p, q; z) = z^2\left[\frac{d^2S_\nu}{dz^2}+\frac{1}{z}\frac{dS_\nu}{dz}+\left(1-\frac{\nu^2}{z^2}\right)S_\nu\right] = \Psi(z, q)-\Psi(z, p), \tag{2.4}$$

where

$$\Psi(z, x) = \frac{z}{\pi i}\left(\frac{z}{2x}\right)^{\nu+1}\exp\left(x-\frac{z^2}{4x}\right). \tag{2.5}$$

Analogously, one can obtain the following recurrence relations

$$S_{\nu-1}+S_{\nu+1}-\frac{2\nu}{z}S_\nu = \frac{2}{z^3}\left[q\Psi(z, q)-p\Psi(z, p)\right]; \tag{2.6}$$

$$S_{\nu-1}-S_{\nu+1}-\frac{2\partial S_\nu}{\partial z} = \frac{2}{z^3}\left[q\Psi(z, q)-p\Psi(z, p)\right], \tag{2.7}$$

from which it follows that

$$z\frac{\partial S_\nu}{\partial z} = \nu S_\nu - zS_{\nu+1}. \tag{2.8}$$

Thus the incomplete cylindrical functions $S_\nu(p, q; z)$ satisfy the non-homogeneous differential equation (2.4) with a uniquely determined right side. This fact permits us to obtain for them a somewhat different representation. By choosing the functions $J_\nu(z)$ and $N_\nu(z)$ as a fundamental set of solutions for the homogeneous equation one obtains by the method of variation of parameters:

$$\begin{aligned} S_\nu(p, q; z) &= \frac{\pi}{2}N_\nu(z)\int_a^z J_\nu(t)\frac{\Psi(t, q)-\Psi(t, p)}{t}dt \\ &\quad -\frac{\pi}{2}J_\nu(z)\int_b^z N_\nu(t)\frac{\Psi(t, q)-\Psi(t, p)}{t}dt, \end{aligned} \tag{2.9}$$

where the constants a and b are to be so chosen that the asymptotic behavior of the right side of (2.9) agrees with the asymptotic behavior of $S_\nu(p, q; z)$, as $z \to 0$. According to (2.2), it is not difficult to see that this condition can be fulfilled by taking $a = 0$, $b = \infty \exp i\lambda$. For this choice, formula (2.9), with the aid of (2.5) becomes

$$S_\nu(p, q; z) = \frac{1}{2i} N_\nu(z) \left[Q_\nu(q, z) - Q_\nu(p, z)\right] + \frac{1}{2i} J_\nu(z) \left[P_\nu(q, z) - P_\nu(p, z)\right], \tag{2.10}$$

where the following notations have been introduced:

$$Q_\nu(x, z) = (2x)^{-\nu-1} e^x \int_0^z t^{\nu+1} J_\nu(t) \exp\left(-\frac{t^2}{4x}\right) dt; \tag{2.11}$$

$$P_\nu(x, z) = (2x)^{-\nu-1} e^x \int_z^{\infty \exp i\lambda} t^{\nu+1} N_\nu(t) \exp\left(-\frac{t^2}{4x}\right) dt. \tag{2.12}$$

Convergence of these integrals at the lower limit is guaranteed if $\operatorname{Re} \nu > -1$, and at infinity if one chooses λ such that for a given x the condition

$$|2\lambda - \arg x| < \pi/2. \tag{2.13}$$

is fulfilled.

Along with the functions just considered, there also occur integrals of the following forms

$$\frac{1}{2\pi i}\left(\frac{z}{2}\right)^\nu \int_p^q t^{-\nu-1} \exp\left(t + \frac{z^2}{4t}\right) dt; \tag{2.14}$$

$$\frac{1}{2}\left(\frac{z}{2}\right)^\nu \int_p^q t^{-\nu-1} \exp\left(-t - \frac{z^2}{4t}\right) dt. \tag{2.15}$$

The first of these for $\bar{p} = q = c + i\infty$ reduces to the modified Bessel function $I_\nu(z)$ while the second for $p = 0$, $q = \infty$ together with the condition $|2 \arg z - \arg x| < \pi/2$ reduces to the MacDonald function $K_\nu(z)$ (see § 1, Chapter I). For arbitrary values of p and q the integrals (2.14) and (2.15) can be simply expressed in terms of the function $S_\nu(p, q; z)$. Thus comparison of (2.2) with the integral (2.14) gives

$$\frac{1}{2\pi i}\left(\frac{z}{2}\right)^\nu \int_p^q t^{-\nu-1} \exp\left(t + \frac{z^2}{4t}\right) dt = S_\nu(p, q; iz)\, e^{-i\nu\pi/2}. \tag{2.16}$$

while comparison of (2.2) with (2.15), after first making the change of variable $t = u \exp(-i\pi)$ yields

$$\frac{1}{2}\left(\frac{z}{2}\right)^\nu \int_p^q t^{-\nu-1} \exp\left(-t - \frac{z^2}{4t}\right) dt = i\pi S_\nu(p e^{i\pi}, q e^{i\pi}; iz)\, e^{i\nu\pi/2}. \tag{2.17}$$

Using (2.10) with $S_\nu(p, q; iz)$ in place of $S_\nu(p, q, z)$ we can write in analogy to (2.11) and (2.12)

$$\tilde{Q}_\nu(x, z) = (2x)^{-\nu-1} e^x \int_0^z t^{\nu+1} I_\nu(t) \exp\left(-\frac{t^2}{4x}\right) dt; \tag{2.18}$$

$$\tilde{P}_\nu(x, z) = (2x)^{-\nu-1} e^x \int_z^{\infty \exp i\lambda} t^{\nu+1} K_\nu(t) \exp\left(-\frac{t^2}{4x}\right) dt, \tag{2.19}$$

which differ from $Q_\nu(x, z)$ and $P_\nu(x, z)$ only in that the ordinary Bessel functions are replaced by the modified ones.

All of these functions will be studied separately in later sections.

3. Incomplete Weber Integrals

It was noted previously that the incomplete cylindrical functions of Sonine-Schlaefli form can be expressed by means of the integrals

$$Q_\nu(x, z) = (2x)^{-\nu-1} e^x \int_0^z t^{\nu+1} J_\nu(t) \exp\left(-\frac{t^2}{4x}\right) dt; \tag{3.1}$$

$$P_\nu(x, z) = (2x)^{-\nu-1} e^x \int_z^{\infty \exp i\lambda} t^{\nu+1} N_\nu(t) \exp\left(-\frac{t^2}{4x}\right) dt; \tag{3.2}$$

$$\tilde{Q}_\nu(x, z) = (2x)^{-\nu-1} e^x \int_0^z t^{\nu+1} I_\nu(t) \exp\left(-\frac{t^2}{4x}\right) dt; \tag{3.3}$$

$$\tilde{P}_\nu(x, z) = (2x)^{-\nu-1} e^x \int_z^{\infty \exp i\lambda} t^{\nu+1} K_\nu(t) \exp\left(-\frac{t^2}{4x}\right) dt, \tag{3.4}$$

where $\operatorname{Re} \nu > -1$ and $|2\lambda - \arg x| < \pi/2$. Here we shall study their characteristics in more detail.

From the following well known relations for the Weber integrals involving the Bessel and Neumann functions (see [10, p. 430] and [11, p. 731])

$$\int_0^{\infty \exp i\lambda} t^{\nu+1} J_\nu(t) \exp\left(-\frac{t^2}{4x}\right) dt = (2x)^{\nu+1} e^{-x}; \tag{3.5}$$

$$\begin{aligned} &\int_0^{\infty \exp i\lambda} t^{\nu+1} N_\nu(t) \exp\left(-\frac{t^2}{4x}\right) dt \\ &= (2x)^{\nu+1} e^{-x} \left[\cot \nu\pi - \frac{\Gamma(\nu)}{\pi x^\nu} F(-\nu, 1-\nu; x)\right] \end{aligned} \tag{3.6}$$

it follows that

$$Q_\nu(x, \infty \exp i\lambda) = 1; \tag{3.7}$$

$$P_\nu(x, 0) = \cot \nu\pi - \frac{\Gamma(\nu)}{\pi x^\nu} F(-\nu, 1-\nu; x), \tag{3.8}$$

where $F(\alpha, \beta; x)$, is the confluent hypergeometric function. Formulae (3.6) and (3.8) retain their validity for $\nu \to n$.

From these relations one sees that $Q_\nu(x, z)$ for $z \to \infty \exp i\lambda$, and $P_\nu(x, z)$ for $z \to 0$ differ only by constant factors from the corresponding Weber integrals. Therefore, in the following discussion $Q_\nu(x, z)$ and $P_\nu(x, z)$ will be referred to as incomplete Weber integrals involving Bessel and Neumann functions, while the corresponding functions $\tilde{Q}_\nu(x, z)$ and $\tilde{P}_\nu(x, z)$, will be called incomplete Weber integrals involving modified Bessel functions. The latter are related to $P_\nu(x, z)$ and $Q_\nu(x, z)$ through a simple change of variable.

Changing the variable of integration in (3.4) by setting $\xi = t \exp(i\pi/2)$ and using the relation

$$K_\nu(\xi e^{-i\pi/2}) = \frac{\pi i}{2} e^{i\nu\pi/2} \mathscr{H}_\nu^{(1)}(\xi) = \frac{\pi i}{2} e^{i\nu\pi/2} [J_\nu(\xi) + iN_\nu(\xi)],$$

we find after simple transformations

$$\tilde{P}_\nu(x, z) = \frac{i\pi}{2} e^{i\nu\pi} e^{2x} [Q_\nu(xe^{i\pi}, \infty \exp i\lambda) - Q_\nu(xe^{i\pi}, iz) + iP_\nu(xe^{i\pi}, iz)].$$

Alternatively, making use of (3.7) we have

$$\tilde{P}_\nu(x, z) = \frac{i\pi}{2} e^{i\nu\pi} e^{2x} [1 - Q_\nu(xe^{i\pi}, iz) + iP_\nu(xe^{i\pi}, iz)]. \tag{3.9}$$

Analogously $Q_\nu(x, z)$ and $\tilde{Q}_\nu(x, z)$ are connected by:

$$\tilde{Q}_\nu(x, z) = e^{2x} Q_\nu(xe^{i\pi}, iz). \tag{3.10}$$

The last relation shows that we need only consider the incomplete Weber integrals containing Bessel and Neumann functions.

The incomplete Weber integrals satisfy simple recursion relations which can be obtained with the aid of the equations:

$$\left.\begin{aligned} t^{\nu+1} J_\nu(t) &= \frac{d}{dt} [t^{\nu+1} J_{\nu+1}(t)]; \\ t^{\nu+1} N_\nu(t) &= \frac{d}{dt} [t^{\nu+1} N_{\nu+1}(t)]; \\ t^{\nu+1} I_\nu(t) &= \frac{d}{dt} [t^{\nu+1} I_{\nu+1}(t)]; \\ t^{\nu+1} K_\nu(t) &= -\frac{d}{dt} [t^{\nu+1} K_{\nu+1}(t)]. \end{aligned}\right\} \tag{3.11}$$

We find

$$Q_{\nu+1}(x, z) - Q_\nu(x, z) = -\left(\frac{z}{2x}\right)^{\nu+1} J_{\nu+1}(z) \exp\left(x - \frac{z^2}{4x}\right); \tag{3.12}$$

$$\tilde{Q}_{\nu+1}(x, z) - \tilde{Q}_\nu(x, z) = -\left(\frac{z}{2x}\right)^{\nu+1} I_{\nu+1}(z) \exp\left(x - \frac{z^2}{4x}\right); \tag{3.13}$$

$$P_{\nu+1}(x, z) - P_\nu(x, z) = +\left(\frac{z}{2x}\right)^{\nu+1} N_{\nu+1}(z) \exp\left(x - \frac{z^2}{4x}\right); \tag{3.14}$$

$$\tilde{P}_{\nu+1}(x, z) + \tilde{P}_\nu(x, z) = \left(\frac{z}{2x}\right)^{\nu+1} K_{\nu+1}(z) \exp\left(x - \frac{z^2}{4x}\right). \tag{3.15}$$

From the above formulae the following more general relations can be obtained:

$$Q_{\nu+n+1}(x, z) = Q_\nu(x, z) - \exp\left(x - \frac{z^2}{4x}\right) \sum_{k=0}^{n} \left(\frac{z}{2x}\right)^{\nu+k+1} J_{\nu+k+1}(z); \quad (3.16)$$

$$\tilde{Q}_{\nu+n+1}(x, z) = \tilde{Q}_\nu(x, z) - \exp\left(x - \frac{z^2}{4x}\right) \sum_{k=0}^{n} \left(\frac{z}{2x}\right)^{\nu+k+1} I_{\nu+k+1}(z); \quad (3.17)$$

$$P_{\nu+n+1}(x, z) = P_\nu(x, z) + \exp\left(x - \frac{z^2}{4x}\right) \sum_{k=0}^{n} \left(\frac{z}{2x}\right)^{\nu+k+1} N_{\nu+k+1}(z); \quad (3.18)$$

$$\begin{aligned} \tilde{P}_{\nu+n+1}(x, z) &= \tilde{P}_\nu(x, z)\,(-1)^{n+1} \\ &+ \exp\left(x - \frac{z^2}{4x}\right) \sum_{k=0}^{n} \left(\frac{z}{2x}\right)^{\nu+k+1} (-1)^{k-n} K_{\nu+k+1}(z). \end{aligned} \quad (3.19)$$

These formulae show that the incomplete Weber integrals for arbitrary index are completely determined by those with index $|\nu| < 1$. In particular, for integer ν, it is sufficient to know their value for $\nu = 0$.

We shall now demonstrate the characteristic behavior of these integrals for large and small values of x. To this end we rewrite formula (3.1) and carry out an integration by parts. The result is

$$\begin{aligned} Q_\nu(x, z) &= (2x)^{-\nu-1} e^x \int_0^z \exp\left(-\frac{t^2}{4x}\right) \frac{d}{dt} \left[t^{\nu+1} J_{\nu+1}(t)\right] dt \\ &= (2x)^{-\nu-1} e^x \Bigg[z^{\nu+1} J_{\nu+1}(z) \exp\left(-\frac{z^2}{4x}\right) \\ &\quad + \frac{1}{2x} \int_0^z t^{\nu+2} J_{\nu+1}(t) \exp\left(-\frac{t^2}{4x}\right) dt \Bigg]. \end{aligned}$$

Carrying out a similar transformation m times, we obtain

$$\begin{aligned} Q_\nu(x, z) &= \exp\left(x - \frac{z^2}{4x}\right) \sum_{k=0}^{m} \left(\frac{z}{2x}\right)^{\nu+1+k} J_{\nu+1+k}(z) \\ &+ (2x)^{-\nu-2-m} e^x \int_0^z t^{\nu+2+m} J_{\nu+1+m}(t) \exp\left(-\frac{t^2}{4x}\right) dt. \end{aligned} \quad (3.20)$$

For $m \to \infty$ the last term tends to zero for any value of x or z and thus for the incomplete Weber integral $Q_\nu(x, z)$: we have the following Neumann series

$$Q_\nu(x, z) = \exp\left(x - \frac{z^2}{x}\right) \sum_{k=0}^{\infty} \left(\frac{z}{2x}\right)^{\nu+1+k} J_{\nu+1+k}(z). \quad (3.21)$$

The same result can be arrived at using the relation of these functions to Lommel's function of two variables (see § 6 below).

By an analogous procedure we obtain for the incomplete Weber integral $\tilde{Q}_\nu(x, z)$,

$$\tilde{Q}_\nu(x, z) = \exp\left(x - \frac{z^2}{4x}\right) \sum_{k=0}^{\infty} \left(\frac{z}{2x}\right)^{\nu+1+k} I_{\nu+1+k}(z). \tag{3.22}$$

The series just obtained are absolutely convergent and are suitable for numerical computation. They also permit asymptotic expansions for large x. The dominant terms of these expansions for $|z/2x| \ll 1$ are

$$Q_\nu(x, z) \approx \left(\frac{z}{2x}\right)^{\nu+1} J_{\nu+1}(z) \exp\left(x - \frac{z^2}{4x}\right); \tag{3.23}$$

$$\tilde{Q}_\nu(x, z) \approx \left(\frac{z}{2x}\right)^{\nu+1} I_{\nu+1}(z) \exp\left(x - \frac{z^2}{4x}\right). \tag{3.24}$$

For small values of x the expansions (3.21) and (3.22) prove to be inconvenient when $|z/2x| \gg 1$. We shall therefore treat this case by a somewhat different method. We rewrite (3.1) in the form

$$Q_\nu(x, z) = (2x)^{-\nu-1} e^x \left[\int_0^{\infty \exp i\lambda} t^{\nu+1} J_\nu(t) \exp\left(-\frac{t^2}{4x}\right) dt - \int_z^{\infty \exp i\lambda} t^{\nu+1} J_\nu(t) \exp\left(-\frac{t^2}{4x}\right) dt \right].$$

Making use of (3.5) we find

$$Q_\nu(x, z) = 1 - (2x)^{-\nu-1} e^x \int_z^{\infty \exp i\lambda} t^{\nu+1} J_\nu(t) \exp\left(-\frac{t^2}{4x}\right) dt. \tag{3.25}$$

Now, integrating by parts m times, and noting that

$$t \exp\left(-\frac{t^2}{4x}\right) = -2x \frac{d}{dt} \exp\left(-\frac{t^2}{4x}\right), \tag{3.26}$$

we obtain

$$\begin{aligned} Q_\nu(x, z) = 1 &- \exp\left(x - \frac{z^2}{4x}\right) \sum_{k=0}^{m} (2x)^{k-\nu} \left(\frac{d}{z\,dz}\right)^k [z^\nu J_\nu(z)] \\ &+ (2x)^{m+1-\nu} e^x \int_z^{\infty \exp i\lambda} \left(\frac{d}{t\,dt}\right)^{m+1} [t^\nu J_\nu(t)] \frac{d}{dt} \exp\left(-\frac{t^2}{4x}\right) dt. \end{aligned} \tag{3.27}$$

and taking advantage of the well known relation

$$\left(\frac{d}{t\,dt}\right)^m [t^\nu J_\nu(t)] = t^{\nu-m} J_{\nu-m}(t), \tag{3.28}$$

we find

$$\begin{aligned} Q_\nu(x, z) = 1 &- \exp\left(x - \frac{z^2}{4x}\right) \sum_{k=0}^{m} \left(\frac{2x}{z}\right)^{k-\nu} J_{\nu-k}(z) \\ &+ (2x)^{m+1-\nu} \int_z^{\infty \exp i\lambda} t^{\nu-m-1} J_{\nu-m-1}(t) \frac{d}{dt} \exp\left(-\frac{t^2}{4x}\right) dt. \end{aligned} \tag{3.29}$$

Assuming that $|2 \arg z - \arg x| < \pi/2$, the integral in the second term of this formula remains bounded for $x \to 0$ ($|z/2x| \gg 1$). Therefore for arbitrary m, formula (3.29) can be written as

$$Q_\nu(x, z) = 1 - \exp\left(x - \frac{z^2}{4x}\right)\left[\sum_{k=0}^{m}\left(\frac{2x}{z}\right)^{k-\nu} J_{\nu-k}(z) + O\left(\left|\frac{2x}{z}\right|^{m+1-\nu}\right)\right]. \tag{3.30}$$

Allowing $m \to \infty$, the resulting Neumann series is convergent only for integer values of ν. Therefore, for arbitrary ν (3.30) can only be regarded as an asymptotic representation for $Q_\nu(x, z)$, valid for sufficiently small x when $|2x/z| \ll 1$. The leading term of this expansion has the form

$$Q_\nu(x, z) \approx 1 - \left(\frac{z}{2x}\right)^\nu J_\nu(z) \exp\left(x - \frac{z^2}{4x}\right) \tag{3.31}$$

and with the aforementioned limitations $|2 \arg z - \arg x| < \pi/2$ we find

$$\lim_{x \to 0} Q_\nu(x, z) = 1. \tag{3.32}$$

The asymptotic expansion for $\tilde{Q}_\nu(x, z)$ is easy to obtain from (3.10) and has the form

$$\tilde{Q}_\nu(x, z) = e^{2x} - \exp\left(x - \frac{z^2}{4x}\right)\left[\sum_{k=0}^{m}\left(\frac{2x}{z}\right)^{k-\nu} I_{\nu-k}(z) + O\left(\left|\frac{2x}{z}\right|^{m+1-\nu}\right)\right]. \tag{3.33}$$

For $m \to \infty$ the resulting series is, as before, only convergent for integer values of ν and is asymptotic otherwise. The leading terms of this expansion, assuming $|2x/z| \ll 1$ have the form

$$\tilde{Q}_\nu(x, z) \approx e^{2x} - \left(\frac{z}{2x}\right)^\nu I_\nu(z) \exp\left(x - \frac{z^2}{4x}\right), \tag{3.34}$$

from which, under the condition $|2 \arg z - \arg x| \leq \pi/2$ we find

$$\underset{x \to 0}{\tilde{Q}_\nu}(x, z) = 1. \tag{3.35}$$

We shall now proceed to the determination of asymptotic expansions for incomplete Weber integrals containing Neumann and MacDonald functions. We shall consider first the case where x is large, and is such that $|z/2x| \ll 1$. Writing formula (3.2) in the form

$$P_\nu(x, z) = (2x)^{-\nu-1} e^x \left[\int_0^{\infty \exp i\lambda} t^{\nu+1} N_\nu(t) \exp\left(-\frac{t^2}{4x}\right) dt - \int_0^z t^{\nu+1} N_\nu(t) \exp\left(-\frac{t^2}{4x}\right) dt\right]. \tag{3.36}$$

it is seen that the first term of this expression is $P_\nu(x, 0)$ which is given by formula (3.8). To evaluate the second term, we first carry out m integrations by parts as we did in obtaining formula (3.20). We find

$$P_\nu(x, z) = \cot \nu\pi - \frac{\Gamma(\nu)}{\pi x^\nu} F(-\nu, 1-\nu; x)$$
$$- \exp\left(x - \frac{z^2}{4x}\right)\left[\sum_{k=0}^{m} \left(\frac{z}{2x}\right)^{\nu+1+k} N_{\nu+1+k}(z) \right. \tag{3.37}$$
$$\left. + O\left(\left|\frac{z}{2x}\right|^{\nu+2+m}\right)\right] - \frac{e^x}{\pi}\left[\sum_{k=0}^{m} \frac{\Gamma(1+\nu+k)}{x^{\nu+1+k}} + O\left(\left|\frac{1}{x}\right|^{\nu+2+m}\right)\right].$$

The resulting series for $m \to \infty$ is non-convergent and can serve only as an asymptotic representation of $P_\nu(x, z)$ for $x \gg 1$. To obtain explicit expressions for the dominant terms in this expression we take advantage of the asymptotic representation of the hypergeometric function (see formula (8.15), Chapter II)

$$F(-\nu, 1-\nu; x) \approx -\frac{\nu e^x}{x} + \Gamma(1-\nu)\, x^\nu e^{\pm i\nu\pi}, \tag{3.38}$$

where the positive sign is taken for $\operatorname{Im} x < 0$, and the negative for $\operatorname{Im} x > 0$. Putting (3.38) into (3.37) and using $\Gamma(1-\nu)\,\Gamma(\nu) = \pi/\sin \nu\pi$, we find, after a simple transformation

$$\underset{|x|\to\infty}{P_\nu(x, z)} \approx \pm i - \left(\frac{z}{2x}\right)^{\nu+1} N_{\nu+1}(z) \exp\left(x - \frac{z^2}{4x}\right), \tag{3.39}$$

where the positive sign corresponds to $\operatorname{Im} x > 0$ and the negative to $\operatorname{Im} x < 0$. For $x \to c \pm i\infty$ only the first term, $(\pm i)$, remains, and therefore

$$P_\nu(x, z) = \pm i; \quad x \to c \pm i\infty.$$

The above arguments were presented with the aim of describing the behavior of the incomplete Weber integral $P_\nu(x, z)$ when x tends to infinity along a line parallel to the imaginary axis. In fact, by taking advantage of the asymptotic expansion of the confluent hypergeometric function, it is seen that the second term is completely compensated for (by the infinite sum in the last term) and so for $|z/2x| \ll 1$ we obtain

$$P_\nu(x, z) \approx \pm i - \left\{\sum_{k=0}^{\infty} \left(\frac{z}{2x}\right)^{\nu+1+k} N_{\nu+1+k}(z)\right\} \exp\left(x - \frac{z^2}{4x}\right). \tag{3.40}$$

Here again only the first term is remained for $x \to c \pm i\infty$.

Assume now that x is small and such that $(2x/z) \ll 1$. Integrating the right side of (3.2) by parts m times we have

$$P_\nu(x, z) \approx \left[\sum_{k=0}^{\infty} \left(\frac{2x}{z}\right)^{k-\nu} N_{\nu-k}(z)\right] \exp\left(x - \frac{z^2}{4x}\right). \tag{3.41}$$

The dominant term of this expression for $(2x/z) \ll 1$ is

$$P_\nu(x, z) \approx \left(\frac{z}{2x}\right)^\nu N_\nu(z) \exp\left(x - \frac{z^2}{4x}\right), \tag{3.42}$$

from which it follows under the condition $|2 \arg z - \arg x| < \pi/2$, that

$$\underset{x \to 0}{P_\nu}(x, z) = 0. \tag{3.43}$$

Similarly, one can obtain asymptotic expansions for the incomplete Weber integral (3.4) by taking advantage of relation (3.9). For large x with $|z/2x| \ll 1$, this relation, together with formulae (3.2) and (3.27), gives

$$\tilde{P}_\nu(x, z) \approx \left[\sum_{k=0}^{\infty} \left(\frac{z}{2x}\right)^{\nu+1+k} (-1)^k K_{\nu+1+k}(z)\right] \exp\left(x - \frac{z^2}{4x}\right). \tag{3.44}$$

Here use has been made of (8.19) and (8.20) of Chapter II:

$$F(-\nu; 1-\nu; -x) = \mathrm{e}^{-x} F(1; 1-\nu; x) = -\nu x^\nu \gamma(-\nu, x) \tag{3.45}$$

and of the asymptotic relation (8.23), Chapter II for the incomplete γ-function.

The dominant term in formula (3.44) is

$$\tilde{P}_\nu(x, z) \approx \left(\frac{z}{2x}\right)^{\nu+1} K_{\nu+1}(z) \exp\left(x - \frac{z^2}{4x}\right). \tag{3.46}$$

Further, relation (3.9) together with formulae (3.30) and (3.41) for $|2x/z| \ll 1$ gives the following asymptotic representation

$$\tilde{P}_\nu(x; z) \approx \left[\sum_{k=0}^{\infty} \left(\frac{2x}{z}\right)^{k-\nu} (-1)^k K_{\nu-k}(z)\right] \exp\left(x - \frac{z^2}{4x}\right). \tag{3.47}$$

The dominant term of which is

$$\tilde{P}_\nu(x, z) \approx \left(\frac{z}{2x}\right)^\nu K_\nu(x) \exp\left(x - \frac{z^2}{4x}\right). \tag{3.48}$$

When the condition $|2 \arg z - \arg x| < \pi/2$ is imposed this gives

$$\lim_{x \to 0} \tilde{P}_\nu(x, z) = 0. \tag{3.49}$$

We have thus studied certain characteristics of incomplete Weber integrals involving cylindrical functions. Recalling now relation (2.10)

$$\begin{aligned} S_\nu(p, q; z) &= \frac{1}{2i} N_\nu(z) \left[Q_\nu(q, z) - Q_\nu(p, z)\right] \\ &\quad + \frac{1}{2i} J_\nu(z) \left[P_\nu(q, z) - P_\nu(p, z)\right], \end{aligned} \tag{3.50}$$

it is easily seen from the basic expressions (3.23) and (3.27) that the behavior of their right sides for small z is precisely the same as that of the function $S_\nu(p, q, z)$ for $z \to 0$, as given by (2.2).

From the formulae (3.23), (3.39) and (3.40) it follows that

$$\left.\begin{aligned} Q_\nu(c+i\infty, z) - Q_\nu(c-i\infty, z) &= 0; \\ P_\nu(c+i\infty, z) - P_\nu(c-i\infty, z) &= 2i \end{aligned}\right\} \tag{3.51}$$

and from the representation (3.50) as follows from the definition (2.2) reduces to the Bessel function $J_\nu(z)$,

$$\lim_{p=\bar{q}\to c-i\infty} S_\nu(p, q; z) = J_\nu(z).$$

Analogously, one can easily see that, in complete agreement with (2.16),

$$\lim_{p=\bar{q}\to c-i\infty} S_\nu(p, q; iz) \exp(-i\nu\pi/2) = I_\nu(z).$$

On the other hand, if the change of variable $z = iz$ is made and if $p \to 0 \exp i\pi$, and $q \to \infty \exp i\pi$ then according to formulae (3.23), (3.27), (3.32), (3.43) and (3.45) we have

$$\begin{aligned} &Q_\nu(\infty \exp i\pi, iz) = 0; \quad Q_\nu(0 \exp i\pi, iz) = 1; \\ &P_\nu(\infty \exp i\pi, iz) = i; \quad P_\nu(0 \exp i\pi, iz) = 0. \end{aligned} \tag{3.52}$$

Therefore for these values of p and q (3.50) gives

$$S_\nu(0 \exp i\pi, \infty \exp i\pi; iz) = \frac{1}{2} \mathscr{H}_\nu^{(2)}(iz), \tag{3.53}$$

which agrees with relation (2.17), as it should.

Thus, the knowledge of the incomplete Weber integrals determines the incomplete cylindrical functions in Sonine Schlaefli form completely, and formula (3.50) can be considered as a new integral representation for these functions.

Conversely, it may be advantageous to express the incomplete Weber integrals in terms of incomplete cylindrical functions. By differentiating equation (3.50) with respect to z

$$\begin{aligned} \frac{dS_\nu(p, q; z)}{dz} &= \frac{1}{2i} N_\nu'(z) \left[Q_\nu(q, z) - Q_\nu(p, z)\right] \\ &+ \frac{1}{2i} J_\nu'(z) \left[P_\nu(q, z) - P_\nu(p, z)\right]. \end{aligned} \tag{3.54}$$

and taking advantage of (2.8) and the corresponding relations for the Bessel functions (see § 1, Chapter II), we find

$$\begin{aligned} S_{\nu+1}(p, q; z) &= \frac{1}{2i} N_{\nu+1}(z) \left[Q_\nu(q, z) - Q_\nu(p, z)\right] \\ &+ \frac{1}{2i} J_{\nu+1}(z) \left[P_\nu(q, z) - P_\nu(p, z)\right]. \end{aligned} \tag{3.55}$$

Solving Eqs. (3.50) and (3.55) we obtain the following expressions for the Weber integrals in terms of incomplete cylindrical functions in Sonine-Schlaefli form:

$$Q_\nu(q, z) - Q_\nu(p, z) = \pi i z [S_\nu(p, q; z) J_{\nu+1}(z) - S_{\nu+1}(p, q; z) J_\nu(z)]; \quad (3.56)$$

$$P_\nu(q, z) - P_\nu(p, z) = -\pi i z [S_\nu(p, q; z) N_{\nu+1}(z) - S_{\nu+1}(p, q; z) N_\nu(z)]. \quad (3.57)$$

In the following section, more detailed analytical expressions for the special values $p = z/2$ and $q = cz/2$ will be obtained.

4. The Connection between the Incomplete Weber Integral, Incomplete Cylindrical Functions, and Lipschitz-Hankel Integrals

In the previous section we obtained a relation between the incomplete Weber integrals and the incomplete cylindrical functions of Sonine-Schlaefli form. We also showed that, for the special values $p = z/2$ and $q = (z/2) \exp w$ there exists a simple connection with the function $\varepsilon_\nu(w, z)$, namely

$$S_\nu\left(\frac{z}{2}, \frac{z}{2} e^w, z\right) = \frac{1}{2} \varepsilon_\nu(w, z). \quad (4.1)$$

On the basis of this relation and formula (2.10) of this Chapter we have

$$\begin{aligned} \varepsilon_\nu(w, z) = {} & i N_\nu(z) \left[Q_\nu\left(\frac{z}{2}, z\right) - Q_\nu\left(\frac{z}{2} e^w, z\right)\right] \\ & + i J_\nu(z) \left[P_\nu\left(\frac{z}{2}, z\right) - P_\nu\left(\frac{z}{2} e^w, z\right)\right]. \end{aligned} \quad (4.2)$$

This formula expresses explicitly the connection between incomplete cylindrical functions of Bessel form and the incomplete Weber integrals. Actually, the solution of the non-homogeneous differential equation (2.8), Chapter III, is expressible in terms of the functions $\varepsilon_\nu(w, z)$, and from this we see that the incomplete Weber integrals play the same role in the construction of incomplete cylindrical functions of Bessel and Sonine-Schlaefli form as the incomplete Lipschitz-Hankel integrals do in the construction of the incomplete cylindrical functions of Poisson form (see § 5, 6 Chapter II).

Formula (4.2) contains the incomplete Weber integrals $Q_\nu(x, z)$ and $P_\nu(x, z)$ for the special values

$$\left.\begin{aligned} x &= \left(z \frac{c}{2}\right); \\ c &= \exp w. \end{aligned}\right\} \quad (4.3)$$

Our present interest lies in studying separately certain characteristics of these integrals:

$$Q_\nu\left(\frac{cz}{2}, z\right) = (cz)^{-\nu-1} e^{cz/2} \int_0^z t^{\nu+1} J_\nu(t) \exp\left(-\frac{t^2}{2cz}\right) dt; \tag{4.4}$$

$$P_\nu\left(\frac{cz}{2}, z\right) = (cz)^{-\nu-1} e^{cz/2} \int_z^{\infty \exp i\lambda} t^{\nu+1} N_\nu(t) \exp\left(-\frac{t^2}{2cz}\right) dt, \tag{4.5}$$

where $\operatorname{Re}\nu > -1$, $|2\lambda - \arg(cz)| < \pi/2$ and the parameter c does not depend on z. The derivative of the first of these functions with respect to z is

$$\begin{aligned} \frac{d}{dz} Q_\nu\left(\frac{cz}{2}, z\right) &= -\frac{\nu+1}{z} Q_\nu\left(\frac{cz}{2}, z\right) + \frac{c}{2} Q_\nu\left(\frac{cz}{2}, z\right) \\ &\quad + c^{-\nu-1} J_\nu(z) \exp\frac{z}{2}\left(c - \frac{1}{c}\right) \\ &\quad - \frac{c}{2}(cz)^{-\nu-2} e^{cz/2} \int_0^z t^{\nu+2} J_\nu(t) \frac{d}{dt} \exp\left(-\frac{t^2}{2cz}\right) dt. \end{aligned} \tag{4.6}$$

Integrating the last term by parts, we have

$$\begin{aligned} -\frac{c}{2}(cz)^{-\nu-2} e^{cz/2} \int_0^z t^{\nu+2} J_\nu(t) \frac{d}{dt} \exp\left(-\frac{t^2}{2cz}\right) dt &= -\frac{J_\nu(z)}{2c^{\nu+1}} \exp\frac{z}{2}\left(c - \frac{1}{c}\right) \\ + \frac{c}{2}(cz)^{-\nu-2} e^{cz/2} \int_0^z t^{\nu+1} [(\nu+2) J_\nu(t) + t J_\nu'(t)] \exp\left(-\frac{t^2}{2cz}\right) dt \\ = -\frac{J_\nu(z)}{2c^{\nu+1}} \exp\frac{z}{2}\left(c - \frac{1}{c}\right) + \frac{\nu+1}{z} Q_\nu\left(\frac{cz}{2}, z\right) - \frac{c}{2} Q_{\nu+1}\left(\frac{cz}{2}, z\right), \end{aligned} \tag{4.7}$$

where the relation $tJ_\nu'(t) = \nu_\nu J(t) - tJ_{\nu+1}(t)$ has been used. Substituting (4.7) into (4.6) we find

$$\frac{d}{dz} Q_\nu\left(\frac{cz}{2}, z\right) = \frac{J_\nu(z)}{2c^{\nu+1}} \exp\frac{z}{2}\left(c - \frac{1}{c}\right) + \frac{c}{2}\left[Q_\nu\left(\frac{cz}{2}, z\right) - Q_{\nu+1}\left(\frac{cz}{2}, z\right)\right]. \tag{4.8}$$

Now, using the recursion relation (3.19) with $x = cz/2$ we obtain

$$\frac{d}{dz} Q_\nu\left(\frac{cz}{2}, z\right) = \frac{1}{2c^{\nu+1}} [J_\nu(z) + cJ_{\nu+1}(z)] \exp\frac{z}{2}\left(c - \frac{1}{c}\right). \tag{4.9}$$

Analogously, for $P_\nu(cz/2, z)$, we have

$$\frac{d}{dz} P_\nu\left(\frac{cz}{2}, z\right) = \frac{-1}{2c^{\nu+1}} [N_\nu(z) + cN_{\nu+1}(z)] \exp\frac{z}{2}\left(c - \frac{1}{c}\right). \tag{4.10}$$

The above equations are easily integrated. Noting that if $\operatorname{Re}\nu > -1$, $Q_\nu(cz/2, z)$ vanishes as $z \to 0$, we find from (4.9)

$$\begin{aligned} Q_\nu\left(\frac{cz}{2}, z\right) &= (cz)^{-\nu-1} e^{cz/2} \int_0^z t^{\nu+1} J_\nu(t) \exp\left(-\frac{t^2}{2cz}\right) dt \\ &= \frac{1}{2c^{\nu+1}} \int_0^z [J_\nu(t) + cJ_{\nu+1}(t)] \exp\frac{t}{2}\left(c - \frac{1}{c}\right) dt. \end{aligned} \tag{4.11}$$

Analogously, from (4.10) we obtain a comparable expression for $P_\nu(cz/2, z)$:

$$P_\nu\left(\frac{cz}{2}, z\right) = (cz)^{-\nu-1} e^{cz/2} \int_z^{\infty \exp i\lambda} t^{\nu+1} N_\nu(t) \exp\left(-\frac{t^2}{2cz}\right) dt \tag{4.12}$$

$$= P_\nu\left(\frac{cz_0}{2}, z_0\right) + \frac{1}{2c^{\nu+1}} \int_z^{z_0} [N_\nu(t) + cN_{\nu+1}(t)] \exp \frac{t}{2}\left(c - \frac{1}{c}\right) dt,$$

where z_0 is any arbitrary value of z for which the function $P_\nu(cz/2, z)$ is defined.

The reference values of z can either be taken from tables or determined analytically. For example if z approaches $z_0 = \infty \exp i\omega$ in such a way that $\operatorname{Re}\,[e^{i\omega}(1/c - c \pm 2i)] \geq 0$, then for this value $P_\nu(cz_0/2, z_0)$ vanishes and (4.12) becomes

$$P_\nu\left(\frac{cz}{2}, z\right) = \frac{1}{2c^{\nu+1}} \int_z^{\infty \exp i\omega} [N_\nu(t) + cN_{\nu+1}(t)] \exp \frac{t}{2}\left(c - \frac{1}{c}\right) dt, \tag{4.13}$$

where

$$\operatorname{Re}\left[e^{i\omega}\left(\frac{1}{c} - c \pm 2i\right)\right] \geq 0.$$

With an appropriate choice of ω, this last condition can be satisfied for any arbitrary fixed value of c. Thus, for all c satisfying $\operatorname{Re}\,(1/c - c) \geq 0$, one can put $\omega = 0$, and if $\operatorname{Re}\,(1/c - c) < 0$, it is sufficient to choose $\omega = \pi$. Appropriate representations of the incomplete Weber integrals in terms of the modified functions $I_\nu(t)$ and $K_\nu(t)$ can be obtained with the aid of relations (3.7)—(3.10) and (4.11)—(4.13) by changing c to $c \exp(i\pi/2)$ and z to iz. However, we shall derive them by a somewhat different method. We consider the functions

$$e^{-cz}\tilde{Q}_\nu\left(\frac{cz}{2}; z\right) = (cz)^{-\nu-1} e^{-cz/2} \int_0^z t^{\nu+1} I_\nu(t) \exp\left(-\frac{t^2}{2cz}\right) dt; \tag{4.14}$$

$$e^{-cz}\tilde{P}_\nu\left(\frac{cz}{2}, z\right) = (cz)^{-\nu-1} e^{-cz/2} \int_z^{\infty \exp i\lambda} t^{\nu+1} K_\nu(t) \exp\left(-\frac{t^2}{2cz}\right) dt \tag{4.15}$$

and take the derivative of the first with respect to z:

$$\frac{d}{dz}\left[e^{-cz}\tilde{Q}_\nu\left(\frac{cz}{2}; z\right)\right] = -\frac{\nu+1}{z}\tilde{Q}_\nu\left(\frac{cz}{2}, z\right) e^{-cz}$$

$$-\frac{ce^{-cz}}{2}\tilde{Q}_\nu\left(\frac{cz}{2}, z\right) + \frac{1}{c^{\nu+1}} I_\nu(z) \exp\left[-\frac{z}{2}\left(c + \frac{1}{c}\right)\right]$$

$$-\frac{c}{2}\frac{e^{-cz/2}}{(cz)^{\nu+2}} \int_0^z t^{\nu+2} I_\nu(t) \frac{d}{dt} \exp\left(-\frac{t^2}{2cz}\right) dt.$$

The procedure is entirely analogous to that employed previously except that the relation $tI'_\nu(t) = \nu I_\nu(t) + tI_{\nu+1}(t)$ is used. After some transformations we obtain

$$\frac{d}{dz}\left[e^{-cz}\tilde{Q}_\nu\left(\frac{cz}{2}, z\right)\right] = \frac{1}{2c^{\nu+1}}\left[I_\nu(z) - cI_{\nu+1}(z)\right] \exp\left[-\frac{z}{2}\left(c + \frac{1}{c}\right)\right], \tag{4.16}$$

from which we finally find the following expression for the incomplete Weber integrals involving modified Bessel functions

$$\begin{aligned}\tilde{Q}_\nu\left(\frac{cz}{2}, z\right) &= (cz)^{-\nu-1} e^{cz/2} \int_0^z t^{\nu+1} I_\nu(t) \exp\left(-\frac{t^2}{2cz}\right) dt \\ &= \frac{e^{cz}}{2c^{\nu+1}} \int_0^z \left[I_\nu(t) - cI_{\nu+1}(t)\right] \exp\left[-\frac{t}{2}\left(c + \frac{1}{c}\right)\right] dt.\end{aligned} \tag{4.17}$$

Similar relations can be obtained for the incomplete Weber integrals containing the MacDonald function

$$\begin{aligned}\tilde{P}_\nu\left(\frac{cz}{2}, z\right) &= (cz)^{-\nu-1} e^{zc/2} \int_z^{\infty \exp i\lambda} t^{\nu+1} K_\nu(t) \exp\left(-\frac{t^2}{2cz}\right) dt \\ &= \frac{e^{cz}}{2c^{\nu+1}} \int_z^{\infty \exp i\omega} \left[K_\nu(t) + cK_{\nu+1}(t)\right] \exp\left[-\frac{t}{2}\left(c + \frac{1}{c}\right)\right] dt.\end{aligned} \tag{4.18}$$

The new integral relations for the incomplete Weber integrals represented by expressions (4.11), (4.13), (4.17) and (4.18) are of independent interest. They are suitable for practical evaluation, since the variable z is not contained in the integrand. In particular, formulae (4.11) and (4.13) with $c = 1$, give

$$Q_\nu\left(\frac{z}{2}, z\right) = \frac{1}{2} \int_0^z \left[J_\nu(t) + J_{\nu+1}(t)\right] dt; \tag{4.19}$$

$$P_\nu\left(\frac{z}{2}, z\right) = \frac{1}{2} \int_z^\infty \left[N_\nu(t) + N_{\nu+1}(t)\right] dt. \tag{4.20}$$

In passing we note some other special cases which result from the previous formulae for $c = \pm 1$ or $c = \pm i$. Setting $c = \pm 1$ gives

$$\int_0^z t^{\nu+1} J_\nu(t) \exp\left(\pm \frac{t^2}{2z}\right) dt = \frac{z^{\nu+1}}{2} e^{\pm z/2} \int_0^z \left[J_\nu(t) \mp J_{\nu+1}(t)\right] dt; \tag{4.21}$$

$$\int_z^{\infty \exp i\lambda} t^{\nu+1} N_\nu(t) \exp\left(\pm \frac{t^2}{2z}\right) dt = \frac{z^{\nu+1}}{2} e^{\pm z/2} \int_z^\infty \left[N_\nu(t) \mp N_{\nu+1}(t)\right] dt; \tag{4.22}$$

$$\int_0^z t^{\nu+1} I_\nu(t) \exp\left(\pm \frac{t^2}{2z}\right) dt = \frac{z^{\nu+1}}{2} e^{\mp z/2} \int_0^z \left[I_\nu(t) \pm I_{\nu+1}(t)\right] e^{\pm t} dt; \tag{4.23}$$

$$\int_z^{\infty \exp i\lambda} t^{\nu+1} K_\nu(t) \exp\left(\pm \frac{t^2}{2z}\right) dt = \frac{z^{\nu+1}}{2} e^{\mp z/2} \int_z^\infty \left[K_\nu(t) \mp K_{\nu+1}(t)\right] e^{\pm t} dt. \tag{4.24}$$

Similarly for $c = i$ we find

$$\int_0^z t^{\nu+1} J_\nu(t) \exp\left(\frac{it^2}{2z}\right) dt = \frac{z^{\nu+1}}{2} e^{-iz/2} \int_0^z [J_\nu(t) + iJ_{\nu+1}(t)] e^{it} dt; \tag{4.25}$$

$$\int_z^{\infty \exp i\lambda} t^{\nu+1} N_\nu(t) \exp\left(\frac{it^2}{2z}\right) dt = \frac{z^{\nu+1}}{2} e^{-iz/2} \int_z^\infty [N_\nu(t) + iN_{\nu+1}(t)] e^{it} dt; \tag{4.26}$$

$$\int_0^z t^{\nu+1} I_\nu(t) \exp\left(\frac{it^2}{2z}\right) dt = \frac{z^{\nu+1}}{2} e^{iz/2} \int_0^z [I_\nu(t) - iI_{\nu+1}(t)] dt; \tag{4.27}$$

$$\int_z^{\infty \exp i\lambda} t^{\nu+1} K_\nu(t) \exp\left(\frac{it^2}{2z}\right) dt = \frac{z^{\nu+1}}{2} e^{iz/2} \int_z^\infty [K_\nu(t) + iK_{\nu+1}(t)] dt. \tag{4.28}$$

while the value for $c = -i$ is obtained simply by replacing i everywhere by $-i$.

The following expressions can also be easily obtained from the last formulae

$$\int_0^z t^{\nu+1} I_\nu(t) \cos\frac{t^2}{2z} dt = \frac{z^{\nu+1}}{2}\left[\cos\frac{z}{2}\int_0^z I_\nu(t) + \sin\frac{z}{2}\int_0^z I_{\nu+1}(t)\, dt\right]; \tag{4.29}$$

$$\int_0^z t^{\nu+1} I_\nu(t) \sin\frac{t^2}{2z} dt = \frac{z^{\nu+1}}{2}\left[\sin\frac{z}{2}\int_0^z I_\nu(t)\, dt - \cos\frac{z}{2}\int_0^z I_{\nu+1}(t)\, dt\right]; \tag{4.30}$$

$$\begin{aligned}&\int_z^{\infty \exp i\lambda} t^{\nu+1} K_\nu(t) \cos\frac{t^2}{2z} dt \\ &= \frac{z^{\nu+1}}{2}\left[\cos\frac{z}{2}\int_z^\infty K_\nu(t)\, dt - \sin\frac{z}{2}\int_z^\infty K_{\nu+1}(t)\, dt\right],\end{aligned} \tag{4.31}$$

$$\begin{aligned}&\int_z^{\infty \exp i\lambda} t^{\nu+1} K_\nu(t) \sin\frac{t^2}{2z} dt \\ &= \frac{z^{\nu+1}}{2}\left[\sin\frac{z}{2}\int_z^\infty K_\nu(t)\, dt - \cos\frac{z}{2}\int_z^\infty K_{\nu+1}(t)\, dt\right].\end{aligned} \tag{4.32}$$

These are well suited to numerical calculations since the integrands on the right sides do not contain z. Most of these integrals are already tabulated [19], [20].

It is also interesting to note that for integer or half-odd values of ν these integrals can be expressed in closed form by means of known and tabulated functions. In fact, for all of these integrals, recurrence relations similar to relation (3.16)—(3.19) can be obtained. For $\nu = 1/2$ the integrals on the right are expressible either in terms of elementary functions or probability integrals, while for $\nu = 0$ they can be expressed either by means of cylindrical functions or by a combination of Struve and cylindrical functions. To see this, we take advantage of the following identities

$$\left.\begin{aligned} [I_0(t) \pm I_1(t)]\, e^{\pm t} &= \frac{d}{dt}[I_0(t)\, e^{\pm t}]; \\ [K_0(t) \mp K_1(t)]\, e^{\pm t} &= \pm \frac{d}{dt}[K_0(t)\, e^{\pm t}]; \\ [J_0(t) + iJ_1(t)\, e^{it} &= -i \frac{d}{dt}[J_0(t)\, e^{it}]; \\ [N_0(t) + iN_1(t)]\, e^{it} &= -i \frac{d}{dt}[N_0(t)\, e^{it}] \end{aligned}\right\} \quad (4.33)$$

and the relations (5.25)—(5.28) of Chapter II for $\nu = 0$.

Conversely, returning to formula (4.2) we can find explicit expressions for the incomplete Weber integrals in terms of incomplete cylindrical functions of Bessel form on substitution of special values of the variable $x = (z/2) \exp w$. It is simpler, however, to obtain these from formulae (3.56) and (3.57) with the aid of (4.1). We have

$$\begin{aligned} Q_\nu\left(\frac{z}{2} e^w, z\right) &= \frac{1}{2} \int_0^z [J_\nu(t) + J_{\nu+1}(t)]\, dt \\ &+ \frac{i\pi z}{2}[\varepsilon_\nu(w, z)\, J_{\nu+1}(z) - \varepsilon_{\nu+1}(w, z)\, J_\nu(z)]; \end{aligned} \quad (4.34)$$

$$\begin{aligned} P_\nu\left(\frac{z}{2} e^w, z\right) &= \frac{1}{2} \int_z^\infty [N_\nu(t) + N_{\nu+1}(t)]\, dt \\ &- \frac{i\pi z}{2}[\varepsilon_\nu(w, z)\, N_{\nu+1}(z) - \varepsilon_{\nu+1}(w, z)\, N_\nu(z)], \end{aligned} \quad (4.35)$$

where $\varepsilon_\nu(w, z)$ is the incomplete cylindrical function of Bessel form:

$$\varepsilon_\nu(w, z) = \frac{1}{\pi i} \int_0^w e^{z \sinh t - \nu t}\, dt,$$

Whose properties have been extensively studied in the previous Chapter. There it was shown that for integer indices ($\nu = n$) these functions can be expressed by means of incomplete cylindrical functions of Poisson form, $E_n^\pm(w, z)$. Therefore, the incomplete Weber integrals, according to Eqs. (4.34), (4.35), can also be expressed in terms of $E_n^\pm(w, z)$. In turn, the incomplete cylindrical functions of Poisson form are connected, as shown in Section 6, Chapter II with incomplete Lipschitz-Hankel integrals. Consequently, for integer $\nu = n$, a connection exists between the incomplete integrals of Weber and Lipschitz-Hankel. The existence of such interrelations is easier to see from the relations (4.11), (4.13), (4.17) and (4.18). The following section will be devoted to the study of these relations.

5. The Connection between Incomplete Integrals of Weber, Lipschitz-Hankel and Incomplete Cylindrical Functions of Poisson form

It was mentioned above that there is a strong connection between the incomplete Weber integrals and the incomplete cylindrical functions of Bessel and Sonine-Schlaefli form. Furthermore the incomplete Lipschitz-Hankel integrals enjoy an analogous connection with incomplete cylindrical functions of Poisson form. From this point of view we can relate these integrals to a class of incomplete cylindrical functions. They play a comparable role in the construction of the corresponding incomplete cylindrical functions and could be used as the basis for their definition. All of these integrals are of independent interest in connection with the solution of many important theoretical and applied problems. As we shall see below, special cases have been studied widely.

In this section we shall investigate the connection between incomplete Weber and Lipschitz-Hankel integrals for integer values of the index $\nu = n$. Since there are simple recurrence relations for the incomplete Weber integrals we need only consider the case of order zero.

According to definition, the incomplete Lipschitz-Hankel integrals with zero index have the form

$$Ie_0(a, z) = \int_0^z I_0(t)\, e^{-at}\, dt; \tag{5.1}$$

$$Je_0(a, z) = \int_0^z J_0(t)\, e^{-at}\, dt; \tag{5.2}$$

$$Ne_0(a, z) = \int_0^z N_0(t)\, e^{-at}\, dt; \tag{5.3}$$

$$Ke_0(a, z) = \int_0^z K_0(t)\, e^{-at}\, dt. \tag{5.4}$$

On the other hand the incomplete Weber integral in terms of $I_0(t)$ has the form

$$\tilde{Q}_0\left(\frac{cz}{2}, z\right) = \frac{1}{cz}\, e^{cz/2} \int_0^z t I_0(t) \exp\left(-\frac{t^2}{2cz}\right) dt \tag{5.5}$$

and according to Eq. (4.17) can be written as

$$\begin{aligned}\tilde{Q}_0\left(\frac{cz}{2}, z\right) = {} & \frac{1}{2c}\, e^{cz} \int_0^z I_0(t) \exp\left[-\frac{t}{2}\left(c + \frac{1}{c}\right)\right] dt \\ & - \frac{1}{2}\, e^{cz} \int_0^z I_1(t) \exp\left[-\frac{t}{2}\left(c + \frac{1}{c}\right)\right] dt.\end{aligned} \tag{5.6}$$

The first term is already in the form of the incomplete Lipschitz-Hankel integral, while the second, following integration by parts using $I_1(z) = I_0'(z)$ can also be written as an integral of this type. The result is

$$\tilde{Q}_0\left(\frac{cz}{2}, z\right) = \frac{1}{2} e^{cz}\left[1 + \frac{1-c^2}{2c} Ie_0(a, z) - I_0(z)\, e^{-az}\right], \qquad (5.7)$$

where $a = (1 + c^2)/2c$.

Analogously, from (4.11) we can find a corresponding expression for the incomplete Weber integral

$$Q_0\left(\frac{cz}{2}, z\right) = \frac{1}{2}\left[1 + \frac{1+c^2}{2c} Je_0(b, z) - J_0(z)\, e^{-bz}\right], \qquad (5.8)$$

where $b = (1 - c^2)/2c$.

To find the corresponding connection between $P_0(cz/2, z)$ and $Ne_0(b,z)$ we first transform the integral (5.3)

$$Ne_0(b, z) = \int_0^{\infty \exp i\omega} N_0(t)\, e^{-bt}\, dt - \int_z^{\infty \exp i\omega} N_0(t)\, e^{-bt}\, dt$$

making use of formulae (5.4), Chapter II, into

$$\int_z^{\infty \exp i\omega} N_0(t)\, e^{-bt}\, dt = -Ne_0(b, z) + \frac{1}{\pi\sqrt{b^2+1}} \ln \frac{\sqrt{b^2+1} - b}{\sqrt{b^2+1} + b}. \qquad (5.9)$$

Now expression (4.13) can be written as

$$\begin{aligned} P_0\left(\frac{cz}{2}, z\right) &= \frac{1}{2c} \int_z^{\infty \exp i\omega} N_0(t) \exp \frac{t}{2}\left(c - \frac{1}{c}\right) dt \\ &\quad - \frac{1}{2} \int_z^{\infty \exp i\omega} \frac{dN_0(t)}{dt} \exp \frac{t}{2}\left(c - \frac{1}{c}\right) dt \\ &= \frac{1}{4}\left(c + \frac{1}{c}\right) \int_z^{\infty \exp i\omega} N_0(t) \exp \frac{t}{2}\left(c - \frac{1}{c}\right) dt \\ &\quad + \frac{1}{2} N_0(z) \exp \frac{z}{2}\left(c - \frac{1}{c}\right) \end{aligned}$$

and, according to (5.9) this gives

$$\begin{aligned} P_0\left(\frac{cz}{2}, z\right) = \frac{1}{4}\frac{c^2+1}{c}\Bigg[&\frac{1}{\pi\sqrt{b^2+1}} \ln \frac{\sqrt{b^2+1} - b}{\sqrt{b^2+1} + b} \\ &- Ne_0(b, z)\Bigg] + \frac{1}{2} N_0(z)\, e^{-bz}, \end{aligned} \qquad (5.10)$$

where, as before, $b = (1 - c^2)/2c$.

Carrying out analogous transformations on Eq. (4.18) and utilizing (5.6), Chapter II:

$$\int_z^{\infty \exp i\omega} K_0(t)\, e^{-at}\, dt = -Ke_0(a, z) - \frac{1}{2\sqrt{a^2-1}} \ln \frac{a - \sqrt{a^2-1}}{a + \sqrt{a^2-1}}, \qquad (5.11)$$

we find the following expression for $\tilde{P}_0(cz/2, z)$

$$\tilde{P}_0\left(\frac{cz}{2}, z\right) = \frac{1}{2} e^{cz}\left[-\frac{1-c^2}{4c\sqrt{a^2-1}} \ln\frac{a-\sqrt{a^2-1}}{a+\sqrt{a^2-1}} - \frac{1-c^2}{2c} Ke_0(a, z) + K_0(z) e^{-az}\right], \tag{5.12}$$

where as previously, $a = (1 + c^2)/2c$.

We shall now give explicit expressions for incomplete Weber integrals with $\nu = 0$ in terms of incomplete cylindrical functions of Poisson form, $E_0^+(w, z)$ and $E_1^+(w, z)$. For this purpose it is sufficient to utilize the corresponding formulae for the incomplete Lipschitz-Hankel integrals obtained in § 6, Chapter II. We have

$$\begin{aligned} Q_0\left(\frac{zc}{2}, z\right) &= \frac{1}{cz} e^{cz/2} \int_0^z t J_0(t) \exp\left(-\frac{t^2}{2cz}\right) dt \\ &= \frac{1}{2}\Big\{1 - \frac{\pi i z}{2}[E_0^+(w, z) J_1(z) - E_1^+(w, z) J_0(z)] \\ &\quad + (z \sin w - 1) J_0(z) e^{iz\cos w}\Big\}; \end{aligned} \tag{5.13}$$

$$\begin{aligned} P_0\left(\frac{zc}{2}, z\right) &= \frac{1}{cz} e^{cz/2} \int_z^{\infty \exp i\lambda} t N_0(t) \exp\left(-\frac{t^2}{2cz}\right) dt \\ &= \frac{1}{2}\Big\{i + \frac{\pi i z}{2}[E_0^+(w, z) N_1(z) - E_1^+(w, z) N_0(z)] \\ &\quad + (1 - z \sin w) N_0(z) e^{iz\cos w}\Big\}, \end{aligned} \tag{5.14}$$

where the variables c and w are connected by

$$c = \exp i\left(\frac{\pi}{2} - w\right). \tag{5.15}$$

Expressions for the other integrals take the form

$$\begin{aligned} \tilde{Q}_0\left(\frac{cz}{2}, z\right) &= \frac{1}{cz} e^{cz/2} \int_0^z t I_0(t) \exp\left(-\frac{t^2}{2cz}\right) dt \\ &= \frac{1}{2} e^{cz}\{1 - \pi i z[F_0^-(w, z) I_1(z) - F_1^-(w, z) I_0(z)] \\ &\quad - I_0(z)(1 + iz \sin w) e^{-z\cos w}\}; \end{aligned} \tag{5.16}$$

$$\begin{aligned} \tilde{P}_0\left(\frac{cz}{2}, z\right) = \frac{1}{2} e^{cz}\{&-\pi i z[F_0^-(w, z) K_1(z) + F_1^-(w, z) K_0(z)] \\ &+ K_0(z)(1 + iz \sin w) e^{-z\cos w}\}. \end{aligned} \tag{5.17}$$

where $F_0^-(w, z)$ and $F_1^-(w, z)$ are incomplete cylindrical functions of Poisson form with purely imaginary argument (iz) (see Section 1, Chapter II), and where c and w are related by

$$c = \exp iw. \tag{5.18}$$

These relations are quite useful in the analytic solution of many problems as well as for purposes of numerical evaluation.

6. The Connection between Incomplete Weber Integrals and Lommel Functions of Two Variables

The Lommel function of two variables $U_\nu(w, z)$ is defined by the following Neumann series (see [10, p. 592])

$$U_\nu(w, z) = \sum_{m=0}^{\infty} (-1)^m \left(\frac{w}{z}\right)^{\nu+2m} J_{\nu+2m}(z). \tag{6.1}$$

Along with this function, it is appropriate to consider, the function $V_\nu(w, z)$, defined by

$$V_\nu(w, z) = U_{-\nu+2}(w, z) + \cos\left(\frac{w}{2} + \frac{z^2}{2w} + \frac{\nu\pi}{2}\right). \tag{6.2}$$

These Lommel functions satisfy the recurrence relations

$$U_{\nu+2}(w, z) + U_\nu(w, z) = \left(\frac{w}{z}\right)^\nu J_\nu(z); \tag{6.3}$$

$$V_{\nu+2}(w, z) + V_\nu(w, z) = \left(\frac{w}{z}\right)^{-\nu} J_{-\nu}(z). \tag{6.4}$$

For purely imaginary values of both variables $w = iw$, $z = iz$ (6.1) becomes

$$i^{-\nu} U_\nu(iw, iz) = Y(w, z) = \sum_{m=0}^{\infty} \left(\frac{w}{z}\right)^{\nu+2m} I_{\nu+2m}(z). \tag{6.5}$$

In the literature, the latter function for $\nu = n$ are called Lommel functions of two imaginary variables (see [6, 21, 22]).

In addition to the representation by Neumann series, there also exist various integral representations. For $\operatorname{Re} \nu > -1$, $U_\nu(w, z)$ can be expressed by (c.f. [10, p. 594])

$$U_{\nu+1}(w, z) = \frac{w^{\nu+1}}{z^\nu} \int_0^1 t^{\nu+1} J_\nu(zt) \cos\left[\frac{1}{2} w(1 - t^2)\right] dt; \tag{6.6}$$

$$U_{\nu+2}(w, z) = \frac{w^{\nu+1}}{z^\nu} \int_0^1 t^{\nu+1} J_\nu(zt) \sin\left[\frac{1}{2} w(1 - t^2)\right] dt. \tag{6.7}$$

Therefore

$$\begin{aligned} U_{\nu+1}(w, z) + iU_{\nu+2}(w, z) &= \frac{w^{\nu+1}}{z^\nu} \int_0^1 t^{\nu+1} J_\nu(zt) \exp\left[\frac{iw}{2}(1 - t^2)\right] dt \\ &= \left(\frac{w}{z^2}\right)^{\nu+1} e^{iw/2} \int_0^z y^{\nu+1} J_\nu(y) \exp\left(-\frac{iw}{2z^2} y^2\right) dy. \end{aligned} \tag{6.8}$$

We shall now show that the incomplete Weber integrals containing the Bessel functions $J_\nu(t)$ and $I_\nu(t)$ can both be expressed by Lommel functions. From the definition we have

$$Q_\nu(x, z) = (2x)^{-\nu-1} e^x \int_0^z t^{\nu+1} J_\nu(t) \exp\left(-\frac{t^2}{4x}\right) dt; \tag{6.9}$$

$$\tilde{Q}_\nu(x, z) = (2x)^{-\nu-1} e^x \int_0^z t^{\nu+1} I_\nu(t) \exp\left(-\frac{t^2}{4x}\right) dt. \tag{6.10}$$

Comparison of these with the integral representation (6.8) leads directly to:

$$Q_\nu(x, z) = i^{\nu+1}\left[U_{\nu+1}\left(\frac{z^2}{2ix}, z\right) + iU_{\nu+2}\left(\frac{z^2}{2ix}, z\right)\right] \exp\left(x - \frac{z^2}{4x}\right); \tag{6.11}$$

$$\tilde{Q}_\nu(x, z) = \left[Y_{\nu+1}\left(\frac{z^2}{2x}, z\right) + Y_{\nu+2}\left(\frac{z^2}{2x}, z\right)\right] \exp\left(x - \frac{z^2}{4x}\right). \tag{6.12}$$

In practical problems we most frequently encounter incomplete Weber integrals of zero order. Their expressions in terms of Lommel's functions are

$$\begin{aligned} Q_0(x, z) &= \frac{e^x}{2x} \int_0^z t J_0(t) \exp\left(-\frac{t^2}{4x}\right) dt \\ &= i\left[U_1\left(\frac{z^2}{2ix}, z\right) + iU_2\left(\frac{z^2}{2ix}, z\right)\right] \exp\left(x - \frac{z^2}{4x}\right); \end{aligned} \tag{6.13}$$

$$\begin{aligned} \tilde{Q}_0(x, z) &= \frac{1}{2x} e^x \int_0^z t I_0(t) \exp\left(-\frac{t^2}{4x}\right) dt \\ &= \left[Y_1\left(\frac{z^2}{2x}, z\right) + Y_2\left(\frac{z^2}{2x}, z\right)\right] \exp\left(x - \frac{z^2}{4x}\right). \end{aligned} \tag{6.14}$$

The latter formula was obtained in [6] in a somewhat different way. There, the following interesting relation for the incomplete Lipschitz-Hankel integral involving the modified Bessel function $I_0(z)$ was derived

$$\begin{aligned} Ie_0(\pm a, z) &= \int_0^z e^{\mp at} I_0(t)\, dt \\ &= \frac{1}{\sqrt{\alpha^2 - 1}}\left\{1 - e^{\mp az}\left[I_0(z) + 2Y_2\left(\frac{z}{c}, z\right) \pm 2Y_1\left(\frac{z}{c}, z\right)\right]\right\}, \end{aligned} \tag{6.15}$$

where the parameters a and c are related by

$$a = \frac{1}{2}\left(c + \frac{1}{c}\right). \tag{6.16}$$

In [17], this formula was obtained directly by replacing $Y_2(z/c, z) \pm Y_1(z/c, z)$ by a series according to Eq. (6.5) and summing. However, these formulae as well as the analogous one for $Je_0(a, z)$

$$\begin{aligned} Je_0(a, z) = \frac{2c}{1 + c^2}\Big\{&-1 + \Big[J_0(z) + 2iU_1\left(\frac{z}{ic}, z\right) \\ &- 2U_2\left(\frac{z}{ic}, z\right)\Big] \exp \frac{z}{2}\left(c - \frac{1}{c}\right)\Big\}, \end{aligned} \tag{6.17}$$

can be obtained very easily from (6.13) and (6.14) by utilizing the relations derived above between the incomplete Weber and Lipschitz-Hankel integrals (see Section 4, this Chapter).

The formulae (6.14) and (6.15) just derived, which express incomplete Weber and Lipschitz-Hankel integrals with Bessel functions of zero order in terms of the Lommel function of two variables, often expedite analytic calculations. Furthermore comprehensive tables of the Lommel functions of two imaginary variables [22] and two real variables [23] permit numerical evaluation of the relevant integrals.

In summary we make the following remarks: In this chapter, as well as in chapters II and III, a connection was established on one hand between incomplete cylindrical functions of Poisson form and incomplete Lipschitz-Hankel integrals, and on the other between incomplete cylindrical functions of Bessel form and incomplete Weber integrals. We have already established a connection between the well known Lommel functions of two variables and the incomplete Lipschitz-Hankel and Weber integrals. Therefore, at first glance it might appear that there exists a close connection between the functions $E_\nu^\pm(w, z)$, $\varepsilon_\nu(w, z)$ and the Lommel functions, and that separate consideration of these is unnecessary. This, however, is not the case, since $E_\nu^\pm(w, z)$, is expressible in terms of $\varepsilon_\nu(w, z)$ for integer ν only. If ν is not an integer, there is no simple relation between these functions (see e.g. Section 4, Chapter III), and they are in general of entirely different types. Only special cases of the incomplete cylindrical functions of Bessel or Sonine-Schlaefli form, namely the incomplete Weber integrals containing $J_\nu(t)$ and $I_\nu(t)$, can be expressed by Lommel functions of two variables. With this restriction, the Lommel functions can also be considered as incomplete cylindrical functions. In general, however, it is impossible to construct incomplete cylindrical functions of Bessel form on this basis. To do this, it would be necessary to introduce Lommel functions defined by expansions of the type (6.1) and (6.5), but containing Neumann or MacDonald functions. However, such series can be shown to be divergent (see Section 3 of this Chapter).

Chapter V

Incomplete Cylindrical Functions of Real Arguments and their Relation to Certain Discontinuous Integrals

1. Incomplete Cylindrical Functions of Real Arguments

For convenience in the following presentation we recall those representations of incomplete cylindrical functions with real arguments of both the Poisson and Bessel form which are most frequently met in applications.

1. Incomplete functions of Bessel and Struve:

$$J_\nu(\alpha, x) = 2\frac{x^\nu}{A_\nu}\int_0^\alpha \cos(x\cos\theta)\sin^{2\nu}\theta\,d\theta; \tag{1.1}$$

$$H_\nu(\alpha, x) = 2\frac{x^\nu}{A_\nu}\int_0^\alpha \sin(x\cos\theta)\sin^{2\nu}\theta\,d\theta, \tag{1.2}$$

whence

$$E_\nu^\pm(\alpha, x) = J_\nu(\alpha, x) \pm iH_\nu(\alpha, x) = 2\frac{x^\nu}{A_\nu}\int_0^\alpha e^{\pm ix\cos\theta}\sin^{2\nu}\theta\,d\theta, \tag{1.3}$$

where $A_\nu = 2^\nu\Gamma(\nu + 1/2)\,\Gamma(1/2)$.

Here as everywhere below, we will take α, x and $\nu + 1/2 > 0$ as real quantities. The variable x can range from 0 to ∞ while the variable α is restricted to the interval $0 \leq \alpha \leq \pi$.

2. Incomplete functions of Anger and Weber:

$$A_\nu(\alpha, x) = \frac{1}{\pi}\int_0^\alpha \cos(x\sin\theta - \nu\theta)\,d\theta; \tag{1.4}$$

$$B_\nu(\alpha, x) = \frac{1}{\pi}\int_0^\alpha \sin(\nu\theta - x\sin\theta)\,d\theta, \tag{1.5}$$

whence

$$\varepsilon_\nu(i\alpha, x) = A_\nu(\alpha, x) - iB_\nu(\alpha, x) = \frac{1}{\pi}\int_0^\alpha e^{i(x\sin\theta - \nu\theta)}\,d\theta. \tag{1.6}$$

3. Modified incomplete functions of Bessel and Struve:

$$I_\nu(\alpha, x) = 2\frac{x^\nu}{A_\nu}\int\limits_0^\alpha \cosh(x\cos\theta)\sin^{2\nu}\theta\,d\theta; \tag{1.7}$$

$$L_\nu(\alpha, x) = 2\frac{x^\nu}{A_\nu}\int\limits_0^\alpha \sinh(x\cos\theta)\sin^{2\nu}\theta\,d\theta; \tag{1.8}$$

so that

$$F_\nu^\pm(\alpha, x) = \frac{1}{2}[I_\nu(\alpha, x) \pm L_\nu(\alpha, x)] = \frac{x^\nu}{A_\nu}\int\limits_0^\alpha e^{\pm x\cos\theta}\sin^{2\nu}\theta\,d\theta. \tag{1.9}$$

4. Incomplete functions of Hankel:

$$\mathscr{H}_\nu(\beta, x) = -\frac{2ie^{-i\nu\pi}}{A_\nu}x^\nu\int\limits_0^\beta e^{ix\cosh u}\sinh^{2\nu}u\,du. \tag{1.10}$$

$$\mathscr{H}_\nu(\beta, x) = -\frac{2ie^{-i\nu\pi}}{A_\nu}x^\nu\int\limits_1^{\cosh\beta} e^{ixt}(t^2-1)^{\nu-1/2}\,dt. \tag{1.11}$$

5. Modified incomplete Hankel function, or incomplete MacDonald functions of Poisson form:

$$K_\nu(\beta, x) = \pi\frac{x^\nu}{A_\nu}\int\limits_0^\beta e^{-x\cosh u}\sinh^{2\nu}u\,du \tag{1.12}$$

and incomplete MacDonald functions of Bessel form:

$$k_\nu(\beta, x) = \frac{1}{2}\int\limits_{-\beta}^\beta e^{-x\cosh t-\nu t}\,dt. \tag{1.13}$$

The concept of incomplete cylindrical functions of Bessel form $j_\nu(w, z)$ and $i_\nu(w, z)$, for real arguments cannot be defined uniquely. This is due to the fact that the definition of these functions introduced earlier (Chapter III, Eqs. (1.4), (1.6))

$$j_\nu(w, x) = \frac{1}{2\pi i}\int\limits_{\overline{w}}^w e^{x\sinh t-\nu t}\,dt; \tag{1.14}$$

$$i_\nu(w, x) = \frac{1}{2\pi i}\int\limits_{\overline{w}}^w e^{x\cosh t-\nu t}\,dt \tag{1.15}$$

involves contour integration, and the reduction to the complete functions must be effected by the selection of an appropriate path in complex plane i.e.

$$\begin{aligned}\lim_{w\to i\pi+\infty} j_\nu(w, z) \to J_\nu(x) &= \frac{1}{\pi}\int\limits_0^\pi \cos(\nu\theta - x\sin\theta)\,d\theta\\ &- \frac{\sin\nu\pi}{\pi}\int\limits_0^\infty e^{-\nu\theta - x\sinh\theta}\,d\theta;\end{aligned} \tag{1.16}$$

$$\begin{aligned}\lim_{w\to i\pi+\infty} i_\nu(w, x) = J_\nu(x) &= \frac{1}{\pi}\int\limits_0^\pi e^{x\cos\theta}\cos\nu\theta\,d\theta\\ &- \frac{\sin\nu\pi}{\pi}\int\limits_0^\infty e^{-x\cosh t-\nu t}\,dt.\end{aligned} \tag{1.17}$$

Incomplete versions of these functions may be introduced either in the first term with integration taken between the limits 0 and α rather than 0 and π, or in the second term with integration from 0 to β rather than from 0 to ∞. In this connection we also consider incomplete integrals of the forms

$$\int_0^\alpha e^{\pm x\cos\theta} \cos \nu\theta \, d\theta; \tag{1.18}$$

$$\int_0^\alpha e^{\pm x\sin\theta} \sin \nu\theta \, d\theta; \tag{1.19}$$

and

$$\int_0^\beta e^{x\sinh t - \nu t} \, dt. \tag{1.20}$$

All of these incomplete functions can be formulated in terms of integrals with complete range if we introduce an appropriate discontinuous auxillary function. We will use such a representation to derive a sequence of properties of the functions heretofore studied.

2. Unit Functions and their Basic Characteristics

In what follows, we will frequently use a discontinuous function of two variables $\Omega(p, x)$ defined by

$$\Omega(p, x) = \begin{Bmatrix} 1 & \text{for } 0 < x < p; \\ 1/2 & \text{for } 0 < x = p > 0; \\ 0 & \text{for } x > p > 0. \end{Bmatrix} \tag{2.1}$$

$\Omega(p, x)$ is called a unit step function and finds wide application in many theoretical and applied problems. Characteristics and various analytical representations of such functions have been studied in some detail in a series of reports [24, p. 76]). The most important representations of unit functions are summarized below. Many of them can be obtained for Re $(\mu) > 1$ from the following discontinuous integral

$$\int_0^\infty J_\mu(pt)\, J_{\mu-1}(qt)\, dt = \begin{Bmatrix} 0; & q > p; \\ \dfrac{q^{\mu-1}}{p^\mu}; & q < p; \\ \dfrac{1}{2p}; & q = p. \end{Bmatrix} \tag{2.2}$$

Comparison of the above with Eq. (2.1) gives immediately

$$\Omega(p, q) = \frac{p^\mu}{q^{\mu-1}} \int_0^\infty J_\mu(pt)\, J_{\mu-1}(qt)\, dt. \tag{2.3}$$

In the case $\mu = 1$ we find an integral representation of Ω in terms of the Bessel functions of order zero and one:

$$\Omega(p, q) = p \int_0^\infty J_1(pt)\, J_0(qt)\, dt. \tag{2.4}$$

On the other hand, putting $\mu = 1/2$ in (2.3) and using the relation

$$\left.\begin{aligned} J_{1/2}(z) &= \sqrt{\frac{2}{\pi z}} \sin z; \\ J_{-1/2}(z) &= \sqrt{\frac{2}{\pi z}} \cos z, \end{aligned}\right\} \tag{2.5}$$

the following well known representation is obtained:

$$\Omega(p, q) = \frac{2}{\pi} \int_0^\infty \frac{\sin pt \cos qt}{t}\, dt = \frac{1}{\pi} \int_{-\infty}^{\infty} \frac{\sin pt}{t}\, e^{iqt}\, dt. \tag{2.6}$$

Note that for $q = |x|$, (2.6) is the usual Fourier transform of the unit function $\Omega(p, |x|)$ while (2.4) becomes the so-called Fourier-Bessel transform.

We now investigate other useful relations for the unit step function. From relation (2.5) we have

$$\frac{\sin z}{z} = \sqrt{\frac{\pi}{2z}}\, J_{1/2}(z).$$

Or, using the integral representation of Sonine for $J_\nu(z)$ (Section 4, Chapter I)

$$J_\nu(z) = \left(\frac{z}{2}\right)^\nu \frac{1}{2\pi i} \int_{c-i\infty}^{c+i\infty} u^{\nu-1} \exp\left(u - \frac{z^2}{4u}\right) du, \tag{2.7}$$

for $\nu = 1/2$, we obtain

$$\frac{\sin z}{z} = \frac{\sqrt{\pi}}{2} \frac{1}{2\pi i} \int_{c-i\infty}^{c+i\infty} u^{-3/2} \exp\left(u - \frac{z^2}{4u}\right) du. \tag{2.8}$$

Substituting this result into Eq. (2.6) and interchanging the order of integration, we find

$$\Omega(p, q) = \frac{1}{\sqrt{\pi}} \frac{p}{2\pi i} \int_{c-i\infty}^{c+i\infty} \frac{du\, e^u}{u^{3/2}} \int_0^\infty e^{-p^2t^2/4u} \cos qt\, dt. \tag{2.9}$$

The inner integral is easily computed and is equal to

$$\int_0^\infty e^{-p^2t^2/4u} \cos qt\, dt = \frac{1}{2} \int_{-\infty}^{\infty} e^{-p^2t^2/4u + iqt}\, dt = \frac{1}{2} \frac{\sqrt{\pi u}}{p}\, e^{-(q/p)^2 u}. \tag{2.10}$$

Thus formula (2.9) gives an expression for the unit function as a contour integral

$$\Omega(p, q) = \frac{1}{2\pi i} \int_{c-i\infty}^{c+i\infty} e^{(1 - q^2/p^2)u} \frac{du}{u}. \tag{2.11}$$

We remark that if $q^2/p^2 > 1$, this integral is equal to zero, to show this we close the contour on the right by a semi-circle in the complex u plane with a infinitely large radius. Then the integral along the complete contour is obviously zero, as long as no singularities exist inside the contour. But the integral along the semi-circle also tends to zero for $q^2/p^2 > 1$, when $R \to \infty$. Consequently, (2.11) is zero. If $q^2 = p^2$, the integral along the semi-circle is equal to πi. Therefore $\Omega(p, p) = 1/2$. Finally, if $q^2/p^2 < 1$, we introduce a change of variable $(1 - q^2/p^2)\, u = t$ so that

$$\Omega(p, q) = \frac{1}{2\pi i} \int_{c-i\infty}^{c+i\infty} \frac{\mathrm{e}t}{t}\, dt.$$

which according to (2.7) is equal to $J_0(0) = 1$, or, $\Omega(p, q) = 1$ for $q^2/p^2 < 1$.

In addition to $\Omega(p, q)$ we will also need its derivative with respect to q:

$$\delta(q - p) = -\frac{d\Omega(p, q)}{dq}. \tag{2.12}$$

According to the definition of $\Omega(p, q)$, $\delta(x)$ is equal to zero everywhere with the exception of the point $x = 0$, where it tends to infinity in such a way that

$$\int_{-\infty}^{\infty} \delta(x)\, dx = 1. \tag{2.13}$$

$\delta(x)$ is called the dirac δ-function, although it is not a function in the usual mathematical sense. The delta function, which is widely used in modern theoretical physics, is characterized primarily by its behavior in certain integration operations. One of these operations is exhibited by (2.13). The delta function has a number of other interesting properties which may be found for example in [24, Ch. V] and [25, Ch. I]. We present here those properties which will be used in the following sections.

The basic property of the Dirac δ-function is

$$\int_a^b f(t)\, \delta(t - x)\, dt = \begin{cases} f(x); & a < x < b; \\ 0; & x < a;\, x > b, \end{cases} \tag{2.14}$$

where $f(t)$ is an arbitrary continuous function.

For a double integral, the following formula holds:

$$\int_{a_1}^{b_1} dt \int_{a_2}^{b_2} f(t, t')\, \delta(t - t')\, dt' = \begin{cases} \int_a^b f(t, t)\, dt; & b > a; \\ 0; & b < a. \end{cases} \tag{2.15}$$

Here, $a = \max(a_1, a_2)$, $b = \min(b_1, b_2)$, and $(a_1 < b_1,\ a_2 < b_2)$.

The following relations connect the δ-function and its derivative:

$$\int_a^b f(t)\,\delta[\varphi(t)]\,dt = \int_a^b f(t) \sum_i \frac{\delta(t-t_i)}{|\varphi'(t_i)|}\,; \tag{2.16}$$

$$\int_a^b f(t)\,\frac{d}{dx}\,\delta(t-x)\,dt = -\frac{\partial f}{\partial x}\,, \tag{2.17}$$

where $a < x < b$ and t_i are the simple roots of $\varphi(t) = 0$ lying in the interval (a, b).

The Dirac function also has analytical representations. From its definition as a derivative we can derive the representation

$$\delta(x) = \lim_{L\to\infty} \frac{\sin Lx}{\pi x}\,. \tag{2.18}$$

In fact, for $x = 0$, the function $(\sin Lx/\pi x)$ equals L/π, and as $|x|$ increases it oscillates with period $2\pi/L$ which decreases when L increases. The integral of $\sin Lx/\pi x$ over $(-\infty, +\infty)$ is unity, independent of the value of L. Thus, the right side of Eq. (2.18) represents the δ-function. A consequence of (2.18) is the useful representation

$$\delta(x) = \frac{1}{2\pi} \int_{-\infty}^{\infty} e^{ikx}\,dk, \tag{2.19}$$

which will aid considerably in calculating certain improper integrals involving the incomplete cylindrical functions.

Along with $\delta(x)$, there is also the less frequently encountered function

$$\delta^{\pm}(x) = \frac{1}{2\pi} \int_0^{\infty} e^{\pm ikx}\,dk, \tag{2.20}$$

which is related to $\delta(x)$ by

$$\delta^{\pm}(x) = \frac{1}{2}\,\delta(x) \pm \frac{i}{2\pi}\,P\,\frac{1}{x}\,. \tag{2.21}$$

Here the symbol P indicates that the integrals containing $\delta^{\pm}(x)$, are to be taken in the Cauchy principal value sense. In other words, to calculate these integrals by quadrature, it is necessary to to delete a neighborhood $(x - \varepsilon, x + \varepsilon)$, of the singularity at $x = 0$ from the range of integration and then proceeding to the limit as $\varepsilon \to 0$.

3. New Integral Representations for Incomplete Cylindrical Functions of Real Arguments

In this section we use the properties of the unit step function obtained above to introduce new representations for the incomplete functions of Bessel, Struve, Hankel and others. We turn first to Eq. (1.3) of this chapter, and for definiteness consider only the function

$$E_\nu^+(\alpha, x) = J_\nu(\alpha, x) + iH_\nu(\alpha, x) = 2\frac{x^\nu}{A_\nu} \int_0^{\alpha} e^{ix\cos\theta} \sin^{2\nu}\theta\,d\theta\,. \tag{3.1}$$

Introducing now the new variable $\cos\theta = u$ and setting $a = \cos\alpha$, we obtain

$$E_\nu^+(\alpha, x) = 2\frac{x^\nu}{A_\nu}\int_a^1 e^{ixu}(1-u^2)^{\nu-1/2}\,du$$
$$= \frac{2x^\nu}{A_\nu}\int_0^1 e^{ixu}(1-u^2)^{\nu-1/2}\,du - \frac{2x^\nu}{A_\nu}\int_0^a e^{ixu}(1-u^2)^{\nu-1/2}\,du. \tag{3.2}$$

Here the first term obviously can be represented in terms of Bessel and Struve functions:

$$\frac{2x^\nu}{A_\nu}\int_0^1 e^{ixu}(1-u^2)^{\nu-1/2}\,du = J_\nu(x) + iH_\nu(x). \tag{3.3}$$

In the second term we transform the interval of integration to (0,1) by introducing the unit step function $\Omega(a, u)$ into the integrand, and obtain

$$E_\nu^+(\alpha, x) = J_\nu(x) + iH_\nu(x) - \frac{2x^\nu}{A_\nu}\int_0^1 \Omega(a, u)\,e^{ixu}(1-u^2)^{\nu-1/2}\,du. \tag{3.4}$$

Substituting for $\Omega(a, u)$ its integral representation of the form (2.6), we have

$$E_\nu^+(\alpha, x) = J_\nu(x) + iH_\nu(x)$$
$$-\frac{2}{\pi}\frac{x^\nu}{A_\nu}\int_0^1 e^{ixu}(1-u^2)^{\nu-1/2}\,du\int_{-\infty}^{\infty}\frac{\sin at}{t}\,e^{iut}\,dt$$
$$= J_\nu(x) + iH_\nu(x) - \frac{x^\nu}{\pi}\int_{-\infty}^{\infty}\frac{\sin at}{t}\left[\frac{2}{A_\nu}\int_0^1 e^{i(t+x)u}(1-u^2)^{\nu-1/2}\,du\right]dt.$$

With the aid of Eq. (3.3) it is not difficult to evaluate the inner integral by means of Bessel and Struve functions, thus

$$E_\nu^+(\alpha, x) = J_\nu(x) + iH_\nu(x) - \frac{1}{\pi}\int_{-\infty}^{\infty}\frac{\sin at}{t}\,\frac{J_\nu(x+t) + iH_\nu(x+t)}{(x+t)^\nu}\,dt. \tag{3.5}$$

Introducing now the new variable of integration $x + t = u$ and separating the real and imaginary parts of the resulting expression, we obtain the following integral representations for the incomplete functions of Bessel and Struve:

$$J_\nu(\alpha, x) = J_\nu(x) - \frac{x^\nu}{\pi}\int_{-\infty}^{\infty}\frac{\sin a(u-x)}{u-x}\,\frac{J_\nu(u)}{u^\nu}\,du; \tag{3.6}$$

$$H_\nu(\alpha, x) = H_\nu(x) - \frac{x^\nu}{\pi}\int_{-\infty}^{\infty}\frac{\sin a(u-x)}{u-x}\,\frac{H_\nu(u)}{u^\nu}\,du. \tag{3.7}$$

In obtaining these equations we assumed $0 \leq a = \cos\alpha < 1$, i.e. $0 < \alpha \leq \pi/2$. We now seek analogous expressions for the case $\pi/2 < \alpha \leq \pi$

$(-1 \le a < 0)$. For this we put $\alpha = \beta + \pi/2$ in Eq. (3.2), so that $a = \cos\alpha = -\sin\beta = -|a|$, $0 \le \beta \le \pi/2$. Then, in using (3.3) we have

$$E_\nu^+(\alpha, x) = J_\nu(x) + iH_\nu(x) + \frac{2x^\nu}{A_\nu} \int_0^{|a|} e^{-ixu}(1 - u^2)^{\nu - 1/2}\, du.$$

Carrying out the same transformations as before, we get

$$\begin{aligned} E_\nu^+(\alpha, x) &= J_\nu(x) + iH_\nu(x) \\ &+ \frac{1}{\pi} \frac{x^\nu}{A_\nu} \int_{-\infty}^{\infty} \frac{\sin|a|\,(u + x)}{u + x} \frac{J_\nu(u + iH_\nu(u)}{u^\nu}\, du. \end{aligned} \tag{3.8}$$

Replacing u with $-u$ and using the relations

$$\left.\begin{aligned} \frac{J_\nu(-z)}{(-z)^\nu} &= \frac{J_\nu(z)}{z^\nu}; \\ \frac{H_\nu(-z)}{(-z)^\nu} &= -\frac{H_\nu(z)}{z^\nu}, \end{aligned}\right\} \tag{3.9}$$

for $\pi/2 \le \alpha \le \pi$ we find

$$J_\nu(\alpha, x) = J_\nu(x) + \frac{x^\nu}{\pi} \int_{-\infty}^{\infty} \frac{\sin|a|\,(u - x)}{u - x} \frac{J_\nu(u)}{u^\nu}\, du; \tag{3.10}$$

$$H_\nu(\alpha, x) = H_\nu(x) - \frac{x^\nu}{\pi} \int_{-\infty}^{\infty} \frac{\sin|a|\,(u - x)}{u - x} \frac{H_\nu(u)}{u^\nu}\, du. \tag{3.11}$$

It is evident that the content of Eqs. (3.6), (3.7) and (3.10), (3.11) can be expressed for all α in the interval $0, \pi$ by

$$J_\nu(\alpha, x) = J_\nu(x) - \frac{x^\nu}{\pi} \int_{-\infty}^{\infty} \frac{\sin a\,(u - x)}{u - x} \frac{J_\nu(u)}{u^\nu}\, du; \tag{3.12}$$

$$H_\nu(\alpha, x) = H_\nu(x) - \frac{x^\nu}{\pi} \int_{-\infty}^{\infty} \frac{\sin|a|\,(u - x)}{u - x} \frac{H_\nu(u)}{u^\nu}\, du. \tag{3.13}$$

Now setting $\alpha = 0$ $(a = 1)$ and noting that $J_\nu(0, x) \equiv 0$ and $H_\nu(0, x) \equiv 0$, we immediately find the following representations for Bessel and Struve functions:

$$J_\nu(x) = \frac{x^\nu}{\pi} \int_{-\infty}^{\infty} \frac{\sin(u - x)}{u - x} \frac{J_\nu(u)}{u^\nu}\, du: \tag{3.14}$$

$$H_\nu(x) = \frac{x^\nu}{\pi} \int_{-\infty}^{\infty} \frac{\sin(u - x)}{u - x} \frac{H_\nu(u)}{u^\nu}\, du. \tag{3.15}$$

The first of these equations is a particular case of the more general relation (cf., e.g. [11, p. 766])

$$\int_{-\infty}^{\infty} \frac{\sin b(t+x)}{(x+t)t^{\nu}} J_{\nu+2n}(t)\,dt = \pi x^{-\nu} J_{\nu+2n}(x). \tag{3.16}$$

Integral representations for Bessel and Struve functions of the form (3.14) and (3.15) can also be obtained from Eq. (3.3) by introducing into the integrand the unit multiplier $\Omega(1, u)$ and carrying out steps analogous to those above.

It is not difficult to find analogous representations for the incomplete modified Bessel function $I_\nu(\alpha, x)$ and for the incomplete modified Struve function $I_\nu(\alpha, x)$. Tracing through the derivations of the above expressions, we see that the desired representations can be obtained by replacing x with $-ix$ in Eq. (3.12) and (3.14). Thus we have

$$I_\nu(\alpha, x) = I_\nu(x) - \frac{x^\nu}{\pi} \int_{-\infty}^{\infty} \frac{\sin a(u+ix)}{u+ix} \frac{J_\nu(u)}{u^\nu}\,du; \tag{3.17}$$

$$L_\nu(\alpha, x) = L_\nu(x) - \frac{ix^\nu}{\pi} \int_{-\infty}^{\infty} \frac{\sin |a|\,(u+ix)}{u+ix} \frac{H_\nu(u)}{u^\nu}\,du. \tag{3.18}$$

By means of elementary transformations we can change the intervals of integration to $(0, \infty)$, whereupon these expressions take the form

$$I_\nu(\alpha, x) = I_\nu(x) - \frac{2x^\nu}{\pi} \int_0^{\infty} \frac{x \sinh ax \cos au + u \sin au \cosh ax}{x^2+u^2} \frac{J_\nu(u)}{u^\nu}\,du; \tag{3.19}$$

$$L_\nu(\alpha, x) = L_\nu(x) - \frac{2x^\nu}{\pi} \int_0^{\infty} \frac{x \cosh |a|\, x \sin |a|\, u - u \sinh |a|\, x \cos |a|\, u}{x^2+u^2} \frac{H_\nu(u)}{u^\nu}\,du. \tag{3.20}$$

We turn now to the consideration of incomplete Hankel functions, defined by Eq. (1.11), which we write in the form

$$\mathscr{H}_\nu(\beta, x) = -\frac{2ie^{-i\nu\pi}x^\nu}{A_\nu} \int_1^{c} e^{iux}(u^2-1)^{\nu-1/2}\,du, \tag{3.21}$$

where $c = \cosh\beta$ $(0 \le \beta < \infty)$.

As before, we introduce the unit multiplier $\Omega(c, u)$ into the integrand, using (2.6), and transform the interval of integration from one to infinity. We find

$$\mathscr{H}_\nu(\beta, x) = -\frac{2ie^{-\nu\pi}x^\nu}{\pi A_\nu} \int_1^{\infty} e^{iux}(u^2-1)^{\nu-1/2}\,du \int_{-\infty}^{\infty} e^{iut} \frac{\sin ct}{t}\,dt.$$

In this expression, as before, we may interchange the order of integration. However, in the present case we must have $|\nu| < 1/2$ in order to perform this operation. Carrying out the interchange, we obtain

$$\mathscr{H}_\nu(\beta, x) = -\frac{2ie^{-\nu\pi}x^\nu}{\pi A^\nu} \int_{-\infty}^{\infty} \frac{\sin ct}{t}\, dt \int_1^{\infty} e^{i(x+t)u} (u^2 - 1)^{\nu-1/2}\, du. \quad (3.22)$$

The inner integral is easily expressed in terms of the Hankel function of the first kind, (cf. Eq. (3.19) of Chapter I) and thus

$$\begin{aligned} \mathscr{H}_\nu(\beta, x) &= \frac{x^\nu}{\pi} \int_{-\infty}^{\infty} \frac{\sin ct}{t} \frac{\mathscr{H}_\nu^{(1)}(t+x)}{(t+x)^\nu}\, dt \\ &= \frac{x^\nu}{\pi} \int_{-\infty}^{\infty} \frac{\sin c(u-x)}{u-x} \frac{\mathscr{H}_\nu^{(1)}(u)}{u^\nu}\, du, \end{aligned} \quad (3.23)$$

Where $c = \cosh\beta$ $(0 \leq \beta < \infty)$ and $-1/2 < \nu < 1/2$. This expression gives another integral representation for the incomplete Hankel function $\mathscr{H}_\nu(\beta, x)$.

Proceeding in an analogous way from Eq. (1.12), we obtain an integral representation for the incomplete MacDonald function:

$$K_\nu(\beta, x) = \frac{x^\nu}{\pi} \int_{-\infty}^{\infty} \frac{\sin ct}{t} \frac{K_\nu(x - it)}{(x - it)^\nu}\, dt. \quad (3.24)$$

Here $c = \cosh\beta$ and $\nu > -1/2$.

As $\beta \to \infty$ $(c \to \infty)$ the functions $\mathscr{H}_\nu(\beta, x)$ and $K_\nu(\beta, x)$ tend by definition to the Hankel function $\mathscr{H}_\nu^{(1)}(x)$ and the MacDonald function $K_\nu(x)$, respectively. In order to recover these results from Eqs. (3.23) and (3.24), we let $c \to \infty$ and make use of Eq. (2.18).

We now apply the methods developed above to obtain new integral representations for incomplete cylindrical functions of the Bessel form. According to (1.6) of this chapter, we have

$$\varepsilon_\nu(i\alpha, x) = A_\nu(\alpha, x) - iB_\nu(\alpha, x) = \frac{1}{\pi} \int_0^a e^{i(x\sin\theta - \nu\theta)}\, d\theta. \quad (3.25)$$

Transforming the interval of integration from 0 to π with the aid of the unit multiplyer

$$\Omega(\alpha, \theta) = \frac{1}{\pi} \int_{-\infty}^{\infty} e^{it\theta} \frac{\sin \alpha t}{t}\, dt,$$

we have

$$\varepsilon_\nu(i\alpha, x) = \frac{1}{\pi} \int_0^\pi e^{i(x\sin\theta - \nu\theta)}\, d\theta \frac{1}{\pi} \int_{-\infty}^{\infty} e^{it\theta} \frac{\sin \alpha t}{t}\, dt$$

and, after interchanging the order of integration, we get

$$\varepsilon_\nu(i\alpha, x) = \frac{1}{\pi} \int_{-\infty}^{\infty} \frac{\sin \alpha t}{t} dt \frac{1}{\pi} \int_0^\pi e^{i(x \sin\theta + t\theta)} d\theta. \tag{3.26}$$

The inner integral can be evaluated from Eq. (3.25) as the function $\varepsilon_{\nu-t}(i\pi, x) = A_{\nu-t}(x) - iB_{\nu-t}(x)$. Therefore, separating real and imaginary parts and introducing the new variable of integration $\xi = \nu - t$, we obtain the following integral representations for the incomplete functions of Anger and Weber:

$$A_\nu(\alpha, x) = \frac{1}{\pi} \int_{-\infty}^{\infty} \frac{\sin \alpha(\xi - \nu)}{\xi - \nu} A_\xi(x)\, d\xi; \tag{3.27}$$

$$B_\nu(\alpha, x) = \frac{1}{\pi} \int_{-\infty}^{\infty} \frac{\sin \alpha(\xi - \nu)}{\xi - \nu} B_\xi(x)\, d\xi, \tag{3.28}$$

where $A_\nu(x) \equiv A_\nu(\pi, x)$ and $B_\nu(x) \equiv B_\nu(\pi, x)$ are, respectively, the Anger and Weber functions. For $\alpha = \pi$ these formulae take the form

$$A_\nu(x) = \frac{1}{\pi} \int_{-\infty}^{\infty} \frac{\sin \pi(\xi - \nu)}{\xi - \nu} A_\xi(x)\, d\xi; \tag{3.29}$$

$$B_\nu(x) = \frac{1}{\pi} \int_{-\infty}^{\infty} \frac{\sin \pi(\xi - \nu)}{\xi - \nu} B_\xi(x)\, d\xi. \tag{3.30}$$

Analogous computations for the function $k_\nu(\beta, x)$ give

$$k_\nu(\beta, x) = \frac{1}{\pi} \int_{-\infty}^{\infty} \frac{\sin \beta t}{t} K_{\nu - it}(x)\, dt. \tag{3.31}$$

By virtue of Eq. (2.18), this formula reduces, as it should, to the MacDonald function as $\beta \to \infty$.

Thus in diverse examples we have seen how the introduction of the unit step function into integral expression permits us to obtain new integral representations for incomplete cylindrical functions. These same methods may be used to obtain new representations for other functions, for example the incomplete Lipschitz-Hankel and Weber integrals.

In conclusion we note that expressions (3.14), (3.15) for the Bessel and Struve functions, as well as Eqs. (3.29), (3.30) for the Anger and Weber functions, may be interpreted as integral equations with symmetric kernels of the form

$$\frac{\sin a(\eta - \lambda)}{\eta - \lambda}. \tag{3.32}$$

In (3.14) and (3.15) the integration is with respect to the variable x for fixed index ν; on the other hand, in representations (3.29) and (3.30), the integration is with respect to the index ν for fixed x.

4. Relation between Weber Integrals of Bessel Functions and Incomplete Cylindrical Functions

We consider the improper integral

$$\int_0^\infty \frac{J_\nu(bt)\,\mathrm{e}^{-p^2t^2}}{t^2+a^2}\,t^{\nu+1}\,dt, \tag{4.1}$$

which appears frequently in applied problems. This integral is continuous with respect to the variables a and b and, as $a^2 \to 0$ it reduces to one of the Weber integrals, which will be studied further in § 2 of Chapter VI:

$$\begin{aligned}\int_0^\infty J_\nu(bt)\,\mathrm{e}^{-p^2t^2}t^{\nu-1}\,dt &= \frac{\Gamma(\nu)}{2p^\nu\Gamma(1-\nu)}\left(\frac{b}{2p}\right)^\nu F\left(\nu;\nu+1;-\frac{b^2}{4p^2}\right)\\ &= 2^{\nu-1}b^{-\nu}\gamma\left(\nu,\frac{b^2}{4p^2}\right),\end{aligned} \tag{4.2}$$

where $\operatorname{Re}\nu > 0$ and $\operatorname{Re}(p^2) > 0$. Here we have used (8.19) of Chapter II which relates the incomplete gamma function $\gamma(\nu, y)$ to the confluent hypergeometric function.

We now show that, for all a different from zero, (4.1) may be expressed in terms of incomplete cylindrical functions. Without loss of generality, we can take the parameters a and b to be real numbers. Substituting the obvious relation

$$\frac{1}{t^2+a^2} = \int_0^\infty \mathrm{e}^{-x(t^2+a^2)}\,dx \tag{4.3}$$

in to the integral (4.1), we find, after interchanging the order of integration

$$\int_0^\infty \frac{J_\nu(bt)\,\mathrm{e}^{-p^2t^2}}{t^2+a^2}\,t^{\nu+1}\,dt = \int_0^\infty \mathrm{e}^{-xa^2}\,dx \int_0^\infty J_\nu(bt)\,\mathrm{e}^{-(p^2+x)t^2}t^{\nu+1}\,dt. \tag{4.4}$$

The inner integral on the right hand side of this equation is precisely the Weber integral (3.5) of Chapter IV, and is equal to

$$\int_0^\infty J_\nu(bt)\,\mathrm{e}^{-(p^2+x)t^2}t^{\nu+1}\,dt = \frac{b^\nu \mathrm{e}^{-b^2/4(p^2+x)}}{[2(p^2+x)]^{\nu+1}}. \tag{4.5}$$

Consequently, the original integral (4.1) may be written in the form

$$\int_0^\infty \frac{J_\nu(bt)\,e^{-p^2t^2}}{t^2+a^2}\,t^{\nu+1}\,dt = \frac{1}{2}\left(\frac{b}{2}\right)^\nu \int_0^\infty e^{-xa^2-b^2/4(p^2+x)}(p^2+x)^{-\nu-1}\,dx, \quad (4.6)$$

or, introducing in the right side the variable of integration t defined by $p^2 + x = t/a^2$, we find

$$\int_0^\infty \frac{J_\nu(bt)\,e^{-p^2t^2}}{t^2+a^2}\,t^{\nu+1}\,dt$$
$$= \frac{1}{2}\left(\frac{ba^2}{2}\right)^\nu e^{a^2p^2} \int_{a^2p^2}^\infty t^{-\nu-1} \exp\left[-t - \frac{(a^2b^2)}{4t}\right] dt. \quad (4.7)$$

Comparison of this expression with formula (2.17) of Chapter IV

$$\left(\frac{z}{2}\right)^\nu \int_c^d t^{-\nu-1} \exp\left(-t - \frac{z^2}{4t}\right) dt = 2\pi i e^{i\nu\pi/2} S_\nu(ce^{i\pi}, de^{i\pi}, iz) \quad (4.8)$$

gives

$$\int_0^\infty \frac{J_\nu(bt)\,e^{-p^2t^2}}{t^2+a^2}\,t^{\nu+1}\,dt = \pi i e^{a^2p^2} a^\nu e^{i\nu\pi/2} S_\nu(p^2a^2e^{i\pi}, \infty\, e^{i\pi}, iab), \quad (4.9)$$

where $S_\nu(p, q, z)$ is an incomplete cylindrical function of the Sonine-Schlaefli form. Using Eq. (3.50) of Chapter IV and taking into account the relations (3.52)

$$Q_\nu(\infty \exp i\pi, iz) = 0; \qquad P_\nu(\infty \exp i\pi, iz) = i, \quad (4.10)$$

and (3.9), (3.10), of the same chapter, we find the following expression for the integral (4.1):

$$\int_0^\infty \frac{J_\nu(bt)\,e^{-p^2t^2}}{t^2+a^2}\,t^{\nu+1}\,dt$$
$$= e^{-a^2p^2} a^\nu \left[K_\nu(ab)\,\tilde{Q}_\nu(a^2p^2, ab) + I_\nu(ab)\,\tilde{P}_\nu(a^2p^2, ab)\right], \quad (4.11)$$

where $\tilde{Q}_\nu(x, z)$ and $\tilde{P}_\nu(x, z)$ are incomplete Weber integrals.

Furthermore, introducing on the right of (4.7) the variable of integration u defined by $t = ab\,(e^u/2)$, we obtain

$$\int_0^\infty \frac{J_\nu(bt)\,e^{-p^2t^2}}{t^2+a^2}\,t^{\nu+1}\,dt = \frac{a^\nu e^{a^2p^2}}{2} \int_{\ln\left(\frac{2ap^2}{b}\right)}^\infty e^{-ab\cosh u - \nu u}\,du. \quad (4.12)$$

or, recalling the definition of the incomplete cylindrical function $\Phi_\nu(b, z)$ (cf. § 6, Chapter III),

$$\Phi_\nu(b, z) = \frac{1}{\pi i} \int_{-\infty+i\lambda}^b e^{z\sinh u - \nu u}\,du, \quad (4.13)$$

we find after an elementary transformation

$$\int_0^\infty \frac{J_\nu(bt)\, e^{-p^2t^2}}{t^2+a^2} t^{\nu+1}\, dt = \pi i \frac{a^\nu e^{a^2p^2}}{2} e^{-i\nu\pi/2} \Phi_{-\nu}(i\pi/2 - \ln 2ap^2/b, iab). \quad (4.14)$$

This equation is valid for $2ap^2/b > 1$. If $2ap^2/b < 1$, it is advantageous to use Eqs. (6.8) of Chapter III to bring (4.14) into the form

$$\int_0^\infty \frac{J_\nu(bt)\, e^{-p^2t^2}}{t^2+a^2} t^{\nu+1}\, dt$$

$$= a^\nu e^{a^2p^2}\left[K_\nu(ab) - \frac{i\pi}{2} e^{i\nu\pi/2} \Phi_\nu\left(\frac{i\pi}{2} - \ln\left(\frac{b}{2ap^2}\right), iab\right)\right], \quad (4.15)$$

where now $b/2ap^2 > 1$. The relation

$$\frac{\pi i}{2} e^{-i\nu\pi/2} \mathscr{H}_\nu^{(1)}(i) = K_\nu(z). \quad (4.16)$$

between the Hankel and MacDonald functions was also used to obtain (4.15).

In view of the detailed study of the asymptotic behavior of $\Phi_\nu(b, z)$ given above, (4.14) and (4.15) are convenient for the analysis of Weber integrals for all values of the parameter $b/2ap^2$ for which $|ab| \gg 1$.

For integer ν, the integral (4.1) can be expressed not only by incomplete cylindrical functions of the Bessel and Sonine-Schlaefli forms, but also in terms of incomplete cylindrical functions of the Poisson form. Thus, for $\nu = 0$ we find from Eq. (4.12):

$$\int_0^\infty \frac{J_0(bt)\, e^{-p^2t^2}}{t^2+a^2} t\, dt = \frac{e^{a^2p^2}}{2}\left[K_0(ab) - K_0\left(\ln\frac{2ap^2}{b}, ab\right)\right], \quad (4.17)$$

where $K_0(\beta, x)$ is the incomplete MacDonald function.

Thus we have shown that the Weber integrals (4.1) are incomplete cylindrical functions and that (4.9), (4.14), and (4.17) may be interpreted as new integral representations for the incomplete cylindrical functions of the Sonine-Schlaefli, Bessel, and Poisson forms, respectively.

We now take advantage of the results thus far obtained to derive an important formula. From relation (4.11) for $\nu = 0$ we have

$$\begin{aligned} &\int_0^\infty \frac{J_0(bt)\, e^{-p^2t^2}}{t^2+a^2} t\, dt \\ &= e^{-a^2p^2}[K_0(ab)\, \tilde{Q}_0(a^2p^2, ab) + I_0(ab)\, \tilde{P}_0(a^2p^2, ab)], \end{aligned} \quad (4.18)$$

where $\tilde{P}_0(x, p)$ and $\tilde{Q}_0(x, p)$ are the incomplete Weber integrals:

$$\tilde{Q}_0(x, \varrho) = \frac{e^x}{2x} \int_0^{\varrho} t I_0(t) \exp\left(-\frac{t^2}{4x}\right) dt; \tag{4.19}$$

$$\tilde{P}_0(x, \varrho) = \frac{e^x}{2x} \int_{\varrho}^{\infty} t K_0(t) \exp\left(-\frac{t^2}{4x}\right) dt. \tag{4.20}$$

The first of these integrals has been well studied, but the second, $\tilde{P}_0(x, p)$, while it appears in many practical problems, has been little studied and is not tabulated. If for simplicity we take $a = 1$, $p^2 = x$, and $b = \varrho$ it is not difficult to obtain the following expression for this integral from (4.17) and (4.18):

$$\tilde{P}_0(x, \varrho) = \frac{e^{2x}}{2I_0(\varrho)} \left[K_0(\varrho) - K_0\left(\ln\frac{2x}{\varrho}, \varrho\right)\right] - \frac{K_0(\varrho)}{I_0(\varrho)} \tilde{Q}_0(x, \varrho). \tag{4.21}$$

The functions appearing on the right hand side of (4.21) have been tabulated for rather wide ranges of their parameters (c.f. Chapter IX). Formula (4.21) is thus completely adequate for the solution of applied problems.

5. Discontinuous Integrals of Gallop and their Relation to Incomplete Cylindrical Functions

In the study of a number of theoretical and applied problems so-called discontinuous integrals, which are analytic expressions for discontinuous functions, appear. A classic example of an integral of this type is given by (2.6) in § 2 of this chapter:

$$\Omega(p, q) = \frac{2}{\pi} \int_0^{\infty} \frac{\sin px \cdot \cos qx}{x} dx = \begin{cases} 1; & q < p; \\ 1/2; & q = p; \\ 0; & q > p, \end{cases} \tag{5.1}$$

Several investigators have studied a more general class of discontinuous integrals of Bessel functions, which, for different values of their parameters, belong to different function classes.

We consider first an integral of the form

$$\Omega_\nu(a, z) = i \int_{-\infty}^{\infty} \frac{1 - e^{ia(t+z)}}{t + z} \frac{J_\nu(t)}{t^\nu} dt, \tag{5.2}$$

where $\operatorname{Re}(\nu) > -3/2$, $J_\nu(t)$ is the Bessel function, and the parameter a will be taken to be a real number. One of the versions of this integral

$$\int_{-\infty}^{\infty} \frac{\sin a(t + z)}{t + z} \frac{J_\nu(t)}{t^\nu} dt \tag{5.3}$$

is, to within a constant multiplier, a particular case of the discontinuous Gallop integral

$$G_{\mu,\nu} = \int_{-\infty}^{\infty} \frac{J_\mu\{a(z+t)\}}{(z+t)^\mu} \frac{J_\nu(t)}{t^\nu} dt. \tag{5.4}$$

Here we have set $\mu = 1/2$, and used the relation

$$J_{1/2}(z) = \sqrt{\frac{2}{\pi z}} \sin z. \tag{5.5}$$

For $\mu = 1/2$ and $\nu = 0$ we can write (5.3) as

$$\int_{-\infty}^{\infty} \frac{\sin a(z+t)}{z+t} J_0(t)\, dt = \begin{cases} \pi J_0(z); & a \geq 1; \\ 2\int_0^a \frac{\cos uz}{\sqrt{1-u^2}} du; & a < 1. \end{cases} \tag{5.6}$$

But, according to the definition of the incomplete cylindrical function of the Poisson form, (c. f. § 1, Chapter II) we have

$$2\int_0^a \frac{\cos uz}{\sqrt{1-u^2}} du = \pi[J_0(z) - J_0(\alpha, z)]. \tag{5.7}$$

Therefore

$$\int_{-\infty}^{\infty} \frac{\sin a(z+t)}{z+t} J_0(t)\, dt = \pi[J_0(z) - \Omega(1, a)\, J_0(\alpha, z)], \tag{5.8}$$

where $\alpha = \arccos a$, $J_0(\alpha, z)$ is the incomplete Bessel function, and $\Omega(1, a)$ is the unit step function, defined by Eq. (5.1).

We show now that the more general integral (5.2) can also be expressed in terms of incomplete cylindrical functions. To this end we shall assume $\operatorname{Re}\nu > -1$, $(c > 0)$, use the following representation for Bessel functions:

$$\frac{J_\nu(t)}{t^\nu} = \frac{1}{2^\nu 2\pi i} \int_{c-i\infty}^{c+i\infty} x^{-\nu-1} \exp\left(x - \frac{t^2}{4x}\right) dx, \tag{5.9}$$

and take note of the identity

$$i\frac{1 - e^{ia(z+t)}}{z+t} = \int_0^a e^{iu(z+t)} du. \tag{5.10}$$

Substituting (5.9) and (5.10) into (5.2), we find

$$\Omega_\nu(a, z) = \frac{1}{2^\nu 2\pi i} \int_0^a e^{iuz}\, du \int_{c-i\infty}^{c+i\infty} x^{-\nu-1} e^x\, dx \int_{-\infty}^{\infty} e^{iut - t^2/4x}\, dt. \tag{5.11}$$

But since $c > 0$, $\operatorname{Re}(x) > 0$, and integration with respect to t gives

$$\int_{-\infty}^{\infty} e^{iut - t^2/4x}\, dt = 2\sqrt{\pi x}\, e^{-u^2 x}, \tag{5.12}$$

Consequently

$$\Omega_\nu(a, z) = \frac{2\sqrt{\pi}}{2^\nu} \int_0^a e^{iuz}\, du \frac{1}{2\pi i} \int_{c-i\infty}^{c+i\infty} x^{-\nu-1/2} e^{x(1-u^2)}\, dx. \tag{5.13}$$

Here the inner integral reduces to zero for $u > 1$, which can be easily seen by closing the contour of integration on the right by a semicircle of infinite radius (c. f. § 2 of this Chapter). If $u < 1$, then by introducing the new variable of integration $\xi = x(1 - u^2)$ we find

$$\frac{2\sqrt{\pi}}{2^\nu 2\pi i} \int_{c-i\infty}^{c+i\infty} x^{-\nu-1/2} e^{x(1-u^2)}\, dx = \frac{2\sqrt{\pi}}{2^\nu} (1 - u^2)^{\nu-1/2} \frac{1}{2\pi i} \int_{c'-i\infty}^{c'+i\infty} \xi^{-\nu-1/2} e^{\xi}\, d\xi,$$

where $c' = c(1 - u^2) > 0$, as before. But by Eq. (4.5) of Chapter I, (or, alternatively, by replacing ν with $\nu - 1/2$ and taking the limit of (5.9) as $t \to 0$):

$$\frac{1}{2\pi i} \int_{c'-i\infty}^{c'+i\infty} \xi^{-\nu-1/2} e^{\xi}\, d\xi = \frac{1}{2\pi i} \int_{-\infty}^{(0+)} \xi^{-\nu-1/2} e^{\xi}\, d\xi = \frac{1}{\Gamma(\nu + 1/2)}. \tag{5.14}$$

Consequently we have

$$\frac{2\sqrt{\pi}}{2^\nu 2\pi i} \int_{c-i\infty}^{c+i\infty} x^{-\nu-/12} e^{x(1-u^2)}\, dx = \frac{2\sqrt{\pi}}{2^\nu \Gamma(\nu + 1/2)} (1 - u^2)^{\nu-1/2}.$$

Taking account of the above, we may now write a single expression for the inner integral of (5.13), valid for all u:

$$\frac{2\sqrt{\pi}}{2^\nu 2\pi i} \int_{c-i\infty}^{c+i\infty} x^{-\nu-1/2} e^{x(1-u^2)}\, dx = \frac{2\pi}{A_\nu} \Omega(1, u)\, (1 - u^2)^{\nu-1/2}, \tag{5.15}$$

where $\Omega(1, u)$ is the unit step function and $A_\nu = 2^\nu \Gamma(\nu + 1/2)\, \Gamma(1/2)$. Substituting this expression in Eq. (5.13), we get

$$\Omega_\nu(a, z) = \frac{2\pi}{z^\nu} \frac{z^\nu}{A_\nu} \int_0^a e^{iuz} \Omega(1, u)\, (1 - u^2)^{\nu-1/2}\, du. \tag{5.16}$$

If $a > 1$, then by definition $\Omega(1, u) = 0$ for all $u > 1$, and (5.16) takes the form

$$\Omega_\nu(a, z) = \frac{2\pi}{z^\nu} \frac{z^\nu}{A_\nu} \int_0^1 e^{iuz} (1 - u^2)^{\nu-1/2}\, du = \frac{\pi}{z^\nu} E_\nu^+\left(\frac{\pi}{2}, z\right). \tag{5.17}$$

If $a < 1$, then $\Omega(1, u) = 1$, since u is always less than unity on the interval of integration. Hence in this case (5.16) may be written as:

$$\begin{aligned} \Omega_\nu(a, z) &= \frac{2\pi}{z^\nu} \frac{z^\nu}{A_\nu} \int_0^a e^{iuz} (1 - u^2)^{\nu-1/2}\, du \\ &= \frac{\pi}{z^\nu} \left[E_\nu^+\left(\frac{\pi}{2}, z\right) - E_\nu^+(\cos^{-1} a, z) \right] \end{aligned} \tag{5.18}$$

where $E_\nu^+(\omega, z)$ is the incomplete cylindrical function of Poisson form. Collecting the above formulae, we have

$$\begin{aligned} \Omega_\nu(a, z) &= i \int_{-\infty}^{\infty} \frac{1 - e^{ia(z+t)}}{z+t} \frac{J_\nu(t)}{t^\nu} dt \\ &= \frac{\pi}{z^\nu}\left[E_\nu^+\left(\frac{\pi}{2}, z\right) - \Omega(1, a) E_\nu^+(\cos^{-1} a, z)\right]. \end{aligned} \tag{5.19}$$

Taking z real, separating the real and imaginary parts, and using the relations

$$E_\nu^+\left(\frac{\pi}{2}, z\right) = J_\nu(z) + iH_\nu(z);$$

$$E_\nu^+(\alpha, z) = J_\nu(\alpha, z) + iH_\nu(\alpha, z),$$

we find

$$\int_{-\infty}^{\infty} \frac{\sin a(z+t)}{z+t} \frac{J_\nu(t)}{t^\nu} dt = \frac{\pi}{z^\nu}[J_\nu(z) - \Omega(1, a) J_\nu(\cos^{-1} a, z)]; \tag{5.20}$$

$$\int_{-\infty}^{\infty} \frac{1 - \cos a(z+t)}{z+t} \frac{J_\nu(t)}{t^\nu} dt = \frac{\pi}{z^\nu}[H_\nu(z) - \Omega(1, a) H_\nu(\cos^{-1} a, z)], \tag{5.21}$$

where $J_\nu(\alpha, z)$ and $H_\nu(\alpha, z)$ are respectively incomplete Bessel and Struve functions. These formulae may be interpreted as new integral representations for the incomplete Bessel and Struve functions.

Up to now we have considered a particular case of the Gallop integral (5.4), for $\mu = 1/2$. We now show that for all half-odd values of μ and for $a \geq 1$ this integral may be represented by a finite sum of Bessel functions and by a similar sum involving incomplete cylindrical-functions when $a < 1$. In fact, if we use (5.9) and the Poisson representation

$$\frac{J_\mu[a(z+t)]}{(z+t)^\mu} = \frac{a^\mu}{A_\mu} \int_0^\pi e^{ia(z+t)\cos\varphi} \sin^{2\mu}\varphi \, d\varphi, \tag{5.22}$$

then, after substitution in the integral (5.4) and carrying out the integrations with respect to t and x, we obtain the following expression for $G_{\mu,\nu}(a, z)$

$$G_{\mu,\nu}(a, z) = \frac{4\pi a^{\mu-1}}{A_\mu A_\nu} \int_0^a \left(1 - \frac{u^2}{a^2}\right)^{\mu-1/2} \Omega(1, u) \cos uz \, dz. \tag{5.23}$$

For this result we introduced the new variable of integration $u = \alpha \cos\varphi$. If now μ is any half-odd-integer, i.e. $\mu - 1/2 = n$, then

$$\left(1 - \frac{u^2}{a^2}\right)^n \equiv \left(\frac{a^2 - 1}{a^2} + \frac{1 - u^2}{a^2}\right)^n$$

can be represented as a polynomial in $(1 - u^2)$ and Eq. (5.23) can be written as:

$$G_{n+1/2,\nu}(a, z) = \sum_{k=0}^{n} M^n_{k,\nu}(a, z) \frac{2z^{k+\nu}}{A_{k+\nu}} \int_0^a \Omega(1, u)\,(1 - u^2)^{k+\nu-1/2} \cos uz\, dz, \tag{5.24}$$

where we have introduced the notation

$$M^n_{k,\nu}(a, z) = \frac{\sqrt{2\pi}\, a^{-n-1/2}(a^2 - 1)^{n-k}}{z^{k+n}} \frac{2^{k+\nu}\Gamma\left(k + \nu + \frac{1}{2}\right)}{k!\,(n-k)!\,\Gamma\left(\nu + \frac{1}{2}\right)}. \tag{5.25}$$

Repeating the same arguments that led to (5.17) and (5.18), we find

$$G_{n+1/2,\nu}(a, z) = \sum_{k=0}^{n} M^{(n)}_{k,\nu}(a, z)\,[J_{k+\nu}(z) - \Omega(1, a)\, J_{k+\nu}(\cos^{-1} a, z)], \tag{5.26}$$

where as before $\Omega(1, a)$ is the unit step function, (5.1). Setting $a = 1$ and noting that $M^n_{k,\nu}(1, z)$ is different from zero only for $k = n$, we obtain the well known result of Hardy (c.f. [10, p. 463])

$$G_{\mu,\nu}(1, z) = \frac{\Gamma(\mu + \nu)\sqrt{2\pi}}{\Gamma\left(\mu + \frac{1}{2}\right)\Gamma\left(\nu + \frac{1}{2}\right)} \frac{J_{\nu+\mu-1/2}(z)}{z^{\nu+\mu-1/2}}. \tag{5.27}$$

In conclusion we consider still another Gallop integral

$$\int_{-\infty}^{\infty} |t| \frac{\sin a(z + t)}{z + t} J_0(bt)\, dt, \tag{5.28}$$

which may be represented by a discontinuous function of the form ([10], p. 464)

$$\int_{-\infty}^{\infty} |t| \frac{\sin a(z + t)}{z + t} J_0(bt)\, dt = \left\{\begin{array}{ll} 0, & a < b; \\ \dfrac{2 \cos az}{\sqrt{a^2 - b^2}} + 2z \displaystyle\int_0^{\cosh^{-1} a/b} \sin(zb \cosh\theta)\, d\theta, & a > b \end{array}\right\}. \tag{5.29}$$

Alternatively, according to the definition of the incomplete Hankel function [c.f. § 1, Chapter II] the last equation may be written:

$$\int_{-\infty}^{\infty} |t| \frac{\sin a(z + t)}{z + t} J_0(bt)\, dt = \left\{\begin{array}{ll} 0, & a < b; \\ \dfrac{2 \cos az}{\sqrt{a^2 - b^2}} + \dfrac{\pi z}{2}\,[\mathcal{H}_0(\beta, bz) - \mathcal{H}_0(\beta, -bz)], & (a > b), \end{array}\right\} \tag{5.30}$$

where $\beta = \cosh^{-1}(a/b)$.

Thus many discontinuous integrals of the Gallop type may be expressed by means of incomplete cylindrical functions of the Poisson form, and may be considered as new integral representations for this class of functions. At the same time, for certain values of their parameters, these integrals may be conveniently evaluated with the aid of the general theory of incomplete cylindrical functions.

6. Sonine's Discontinuous Integral and its Connection with Incomplete Bessel Functions[1]

Sonine, as is well known, studied a whole class of discontinuous improper integrals containing Bessel functions. In particular, he showed (c.f. e.g. [10, p. 458]) that the integral

$$Z_{\mu,\nu}(a, z) = \int_0^\infty J_{\mu+1}(bt) \frac{J_\nu(a\sqrt{z^2 + t^2})}{(z^2 + t^2)^{\nu/2}} t^\mu \, dt \tag{6.1}$$

for $\operatorname{Re}(\nu + 1) > \operatorname{Re}\mu > -1$ could, when $a < b$, be expressed in terms of Bessel functions by:

$$Z_{\mu,\nu}(a, z) = \frac{2^\mu \Gamma(\mu + 1)}{b^{\mu+1}} \frac{J_\nu(az)}{z^\nu}. \tag{6.2}$$

We will study these integrals in detail for arbitrary positive values of the parameters a and b. Without loss of generality, we may take $b = 1$. Equations (6.1), (6.2) then take the form

$$Z_{\mu,\nu}(a, z) = \int_0^\infty J_{\mu+1}(t) \frac{J_\nu(a\sqrt{z^2 + t^2})}{(z^2 + t^2)^{\nu/2}} t^\mu \, dt; \tag{6.3}$$

$$Z_{\mu,\nu}(a, z) = 2^\mu \Gamma(\mu + 1) \frac{J_\nu(az)}{z^\nu} \tag{6.4}$$

for $a < 1$.

For further transformations of this integral we use the second of the Sonine definite integrals [10, p. 410])

$$\int_0^{\pi/2} J_\mu(z \sin\theta) J_\lambda(y \cos\theta) \sin^{\mu+1}\theta \cos^{\lambda+1}\theta \, d\theta = \frac{z^\mu y^\lambda J_{\mu+\lambda+1}(\sqrt{z^2 + y^2})}{(y^2 + z^2)^{(\mu+\lambda+1)/2}},$$

which for $\lambda = -1/2$ and $\mu = \nu - 1/2$ may be written as

$$\begin{aligned} \frac{J_\nu(a\sqrt{z^2 + t^2})}{(z^2 + t^2)^{\nu/2}} &= \frac{az^{1/2}}{t^{\nu-1/2}} \int_0^{\pi/2} J_{\nu-1/2}(at \sin\theta) \\ &\times J_{-1/2}(az \cos\theta) \sin^{\nu+1/2}\theta \cos^{1/2}\theta \, d\theta. \end{aligned} \tag{6.5}$$

Substituting this in Eq. (6.3) and changing the order of integration, we obtain

$$\begin{aligned} Z_{\mu,\nu}(a, z) &= \sqrt{\frac{2a}{z}} \int_0^{\pi/2} \cos(az \cos\theta) \sin^{\nu+1/2}\theta \, d\theta \\ &\times \int_0^\infty J_{\mu+1}(t) \frac{J_{\nu-1/2}(at \sin\theta)}{t^{\nu-1/2}} t^\mu \, dt, \end{aligned} \tag{6.6}$$

[1] The results of this section were obtained jointly with M. M. Rikenglaz.

where we have used the fact that $J_{-1/2}(z) = \sqrt{z/\pi z} \cos(z)$. It is not difficult to verify that the inner integral in this equation is equal to a Sonine integral for $z = 0$, i.e.

$$Z_{\mu,\nu-1/2}(a \sin \theta, 0) = \int_0^\infty J_{\mu+1}(t) \frac{J_{\nu-1/2}(at \sin \theta)}{t^{\nu-1/2}} t^\mu \, dt. \tag{6.7}$$

On the other hand, if we use the results of [10, p. 458], then (6.3) may also be represented as

$$Z_{\mu,l}(y, z) = \frac{1}{yl} \int_0^y u^{2\mu+1} \Omega(1, u) \left(\frac{\sqrt{y^2 - u^2}}{z}\right)^{l-\mu-1} J_{l-\mu-1}\left(z\sqrt{y^2 - u^2}\right) du. \tag{6.8}$$

Here we have inserted the unit step function $\Omega(1, u)$ into the integrand to reflect the fact that the upper limit of integration is equal to unity for all $y > 1$ and equal to y for $y \leq 1$. Substituting $l = \nu - 1/2$, and $y = a \sin \theta$ in (6.8) and passing to the limit as $z \to 0$, we find after some manipulation

$$Z_{\mu,\nu-1/2}(a \sin \theta, 0) = \\ = \begin{cases} 0 & ; \quad a^2 \sin^2 \theta < 1; \\ \dfrac{2^{\mu+3/2-\nu}}{\Gamma\left(\nu - \mu - \frac{1}{2}\right)} \dfrac{(a^2 \sin^2 \theta - 1)^{\nu-\mu-3/2}}{(a \sin \theta)^{\nu-1/2}} & ; \quad a^2 \sin^2 \theta > 1. \end{cases} \tag{6.9}$$

With this formula and (6.7) we obtain the following expression for $Z_{\mu,\nu}(a, z)$:

$$Z_{\mu,\nu}(a, z) = \frac{2^{\mu+1} a^\nu}{2^\nu \Gamma\left(\nu - \mu - \frac{1}{2}\right) \Gamma\left(\frac{1}{2}\right)} \int_0^{\pi/2} \cos(az \cos \theta) \sin^{2\nu} \theta L(\theta) \, d\theta, \tag{6.10}$$

where we have introduced the notation

$$L(\theta) = \int_0^{1/(a^2 \sin^2 \theta)} t^\mu (1 - t)^{\nu-\mu-3/2} \Omega(1, t) \, dt. \tag{6.11}$$

If $a < 1$, the upper limit of integration in this formula is larger than one for all θ, and according to the definition of $\Omega(1, t)$ we have

$$L(\theta) = \int_0^1 t^\mu (1 - t)^{\nu-\mu-3/2} \, dt = \frac{\Gamma(\mu + 1)\, \Gamma\left(\nu - \mu - \frac{1}{2}\right)}{\Gamma\left(\nu + \frac{1}{2}\right)}. \tag{6.12}$$

In this case $L(\theta)$ does not depend on the parameter θ, and (6.10) reduces immediately to (6.4). If $a > 1$, then it is necessary to divide the interval of integration in (6.10) into two intervals $(0, \theta_0)$ and $(\theta_0, \pi/2)$, where θ_0 is defined by

$$a^2 \sin^2 \theta_0 = 1. \tag{6.13}$$

In the first of these integrals, $L(\theta)$ is, as before, independent of the parameter θ and satisfies formula (6.12); in the second integral, $\Omega(1, t) \equiv 1$, so that for $\theta_0 < \theta < \pi/2$,

$$L(\theta) = \int_0^{1/(a^2 \sin^2\theta)} t^\mu (1-t)^{\nu-\mu-3/2}\, dt. \tag{6.14}$$

Thus the Sonine integral for $a > 1$ may be written as

$$\begin{aligned} Z_{\mu,\nu}(a, z) &= \frac{2^\mu \Gamma(\mu+1)}{z^\nu} J_\nu(\theta_0, az) \\ &+ \frac{2^{\mu+1-\nu} a^\nu}{\Gamma\left(\nu-\mu-\frac{1}{2}\right)\Gamma\left(\frac{1}{2}\right)} \int_{\theta_0}^{\pi/2} \cos(az\cos\theta) \sin^{2\nu}\theta L(\theta)\, d\theta, \end{aligned} \tag{6.15}$$

where $J_\nu(\theta_0, az)$ is the incomplete Bessel function of Poisson form and $L(\theta)$ is defined by Eq. (6.14).

For convenience in the following analysis we carry out some preliminary transformations on the function $L(\theta)$. Making the substitution $(1-u) = ta^2 \sin^2\theta$, in Eq. (6.14) we get

$$\begin{aligned} L(\theta) &= (a\sin\theta)^{1-2\nu} x^{-\nu+\mu+3/2} \int_0^1 (1-u)^\mu (1+ux)^{\nu-\mu-3/2}\, du \\ &= (a\sin\theta)^{1-2\nu} x^{-\nu+\mu+3/2} F(-\nu+\mu+3/2, 1; \mu+2; -x), \end{aligned} \tag{6.16}$$

where we have used (8.8) of Chapter II and, for brevity, set

$$\frac{1}{x} = a^2 \sin^2\theta - 1 > 0. \tag{6.17}$$

We consider now some particular cases. Let μ and ν be such that

$$\nu - \mu - \frac{3}{2} = n, \tag{6.18}$$

where n is any integer. In this case the hypergeometric function in (6.16) may be represented as a polynomial of degree n in x, having the form

$$\begin{aligned} &\frac{1}{\mu+1} F(-n, 1; \mu+2; -x) \\ &= \Gamma(\mu+1)\Gamma(n+1) \sum_{k=0}^{n} \frac{x^k}{\Gamma(\mu+2+k)\Gamma(n-k+1)}. \end{aligned} \tag{6.19}$$

Substituting this in (6.16) and subsequently (6.16) in (6.15), and using there the substitution

$$\sqrt{a^2-1}\cos\psi = a\cos\theta, \tag{6.20}$$

which maps the interval of integration $(\theta_0, \pi/2)$ onto the interval $(0, \pi/2)$, we find

$$Z_{\mu,\nu}(a, z) = \frac{2^\mu \Gamma(\mu+1)}{z^\nu} J_\nu(\theta_0, az)$$

$$+ \frac{2^{\mu-\nu}\Gamma(\mu+1)}{a^\nu \sqrt{\pi}} \sum_{k=0}^{n} \frac{(a^2-1)^{n-k+1/2}}{\Gamma(\mu+2+k)\,\Gamma(n-k+1)}$$

$$\times 2 \int_0^{\pi/2} \cos\left(\sqrt{a^2-1}\, z \cos\psi\right) \sin^{2(n-k+1/2)} \psi \, d\psi .$$

But the last integral is easy to express in terms of Bessel functions. Therefore, introducing the new summation index $l = n - k$, we obtain finally

$$Z_{\mu,\nu}(a, z) = \frac{2^\mu \Gamma(\mu+1)}{z^\nu} J_\nu\left[\sin^{-1}\left(\frac{1}{a}\right), az\right]$$

$$+ \frac{2^{\mu-\nu}\Gamma(\mu+1)}{a^\nu} \Omega(a, 1) \sum_{l=0}^{n} \frac{(a^2-1)^{l/2+1/4}}{\Gamma\left(\nu+\frac{1}{2}-l\right)} \left(\frac{2}{z}\right)^{l+1/2} \tag{6.21}$$

$$\times J_{l+1/2}\left(z\sqrt{a^2-1}\right),$$

where $\nu - \mu - 3/2 = n$, and $\Omega(a, 1)$ is the unit step function given by Eq. (5.1).

Thus it has been shown that for $a > 1$ and $\nu - \mu - 3/2 = n$, the Sonine integral $Z_{\mu,\nu}(a, z)$ may be expressed in terms of an incomplete Bessel function and a finite sum of Bessel functions with half-odd indices, which, in turn, may be expressed in terms of elementary functions. In particular, taking $\nu = \mu + 3/2$ $(n = 0)$, we obtain

$$Z_{\mu,\mu+3/2}(a, z) = \int_0^\infty J_{\mu+1}(t) \frac{J_{\mu+3/2}\left(a\sqrt{z^2+t^2}\right)}{\left(\sqrt{z^2+t^2}\right)^{\mu+3/2}} t^\mu \, dt$$

$$= \frac{2^\mu \Gamma(\mu+1)}{z^{\mu+3/2}} J_{\mu+3/2}\left[\sin^{-1}\left(\frac{1}{a}\right), az\right] \tag{6.22}$$

$$+ \frac{a^{-\mu-3/2}}{(1+\mu)\sqrt{2\pi}} \Omega(a, 1) \frac{\sin z\sqrt{a^2-1}}{z} .$$

Here μ is any number satisfying the condition $\operatorname{Re}(\mu) > -1$.

This equation shows that a quite wide class of discontinuous Sonine integrals can be expressed simply in terms of incomplete cylindrical functions of the Poisson form.

We now take the index μ of the Sonine integral $Z_{\mu,\nu}$ to be an integer, and allow the second index to be arbitrary, subject to the requirement for convergence, $\operatorname{Re}(\nu+1) > \operatorname{Re}(\mu) > -1$. We shall show that in this case the Sonine integral can be expressed in closed form by means of Bessel functions. With this goal in view we make use of Eq. (8.11) of Chapter II and represent the hypergeometric function appearing in equation (6.16), as:

$$F\left(-\nu+\mu+\frac{3}{2},\ 1;\ \mu+2;\ -x\right)$$
$$=\frac{\Gamma(\mu+2)\,\Gamma\left(\nu-\mu-\frac{1}{2}\right)}{\Gamma\left(\nu+\frac{1}{2}\right)}x^{\nu-\mu-3/2}$$
$$\times F\left(-\nu+\mu+\frac{3}{2},\ -\nu+\frac{1}{2};\ -\nu+\mu+\frac{3}{2};\ -\frac{1}{x}\right)$$
$$+\frac{\mu+1}{x\left(-\nu+\mu+\frac{1}{2}\right)}F\left(1,\ -\mu;\ \frac{1}{2}+\nu-\mu;\ -\frac{1}{x}\right),$$

or, applying to the hypergeometric function in the initial term, the second relation of (8.9) in Chapter II, and representing the hypergeometric function in the second term as a series, we obtain with use of (6.17) the following expression for $L(\theta)$:

$$L(\theta=\frac{\Gamma(\mu+1)\,\Gamma\left(\nu-\mu-\frac{1}{2}\right)}{\Gamma\left(\nu+\frac{1}{2}\right)}$$
$$-\frac{\Gamma(\mu+1)\,\Gamma\left(\nu-\mu+\frac{1}{2}\right)}{\nu-\mu-\frac{1}{2}}(a\sin\theta)^{1-2\nu}$$
$$\times\sum_{k=0}^{\infty}\frac{(a^2\sin^2\theta-1)^{\nu-\mu+k-1/2}}{\Gamma(\mu-k+1)\,\Gamma\left(\nu+k-\mu+\frac{1}{2}\right)},$$

Substituting this expression into (6.15) we get

$$Z_{\mu,\nu}(a,z)=\frac{2^{\mu}\,\Gamma(\mu+1)}{z^{\nu}}J_{\nu}(\theta_0,az)$$
$$+\frac{2^{\mu+1}\,\Gamma(\mu+1)\,(az)^{\nu}}{z^{y}\,2^{\nu}\,\Gamma\left(\nu+\frac{1}{2}\right)\Gamma\left(\frac{1}{2}\right)}\int_{\theta_0}^{\pi/2}\cos(az\cos\theta)\sin^{2\nu}\theta\,d\theta$$
$$-\frac{2^{\mu+1-\nu}\,\Gamma(\mu+1)}{\Gamma\left(\frac{1}{2}\right)}\sum_{k=0}^{\infty}\frac{1}{\Gamma(\mu+1-k)\;\Gamma\left(\nu-\mu+k+\frac{1}{2}\right)}$$
$$\times\int_{\theta_0}^{\pi/2}\cos(az\cos\theta)\,(a^2\sin^2\theta-1)^{\nu-\mu+k-1/2}\,a\sin\theta\,d\theta.$$

It is not difficult to verify that the sum of the first two terms of this equation is

$$\frac{2^\mu \Gamma(\mu+1)}{z^\nu} J_\nu(az).$$

In order to simplify the third term, we use (6.20), and note that the result is an integral for the Bessel function. We obtain

$$Z_{\mu,\nu}(a,z) = \frac{2^\mu \Gamma(\mu+1)}{z^\nu} J_\nu(az) \\ - \frac{\Gamma(\mu+1)}{a^\nu} \Omega(a,1) \sum_{k=0}^{\infty} \frac{2^k}{\Gamma(\mu+1-k)} \left(\frac{\sqrt{a^2-1}}{z}\right)^{\nu-\mu+k} J_{\nu-\mu+k}\left(z\sqrt{a^2-1}\right). \tag{6.23}$$

For $\mu = m$, an integer the series terminates with the m-th term, and if we introduce the new summation index $l = m - k$, the last formula may be written as

$$Z_{m,\nu}(a,z) = \frac{2^m \Gamma(m+1)}{z^\nu} J_\nu(az) \\ - \frac{2^m \Gamma(m+1)}{a^\nu} \Omega(a,1) \sum_{l=0}^{m} \frac{2^{-l}}{l!} \left(\frac{\sqrt{a^2-1}}{z}\right)^{\nu-l} J_{\nu-l}\left(z\sqrt{a^2-1}\right). \tag{6.24}$$

Thus a wide class of discontinuous Sonine integrals $Z_{m,\nu}$ can be expressed in closed form by means of Bessel functions for all positive values of a. In particular, for $m = 0$ and $\operatorname{Re}(\nu+1) > 0$ we have

$$Z_{0,\nu}(a,z) = \int_0^\infty J_1(t) \frac{J_\nu\left(a\sqrt{z^2-t^2}\right)}{\left(\sqrt{z^2+t^2}\right)^\nu} dt \\ = \begin{cases} \dfrac{J_\nu(az)}{z^\nu}; & a \le 1; \\ \dfrac{J_\nu(az)}{z^\nu} - \left(\dfrac{\sqrt{a^2-1^\nu}}{a}\right) \dfrac{J_\nu\left(z\sqrt{a^2-1}\right)}{z^\nu}; & a > 1. \end{cases} \tag{6.25}$$

In connection with the results just obtained it is interesting to note that a remark in the monograph [10, p. 458] that the Sonine integral $Z_{\mu,\nu}(a,z)$ for $a > 1$ "evidently cannot be evaluated" is far too pessimistic. In fact, a large class of these integrals for $\mu = m$ and all $\operatorname{Re}(\nu+1) > m \ge 0$, can, according to (6.24), be expressed in closed form for $a < 1$ as well as for $a > 1$. Moreover, the general theory of incomplete cylindrical functions together with associated tables permits still further widening of the class of Sonine integrals which can be related to known and tabulated functions.

Chapter VI

Integrals Involving Incomplete Cylindrical Functions

In the first paragraphs of this Chapter we shall consider some improper integrals containing incomplete cylindrical functions in their integrands. The essential part of the method of their evaluation will consist in replacing the incomplete cylindrical function by integral representations and subsequently changing the order of integration. Such improper integrals are of interest not only from the purely mathematical point of view, but also because they are of considerable importance in many areas of mathematical physics.

1. Improper Integral of Lipschitz-Hankel

We consider first the following improper integral, which contains the incomplete cylindrical functions of Poisson form

$$\int_0^\infty e^{-cx} E_\nu^+(\alpha, bx)\, x^\lambda d\, x. \tag{1.1}$$

To guarantee convergence at the upper and lower limits, it is necessary to take $\operatorname{Re}(c \pm ib) > 0$, $\operatorname{Re}(c + ib\cos\alpha) > 0$; $\operatorname{Re}(\lambda + \nu) > -1$.

Substituting the expression for $E_\nu^+(a, bx)$ from (1.4) of Chapter II into (1.1) and interchanging the order of integration, we obtain

$$\int_0^\infty e^{-cx} x^\lambda E_\nu^+(\alpha, bx)\, dx = \frac{2b^\nu}{A_\nu} \int_0^a \sin^{2\nu}\theta\, d\theta \int_0^\infty e^{-x(c-ib\cos\theta)} x^{\nu+\lambda}\, dx,$$

where $A_\nu = 2^\nu \Gamma(\nu + 1/2)\, \Gamma(1/2)$. The inner integral, after introduction of the new varibale of integration $u = x(c - ib\cos\theta)$ takes the form

$$\frac{1}{(c - ib\cos\theta)^{\nu+\lambda+1}} \int_0^\infty e^{-u} u^{\nu+\lambda}\, du = \frac{\Gamma(\nu + \lambda + 1)}{(c - ib\cos\theta)^{\nu+\lambda+1}}.$$

Finally we obtain

$$\int_0^\infty e^{-cx} x^\lambda E_\nu^+(\alpha, bx)\, dx = \frac{2b^\nu}{A_\nu} \Gamma(\nu + \lambda + 1) \int_0^a \frac{\sin^{2\nu}\theta\, d\theta}{(c - ib\cos\theta)^{\nu+\lambda+1}}. \tag{1.2}$$

In this way the improper integral (1.1) is reduced to a proper integral, which in each specific case may be evaluated without difficulty. Moreover, for half-odd and integer values of the parameters λ and ν it can be expressed in terms of elementary functions. For example, taking $\lambda = \nu = 0$, we get

$$\int_0^\infty e^{-cx} E_0^+(\alpha, bx)\, dx = \frac{2}{\pi} \int_0^\alpha \frac{d\theta}{c - ib\cos\theta}$$
$$= \frac{2}{\pi\sqrt{c^2+b^2}} \tan^{-1} \frac{c\tan\alpha}{\sqrt{c^2+b^2}} + i \frac{1}{\pi\sqrt{c^2+b^2}} \ln \frac{\sqrt{c^2+b^2} + b\sin\alpha}{\sqrt{c^2+b^2} - b\sin\alpha}. \tag{1.3}$$

Using the relation $E_0^+(\alpha, x) = J_0(\alpha, x) + iH_0(\alpha, x)$, we can evaluate the following special cases involving incomplete Bessel and Struve functions

$$\int_0^\infty e^{-cx} J_0(\alpha, bx)\, dx = \frac{2}{\pi\sqrt{c^2+b^2}} \tan^{-1} \frac{c\tan\alpha}{\sqrt{c^2+b^2}}; \tag{1.4}$$

$$\int_0^\infty e^{-cx} H_0(\alpha, bx)\, dx = \frac{1}{\pi\sqrt{c^2+b^2}} \cdot \ln \frac{\sqrt{c^2+b^2} + b\sin\alpha}{\sqrt{c^2+b^2} - b\sin\alpha}. \tag{1.5}$$

Taking now $\alpha = \pi/2$, we obtain the well known integrals of Lipschitz involving Bessel and Struve functions:

$$\int_0^\infty e^{-cx} J_0(bx)\, dx = \frac{1}{\sqrt{c^2+b^2}}; \tag{1.6}$$

$$\int_0^\infty e^{-cx} H_0(bx)\, dx = \frac{2}{\pi} \frac{1}{\sqrt{c^2+b^2}} \ln \frac{\sqrt{c^2+b^2} + b}{c}. \tag{1.7}$$

The corresponding integrals of Lipschitz-Hankel involving incomplete cylindrical functions of the Bessel form will have the form

$$\int_0^\infty e^{-cx} x^\lambda \varepsilon_\nu(i\alpha, bx)\, dx = \frac{1}{\pi} \int_0^\infty e^{-cx}\, dx \int_0^\alpha e^{ixb\sin\theta - i\nu\theta} x^\lambda\, d\theta$$
$$= \frac{1}{\pi} \int_0^\alpha e^{-i\nu\theta}\, d\theta \int_0^\infty e^{-cx + ixb\sin\theta} x^\lambda\, dx = \frac{\Gamma(\lambda+1)}{\pi} \int_0^\alpha \frac{e^{-i\nu\theta}\, d\theta}{(c - ib\sin\theta)^{\lambda+1}}. \tag{1.8}$$

Here, to guarantee convergence, it is sufficient to take $\operatorname{Re}\lambda > -1$ and $\operatorname{Re}(c - ib\sin\theta) > 0$. From these equations it is not difficult to obtain expressions for integrals involving incomplete Anger functions and incomplete Weber functions. To this end it is necessary in (1.8) to separate real and imaginary parts, taking parameters c and b to be real, and to use the relation $\varepsilon_\nu(\alpha, z) = A_\nu(\alpha, z) - iB_\nu(\alpha, z)$. Thus, for $\lambda = 0$ we have

$$\int_0^\infty e^{-cx} A_\nu(\alpha, bx)\, dx = \frac{1}{\pi} \int_0^\alpha \frac{c\cos\nu\theta + b\sin\theta\sin\nu\theta}{c^2 + b^2\sin^2\theta}\, d\theta; \tag{1.9}$$

$$\int_0^\infty e^{-cx} B_\nu(\alpha, bx)\, dx = \frac{1}{\pi} \int_0^\alpha \frac{c\sin\nu\theta - b\sin\theta\cos\nu\theta}{c^2 + b^2\sin^2\theta}\, d\theta. \tag{1.10}$$

The results obtained in (1.8)—(1.10) show that improper Lipschitz-Hankel integrals involving incomplete cylindrical functions of the Bessel form may also be expressed as proper integrals which, for integral values of ν and λ, can be reduced to elementary functions.

Following the methods of this section, it is not difficult to obtain expressions for analogous improper integrals of all the other incomplete cylindrical functions. It is also not difficult to verify the following relations: for $\operatorname{Re}(\nu + \mu + \lambda + 1) > 0$, $\operatorname{Re}(c - ib - id) > 0$ and $\operatorname{Re}(c - ib\cos\alpha - id\cos\alpha) > 0$

$$\int_0^\infty e^{-cx} E_\nu^+(\alpha, bx)\, E_\mu^+(\alpha, dx)\, x^\lambda\, dx$$
$$= \frac{4\Gamma(\nu + \mu + \lambda + 1)\, b^\nu d^\mu}{A_\nu A_\mu} \int_0^\alpha \int_0^\alpha \frac{\sin^{2\nu}\theta \sin^{2\mu}\psi\, d\theta\, d\psi}{(c - ib\cos\theta - id\cos\psi)^{\nu+\lambda+\mu+1}}; \tag{1.11}$$

for $\operatorname{Re}\lambda > 0$, $\operatorname{Re}(c - ib - id) > 0$ and $\operatorname{Re}(c - ib\sin\alpha - id\sin\alpha) > 0$

$$\int_0^\infty e^{-cx} \varepsilon_\nu(\alpha, bx)\, \varepsilon_\mu(\alpha, dx)\, x^\lambda\, dx$$
$$= \frac{\Gamma(\lambda + 1)}{\pi^2} \int_0^\alpha \int_0^\alpha \frac{e^{-i\nu\theta - i\mu\psi} d\theta\, d\psi}{(c - ib\sin\theta - id\sin\psi)^{\lambda+1}}. \tag{1.12}$$

2. Integrals of Weber Type Involving Incomplete Cylindrical Functions

In the preceeding section we considered improper integrals of incomplete cylindrical functions with weighting finction $x^\lambda \exp(-cx)$. We turn now to integrals of a similar form, but with weighting function $x^\lambda \exp(-p^2x^2)$

$$\int_0^\infty e^{-p^2x^2} E_\nu^+(\alpha, bx)\, x^\lambda\, dx. \tag{2.1}$$

In order to guarantee convergence we must take $\operatorname{Re}(p^2) > 0$, $\operatorname{Re}(\lambda + \nu) > -1$. Integrals of this form for cylindrical functions are called exponential Weber integrals, and in general can be represented as hypergeometric series. As an example, we have for Besssel functions (c.f. [10, p. 431])

$$\int_0^\infty e^{-p^2x^2} J_\nu(bx)\, x^\lambda\, dx$$
$$= \frac{\Gamma\left(\frac{\nu + \lambda + 1}{2}\right)}{2p^{\lambda+1}\, \Gamma(\nu + 1)} (b/2p)^\nu F\left(\frac{\nu + \lambda + 1}{2}, \nu + 1; -b^2/4p^2\right). \tag{2.2}$$

For certain values of the parameters ν and λ this integral reduces to elementary functions.

We shall now show that in certain cases, integrals of the form (2.1) may also be reduced to known functions. For the incomplete Bessel function (2.1) is

$$\int_0^\infty e^{-p^2x^2} J_\nu(\alpha, bx)\, x^\lambda\, dx. \tag{2.3}$$

Using the integral representation for $J_\nu(\alpha, bx)$, and changing the order of integration, we have

$$\int_0^\infty x^\lambda e^{-p^2x^2} J_\nu(\alpha, bx)\, dx = \frac{2b^\nu}{A_\nu} \int_0^\alpha \sin^{2\nu}\theta\, d\theta \int_0^\infty e^{-p^2x^2} x^{\lambda+\nu} \cos(bx\cos\theta)\, dx. \tag{2.4}$$

If $\lambda + \nu$ is an integer, then the inner integral is easily evaluated as

$$\left(-\frac{d}{dp^2}\right)^k \int_0^\infty e^{-p^2x^2} \cos(bx\cos\theta)\, dx = \frac{\sqrt{\pi}}{2} \left(-\frac{d}{dp^2}\right)^k \left[\frac{1}{p} \exp\left(-\frac{b^2\cos{}^2\theta}{4p^2}\right)\right], \tag{2.5}$$

for $\lambda + \nu = 2k$ even, or as

$$\begin{aligned} &\frac{d}{\cos\theta\, db}\left[\left(-\frac{d}{dp^2}\right)^k \int_0^\infty e^{-p^2x^2} \sin(bx\cos\theta)\, dx\right] \\ &= \frac{d}{\cos\theta\, db}\left\{\left(-\frac{d}{dp^2}\right)^k \left[\frac{1}{p} W\left(\frac{\mathrm{b}\cos\theta}{2p}\right)\right]\right\} \qquad (2.6) \\ &= \left(-\frac{d}{dp^2}\right)^k \left[\frac{1}{2p^2} - \frac{b\cos\theta}{2p^3} W\left(\frac{b\cos\theta}{2p^3}\right)\right], \end{aligned}$$

for $\lambda + \mu = 2k + 1$ odd. Here $W(z)$ is the integral;

$$W(z) = e^{-z^2} \int_0^z e^{x^2}\, dx. \tag{2.7}$$

Thus, for $\nu + \lambda = 2k$ we have:

$$\begin{aligned} &\int_0^\infty e^{-p^2x^2} J_\nu(\alpha, bx)\, x^\lambda\, dx \qquad (2.8) \\ &= \frac{b^\nu}{A_\nu} \sqrt{\pi} \left(-\frac{d}{dp^2}\right)^k \left[\frac{1}{p} \int_0^\alpha \sin^{2\nu}\theta \exp\left(-\frac{b^2\cos^2\theta}{4p^2}\right) d\theta\right]; \end{aligned}$$

and for $\nu + \lambda = 2k + 1$

$$\begin{aligned} &\int_0^\infty e^{-p^2x^2} J_\nu(\alpha, bx)\, x^\lambda\, dx \qquad (2.9) \\ &= \frac{b^\nu}{A_\nu} \left\{\left(-\frac{d}{dp^2}\right)^k \left[\frac{1}{p^2} \int_0^\alpha \sin^{2\nu}\theta\, d\theta - \frac{b}{p^3} \int_0^\alpha \sin^{2\nu}\theta \cos\theta\, W\left(\frac{b\cos\theta}{2p}\right) d\theta\right]\right\}. \end{aligned}$$

These formulae reduce improper Weber integrals to proper integrals for integer values of $\lambda + \nu$. Taking $\nu = \lambda = 0$ ($k = 0$) in Eq. (2.8), we obtain

$$\int_0^\infty e^{-p^2x^2} J_0(\alpha, bx)\, dx = \frac{1}{\sqrt{\pi}p} \int_0^\alpha \exp\left(-\frac{b^2 \cos^2\theta}{4p^2}\right) d\theta.$$

This last integral is not difficult to evaluate by means of the incomplete modified function $F_0^-(w, z)$, (c.f. § 1, Chapter II) if one uses the identity $2\cos^2\theta \equiv 1 + \cos 2\theta$ and makes an obvious change of the variable of integration. We find

$$\int_0^\infty e^{-p^2x^2} J_0(\alpha, bx)\, dx = \frac{\sqrt{\pi}}{2p} e^{-b^2/8p^2} F_0^-(2\alpha, b^2/8p^2). \qquad (2.10)$$

Setting $\alpha = \pi/2$ and recalling that $J_0(\pi/2, bx) \equiv J_0(bx)$, $F_0^-(\pi, z) = I_0(z)$ we obtain the well-known Weber integral for Bessel functions, [10, p. 432]

$$\int_0^\infty e^{-p^2x^2} J_0(bx)\, dx = \frac{\sqrt{\pi}}{2p} e^{-b^2/8p^2} I_0(b^2/8p^2). \qquad (2.11)$$

If we set $\nu = 0$ and $\lambda = 1$ ($k = 0$) in (2.9), we have

$$\int_0^\infty e^{-p^2x^2} J_0(\alpha, bx)\, x\, dx = \frac{1}{\pi}\left[\frac{1}{p^2} - \frac{b}{p^3}\int_0^\alpha \cos\theta W\left(\frac{b\cos\theta}{2p}\right) d\theta\right]. \qquad (2.12)$$

For $\alpha = \pi/2$, as one would expect, this equation reduces to the well known result of Weber

$$\int_0^\infty J_0(bx) \exp(-p^2x^2)\, x\, dx = \frac{1}{2p^2}\exp\left(-\frac{b^2}{4p^2}\right). \qquad (2.13)$$

To verify this, it is sufficient to express the probability integral as an infinite series, i.e.

$$W(z) = \frac{\sqrt{\pi}}{2}\sum_{n=0}^\infty (-1)^n \frac{z^{2n+1}}{\Gamma(n + 3/2)} \qquad (2.14)$$

and carry out the integration of (2.12) for $\alpha = \pi/2$.

It is also of interest to have an expression for (2.8) when $\lambda = 0$, $\nu = 2$, $k = 1$:

$$\int_0^\infty e^{-p^2x^2} J_2(\alpha, bx)\, dx = -\frac{b^2}{6\sqrt{\pi}p}\frac{d}{dp}\left[\frac{1}{p}\int_0^\alpha \sin^4\theta \exp(-b^2\cos^2\theta/4p^2)\, d\theta\right]$$

$$= \frac{b^2}{6\sqrt{\pi}}\frac{1}{p^3}\left[\int_0^\alpha \sin^4\theta \exp(-2y\cos^2\theta)\, d\theta \right. \qquad (2.14a)$$

$$\left. + 2y\int_0^\alpha \sin^3\theta\cos\theta \exp(-2y\cos^2\theta)\, d(\cos^2\theta)\right],$$

where $y = b^2/8p^2$. Carrying out an integration by parts in the last terms, we get

$$\int_0^\infty e^{-p^2x^2} J_2(\alpha, bx)\, dx = \frac{b^2}{6p^3\sqrt{\pi}} \left[3\int_0^\alpha \sin^2\theta \cos^2\theta \exp(-2y\cos^2\theta)\, d\theta \right.$$
$$\left. - \sin^3\alpha \cdot \cos\alpha \exp(-2y\cos^2\alpha) \right] \tag{2.14b}$$

and, introducing the new variable of integration $\psi = 2\theta$, we find with little difficulty that:

$$\int_0^\infty e^{-p^2x^2} J_2(\alpha, bx)\, dx = \frac{\sqrt{\pi}}{2p} e^{-b^2/8p^2} F_1^-\left(2\alpha, \frac{b^2}{8p^2}\right)$$
$$- \frac{b^2}{6p^3\sqrt{\pi}} \sin^3\alpha \cos\alpha \exp\left(\frac{-b^2\cos^2\alpha}{4p^2}\right), \tag{2.15}$$

where $F_1^-(2\alpha, z)$ is an incomplete modified function of first order. Taking now $\alpha = \pi/2$, we have

$$\int_0^\infty e^{-p^2x^2} J_2(bx)\, dx = \frac{\sqrt{\pi}}{2p} e^{-b^2/8p^2} I_1\left(\frac{b^2}{8p^2}\right). \tag{2.16}$$

Which for $\nu = 1$ is a particular case of the well-known result

$$\int_0^\infty e^{-p^2x^2} J_{2\nu}(bt)\, dt = \frac{\sqrt{\pi}}{2p} e^{-b^2/8p^2} I_\nu\left(\frac{b^2}{8p^2}\right). \tag{2.17}$$

[10, p. 432]. Comparing (2.10) with (2.11) and (2.15) with (2.16), we note that the expressions for exponential Weber integrals of incomplete Bessel functions have the same structure as the corresponding integrals of Bessel functions. In the integrals involving incomplete Bessel functions, additional terms arise due to the character of the recursion properties of incomplete cylindrical functions.

Evaluation of exponential Weber integrals involving incomplete Struve functions may be accomplished similarly. These integrals also simplify if $\lambda + \nu$ is an integer. Omitting the details, we give only the final results, namely

$$\int_0^\infty e^{-p^2x^2} H_\nu(\alpha, bx)\, x^\lambda\, dx$$
$$= \frac{2b^\nu}{A_\nu}\left(-\frac{d}{dp^2}\right)^k \left[\frac{1}{p}\int_0^\alpha \sin^{2\nu}\theta W(b\cos\theta/2p)\, d\theta\right], \tag{2.18}$$

when $\nu + \lambda = 2k$ and

$$\int_0^\infty e^{-p^2x^2} H_\lambda(\alpha, bx)\, x^\lambda\, dx$$
$$= \frac{\sqrt{\pi}\, b^{\nu+1}}{2A_\nu}\left(-\frac{d}{dp^2}\right)^k \left[\frac{1}{p^3}\int_0^\alpha \sin^{2\nu}\theta \cos\theta \exp\left(-\frac{b^2\cos^2\theta}{4p^2}\right) d\theta\right], \tag{2.19}$$

when $\nu + \lambda = 2k + 1$. From these formulae a series of results for special cases can be obtained. For example, taking $\nu = 0$ and $\lambda = 1$ in (2.19), we find an expression for the Weber integral of the incomplete Struve function of zero order

$$\begin{aligned} \int_0^\infty e^{-p^2x^2} H_0(\alpha, bx)\, x\, dx &= \frac{1}{2\sqrt{\pi}} \frac{b}{p^3} \int_0^\alpha \cos\theta \exp\left(-\frac{b^2\cos^2\theta}{4p^2}\right) d\theta \\ &= \frac{1}{\sqrt{\pi}} \frac{1}{p^2} e^{-b^2\cos^2\alpha/4p^2}\, W\left(\frac{b \sin\alpha}{2p}\right). \end{aligned} \tag{2.20}$$

The above methods may also be applied to evaluate exponential integrals of Weber type involving incomplete cylindrical functions of the Bessel and Sonine-Schlaefli forms. In many cases expressions in terms of proper integrals of the probability function can be obtained.

We note that on the interval $(-\infty, \infty)$ these integrals are expressed by incomplete cylindrical functions of the same form. As an illustration, for the functions

$$j_\nu(w, z) = \frac{1}{2\pi i} \int_{\bar{w}}^{w} e^{z\sinh t - \nu t}\, dt,$$

$$i_\nu(w, z) = \frac{1}{2\pi i} \int_{\bar{w}}^{w} e^{z\cosh t - \nu t}\, dt;$$

we have

$$\int_{-\infty}^{\infty} e^{-p^2x^2} j_\nu(w, bx)\, dx = \int_{-\infty}^{\infty} e^{-p^2x^2}\, dx\, \frac{1}{2\pi i} \int_{\bar{w}}^{w} e^{bx\sinh t - \nu t}\, dt,$$

where $\operatorname{Re}(p^2) > 0$. Interchanging the order of integration and taking account of the relation

$$\int_{-\infty}^{\infty} e^{-p^2x^2 + bx\sinh t}\, dx = \frac{\sqrt{\pi}}{p} e^{b^2/4p^2 \sinh^2 t},$$

we find

$$\int_{-\infty}^{\infty} e^{-p^2x^2} j_\nu(w, bx)\, dx = \frac{\sqrt{\pi}}{p} \frac{1}{2\pi i} \int_{\bar{w}}^{w} e^{-\nu t + b^2\sinh^2 t/4p^2}\, dt.$$

Introducing the new variable of integration $2t = u$ and using the definition of $i_\nu(w, z)$, we have finally

$$\int_{-\infty}^{\infty} e^{-p^2x^2} j_\nu(w, bx)\, dx = \frac{\sqrt{\pi}}{2p} e^{-b^2/8p^2}\, i_{\nu/2}(2w, b^2/8p^2). \tag{2.21}$$

3. Improper Integrals of Certain Incomplete Cylindrical Functions with Respect to Index

Until now we have considered improper integrals of the incomplete cylindrical functions $E_\nu(\alpha, bx)$ and $\varepsilon_\nu(\alpha, bx)$ with respect to the variable x.

It is also of interest to consider certain improper integrals of products of incomplete cylindrical functions with respect to the index. Such integrals are sometimes called Ramanujan integrals (c.f. e.g. [10, p. 494], and also [26]). While we prefer to evaluate some integrals of this type by introducing the integral representations for incomplete Anger functions and Weber functions of § 3, Chapter V, it should be noted that the evaluation of improper integrals of cylindrical functions with respect to index is in most cases substantially simplified if the Bessel-Schlaefli integral representation is used. This is due to the fact in this representation the index appears in the integrand only in an exponential term. With this consideration in mind, we now consider some improper integrals of products of incomplete cylindrical functions of the Bessel and Sonine-Schlaefli forms. As we shall see below, these integrals have a structure identical to that of the corresponding integrals of complete cylindrical functions; we therefore shall limit ourselves to selected important examples to illustrate the methods of computation.

We consider the following integral of the product of incomplete cylindrical functions in the Bessel form:

$$\int_{-\infty}^{\infty} \varepsilon_{\nu-\xi}(i\alpha, x)\, \varepsilon_{\mu+\xi}(i\alpha, y)\, d\xi. \tag{3.1}$$

Using the integral representation

$$\varepsilon_\lambda(i\alpha, x) = \frac{1}{\pi} \int_0^{\alpha} e^{ix\sin\theta - i\lambda\theta}\, d\theta, \tag{3.2}$$

we have

$$\begin{aligned}
&\int_{-\infty}^{\infty} \varepsilon_{\nu-\xi}(i\alpha, x)\, \varepsilon_{\mu+\xi}(i\alpha, y)\, d\xi \\
&= \frac{1}{\pi^2} \int_{-\infty}^{\infty} d\xi \int_0^{\alpha} e^{ix\sin\theta - i(\nu-\xi)\theta}\, d\theta \int_0^{\alpha} e^{iy\sin\psi - i(\mu+\xi)\psi}\, d\psi \\
&= \frac{1}{\pi^2} \int_0^{\alpha}\int_0^{\alpha} e^{i(x\sin\theta + y\sin\psi) - i(\nu\theta + \mu\psi)}\, d\theta\, d\psi \int_{-\infty}^{\infty} e^{i(\theta-\psi)\xi}\, d\xi
\end{aligned}$$

But the inner integral is equal to $2\pi\delta(\theta - \varphi)$, where $\delta(z)$ is the delta function. Taking accounting of the properties of $\delta(z)$ expressed in (2.15) of Chapter V, we finally obtain

$$\begin{aligned}
&\int_{-\infty}^{\infty} \varepsilon_{\nu-\xi}(i\alpha, x)\, \varepsilon_{\mu+\xi}(i\alpha, y)\, d\xi \\
&= \frac{2}{\pi} \int_0^{\alpha} e^{i(x+y)\sin\theta - i(\nu+\mu)\theta}\, d\theta = 2\varepsilon_{\nu+\mu}(i\alpha, x+y).
\end{aligned} \tag{3.3}$$

It is interesting to consider the following particular case of this formula. Taking $x = -y$, we find

$$\int_{-\infty}^{\infty} \varepsilon_{\nu-\xi}(i\alpha, x)\, \varepsilon_{\mu+\xi}(i\alpha, -x)\, d\xi = \frac{2\left(1 - e^{-i(\nu+\mu)\alpha}\right)}{\pi i(\nu+\mu)}. \tag{3.4}$$

Analogous formulae may be obtained for the other incomplete cylindrical functions in the Bessel form. In particular, for the incomplete MacDonald function

$$k_\nu(\beta, x) = \frac{1}{2} \int_{-\beta}^{\beta} e^{-x\cosh t - \nu t}\, dt \tag{3.5}$$

we have

$$\int_{-\infty}^{\infty} k_{i\xi+i\nu}(\beta, x)\, k_{i\xi+i\mu}(\beta, y)\, d\xi = \pi k_{i(\nu-\mu)}(\beta, x+y); \tag{3.6}$$

$$\int_0^{\infty} \cos(y\xi)\, k_{i\xi}(\beta, x)\, d\xi = \begin{Bmatrix} \pi/2\, e^{-x\cosh y}, & |y| \le \beta; \\ 0 & |y| > \beta, \end{Bmatrix}. \tag{3.7}$$

The latter, for $\beta \to \infty$ goes into the corresponding formula for the MacDonald function [11, p. 787]

$$\int_0^{\infty} \cos(y\xi)\, K_{i\xi}(x)\, d\xi = \frac{\pi}{2} e^{-x\cosh y}, \tag{3.8}$$

since $k\nu(\beta, x)$ reduces to $K_\nu(x)$ as $\beta \to \infty$.

The preceeding examples are fairly simple. We now give an example of the evaluation of a more complicated improper integral of incomplete cylindrical functions with respect to index. We consider an integral of the form

$$\int_{-\infty}^{\infty} e^{iy\xi}\, k_{\nu+i\xi}(\beta, x)\, k_{\nu-i\xi}(\beta, y)\, d\xi. \tag{3.9}$$

Using the expression for k_ν given in (3.5) and interchanging the order of integration, we have

$$\begin{aligned} &\int_{-\infty}^{\infty} e^{i\gamma\xi}\, k_{\nu+i\xi}(\beta, x)\, k_{\nu-i\xi}(\beta, y)\, d\xi \\ &= \frac{\pi}{2} \int_{-\beta-\gamma}^{\beta-\gamma} e^{-x\cosh(u+\gamma)-(\gamma+u)\nu}\, du \int_{-\beta}^{\beta} e^{-y\cosh t' - \nu t'}\, \delta(t'-u)\, dt', \end{aligned} \tag{3.10}$$

where for convenience we have introduced the new variable of integration $u = t - \gamma$, and $\delta(z)$ as before denotes the Dirac delta function. By the property of the delta function given in (2.14) of Chapter V, the last integral is equal to

$$\int_{-\beta}^{\beta} e^{-y\cosh t' - \nu t'}\, \delta(t'-u)\, dt' = \begin{Bmatrix} e^{-y\cosh u - \nu u}, & -\beta < u < \beta; \\ 0, & |u| > \beta. \end{Bmatrix}. \tag{3.11}$$

From this we see that the lower limit of the integration with respect to u in the right side of (3.10) must be $-\beta$, if $\gamma > 0$, while the upper limit remains the same, provided $\gamma < 2\beta$. In case $\gamma \geq 2\beta$ the integral (3.10) is equal to zero. Thus we have

$$\begin{aligned}&\int_{-\infty}^{\infty} e^{i\gamma\xi}\, k_{\nu+i\xi}(\beta, x)\, k_{\nu-i\xi}(\beta, y)\, d\xi \\ &= \frac{\pi}{2} e^{-\nu\gamma} \int_{-\beta}^{\beta-\gamma} e^{-x\cosh(u+\gamma) - y\cosh u - 2\nu u}\, du.\end{aligned} \tag{3.12}$$

This expression may be simplified somewhat, if the quantities r and ψ are introduced by means of the relations

$$\left.\begin{aligned} x\cosh\gamma + y &= r\cosh\psi; \\ x\sinh\gamma &= r\sinh\psi \end{aligned}\right\} \tag{3.13}$$

and the new parameter of integration $\theta = u + i\psi$ is used. As a result we have

$$\begin{aligned}&\int_{-\infty}^{\infty} e^{i\gamma\xi}\, k_{\nu+i\xi}(\beta, x)\, k_{\nu-i\xi}(\beta, y)\, d\xi \\ &= \frac{\pi}{2}\left(\frac{x + ye^{\gamma}}{xe^{\gamma} + y}\right)^{-\nu} \int_{-\beta+\psi}^{\beta+\psi-\gamma} e^{-r\cosh\theta - 2\nu\theta}\, d\theta \\ &= \frac{i\pi^2}{2} e^{i\nu\pi}\left(\frac{x + ye^{\gamma}}{xe^{\gamma} + y}\right)^{-\nu} [\varepsilon_{2\nu}(\beta + \psi - \gamma + i\pi/2, ir) - \varepsilon_{2\nu}(-\beta + \psi + i\pi/2, ir)].\end{aligned} \tag{3.14}$$

From this example it is evident that integrals of the type (3.9) involving incomplete MacDonald functions cannot be directly expressed in terms of complete MacDonald functions. However, letting β tend to infinity in (3.14) we recover the corresponding relation for MacDonald functions,

$$\int_{-\infty}^{\infty} e^{i\gamma\xi}\, K_{\nu+i\xi}(x)\, K_{\nu-i\xi}(y)\, d\xi = \frac{\pi}{2}\left(\frac{x + ye^{\gamma}}{xe^{\gamma} + y}\right)^{-\nu} K_{2\nu}(r). \tag{3.15}$$

It is worthwhile to note the following particular case of (3.14). If we set $x = y$, then Eqs. (3.13) imply $\psi = \gamma/2$, and $r = 2x\cosh(\gamma/2)$. Therefore

$$\int_{-\infty}^{\infty} e^{i\gamma\xi}\, k_{\nu+i\xi}(\beta, x)\, k_{\nu-i\xi}(\beta, x)\, d\xi = \pi k_{2\nu}[\beta - \gamma/2, 2x\cosh(\gamma/2)]. \tag{3.16}$$

The methods presented here may be applied without significant change to the evaluation of many other improper integrals of incomplete cylindrical functions with index as the variable of integration, such as those studied by various authors and collected in [11, pp. 784—789].

4. Some Discontinuous Integrals of Weber-Schafheitlin Type Involving Incomplete Cylindrical Functions

In the theory of cylindrical functions one considers improper integrals of the form

$$\int_0^\infty \frac{J_\mu(at)\,J_\nu(bt)}{t^\lambda}\,dt,$$

where for convergence it is necessary to have $a > 0$, $b > 0$ and

$$\operatorname{Re}(\mu+\nu+1) > \operatorname{Re}(\lambda) > -1, \qquad a \neq b;$$

$$\operatorname{Re}(\mu+\nu+1) > \operatorname{Re}(\lambda) > 0, \qquad a = b.$$

These integrals, as a rule, are discontinuous and are customarily called discontinuous Weber-Schafheitlin integrals. In general they may be expressed in terms of hypergeometric series (c.f. [10, p. 436—444]):

$$\int_0^\infty \frac{J_\mu(at)\,J_\nu(bt)}{t^\lambda}\,dt = \frac{\Gamma\left(\frac{\mu+\nu-\lambda+1}{2}\right)}{2^\lambda} \times \begin{cases} \dfrac{b^\nu a^{\lambda-1-\nu}}{\Gamma(1+\nu)\,\Gamma\left(\frac{\lambda+\mu+1-\nu}{2}\right)} \\ \quad \times F\left(\frac{\mu+\nu-\lambda+1}{2}, \frac{\nu-\lambda-\mu+1}{2}; 1+\nu; \frac{b^2}{a^2}\right); & b < a; \\ \dfrac{a^{\lambda-1}\,\Gamma(\lambda)}{\Gamma\left(\frac{\lambda+\nu-\mu+1}{2}\right)\Gamma\left(\frac{\lambda+\mu+\nu+1}{2}\right)\Gamma\left(\frac{\lambda+\mu-\nu+1}{2}\right)}; & b = a; \\ \dfrac{a^\nu b^{\lambda-1-\nu}}{\Gamma(1+\mu)\,\Gamma\left(\frac{\lambda-\mu+\nu+1}{2}\right)} \\ \quad \times F\left(\frac{\mu+\nu-\lambda+1}{2}, \frac{\mu-\nu-\lambda+1}{2}; 1+\mu; \frac{a^2}{b^2}\right), & b > a, \end{cases} \tag{4.1}$$

which for certain values of the parameters μ, ν, and λ reduce to elementary functions. An example of such an integral, (where $\lambda = 0$ and $\mu - 1 = \nu$) is the integral representation (2.3), (2.6) of Chapter V for the unit step function

$$\Omega(a,b) = \frac{a^\mu}{b^{\mu-1}} \int_0^\infty J_\mu(at)\,J_{\mu-1}(bt)\,dt = \begin{cases} 1, & b < a; \\ 1/2, & b = a; \\ 0, & b > a. \end{cases}$$

Furthermore, for $\mu = 0$, $\nu = \pm 1/2$ and $\lambda = -1/2$ we have the following relations:

$$\int_0^\infty J_0(at) \sin bt \, dt = \frac{1}{\sqrt{b^2 - a^2}} \Omega(b, a); \tag{4.2}$$

$$\int_0^\infty J_0(at) \cos bt \, dt = \frac{1}{\sqrt{a^2 - b^2}} \Omega(a, b), \tag{4.3}$$

where $\Omega(b, a)$ and $\Omega(a, b) = 1 - \Omega(b, a)$ are unit step functions.

Considering analogous integrals for incomplete Bessel functions of the Poisson form

$$J_0(\alpha, at) = \frac{2}{\pi} \int_0^\alpha \cos(at \cos\theta) \, d\theta, \tag{4.4}$$

we have

$$\begin{aligned} \int_0^\infty J_0(\alpha, at) \sin bt \, dt &= \frac{2}{\pi} \int_0^\alpha d\theta \int_0^\infty \cos(at \cos\theta) \sin bt \, dt \\ &= \frac{1}{2\pi i} \int_0^\alpha d\theta \int_0^\infty (e^{i(a\cos\theta + b)t} - e^{-i(a\cos\theta + b)t} \\ &\quad - e^{i(a\cos\theta - b)t} + e^{-i(a\cos\theta - b)t}) \, dt. \end{aligned} \tag{4.5}$$

For the sake of definiteness we shall take α, a, and b to be positive. From the fundamental formulae (2.20) and (2.21) of Chapter V we have

$$\int_0^\infty J_0(\alpha, at) \sin bt \, dt = \frac{1}{\pi} P\left(\int_0^\alpha \frac{d\theta}{a\cos\theta + b} - \int_0^\alpha \frac{d\theta}{a\cos\theta - b}\right), \tag{4.6}$$

where the symbol P indicates that the integrals to the right are to be taken in the sense of their principal values. The first integral for $a > 0$, $b > 0$, $\alpha < \pi/2$ can therefore be evaluated by the usual methods. The evaluation of the second integral depends on the relationship among the parameters α, a and b. For $a < b$, and also for $a > b$ and $a\cos\alpha - b > 0$, its evaluation is straightforward. If, however, $a > b$, but $a\cos\alpha - b < 0$, this integral must be evaluated according to:

$$P\int_0^\alpha \frac{d\theta}{a\cos\theta - b} = \lim_{\varepsilon \to 0}\left(\int_0^{\theta_0 - \varepsilon} \frac{d\theta}{y\cos\theta - b} + \int_{\theta_0 + \varepsilon}^\alpha \frac{d\theta}{a\cos\theta - b}\right),$$

where $\theta_0 = \arccos(b/a)$, $b < a$. Carrying out all the indicated operations, we find after a series of simplifications

$$\int_0^\infty J_0(\alpha, at) \sin bt \, dt$$

$$= \begin{cases} \dfrac{1}{\pi\sqrt{a^2-b^2}} \ln \dfrac{|\sin(\alpha - \theta_0)|}{\sin(\theta_0 + \alpha)}, & a > b; \\ \infty, & a = b; \\ \dfrac{2}{\pi\sqrt{b^2-a^2}} \cot^{-1}\left(\dfrac{b}{\sqrt{b^2-a^2}} \tan\alpha\right), & a < b. \end{cases} \tag{4.7}$$

For $\alpha = \pi/2$, we obtain (4.2) again.

Considering now the other improper integral (4.3), we have

$$\int_0^\infty J_0(\alpha, at) \cos bt \, dt = \frac{2}{\pi} \int_0^\alpha d\theta \int_0^\infty \cos(at\cos\theta) \cos bt \, dt$$

$$= \frac{1}{2\pi} \int_0^\alpha d\theta \int_0^\infty (e^{i(a\cos\theta + b)t} + e^{-i(a\cos\theta + b)t} + e^{i(a\cos\theta - b)t} + e^{-i(a\cos\theta - b)t}) \, dt.$$

Applying again (2.20) and (2.21) of Chapter V, we find

$$\int_0^\infty J_0(\alpha, at) \cos bt \, dt = \int_0^\alpha \delta(a\cos\theta + b) \, d\theta + \int_0^\alpha \delta(a\cos\theta - b) \, d\theta, \tag{4.8}$$

where $\delta(z)$ is the Dirac delta function. By virtue of the properties of δ, the right side of (4.8) is identically zero for $a < b$. If $a > b$, then by the property of the delta function given in (2.16) of Chapter V the right hand side of (4.8) may be written as:

$$\int_0^\infty J_0(\alpha, at) \cos bt \, dt = \frac{1}{\sqrt{a^2-b^2}} \left[\int_0^\alpha \delta(\theta - \theta_0) \, d\theta + \int_0^\alpha \delta(\theta + \theta_0 - \pi) \, d\theta \right],$$

where $\theta_0 = \arccos(b/a)$. If we consider only those cases where $a > 0$ and $b > 0$, then $\theta_0 < \pi/2$, and the second term in this equation differs from zero only if $\alpha > \pi/2$. Therefore, restricting ourselves to the interval $0 \leq \alpha \leq \pi/2$ as usual, we have

$$\int_0^\infty J_0(\alpha, at) \cos bt \, dt = \frac{1}{\sqrt{a^2-b^2}} \Omega(a, b) \, \Omega\left(\frac{b}{a}, \cos\alpha\right), \tag{4.9}$$

where $\Omega(a, b)$ and $\Omega(b/a, \cos\alpha)$ are unit step functions. In particular

$$\Omega\left(\frac{b}{a}, \cos\alpha\right) = \begin{cases} 1, & \cos\alpha < \dfrac{b}{a}; \\ 1/2 & \cos\alpha = \dfrac{b}{a}; \\ 0, & \cos\alpha > \dfrac{b}{a}. \end{cases} \tag{4.10}$$

Taking here $\alpha = \pi/2$, we obtain the well known result (4.3) for Bessel functions. From these results it is evident that the integrals we have considered involving incomplete Bessel functions correspond entirely with the analogous integrals of complete Bessel functions in the region $0 \leq a \leq b/\cos(\alpha)$ for fixed b and a. However, $a > b/\cos\alpha$, integral (4.9) is identically zero. The difference between these integrals is shown clearly in Fig. 8, where the solid line represents the region of equality of integrals (4.9) and (4.3).

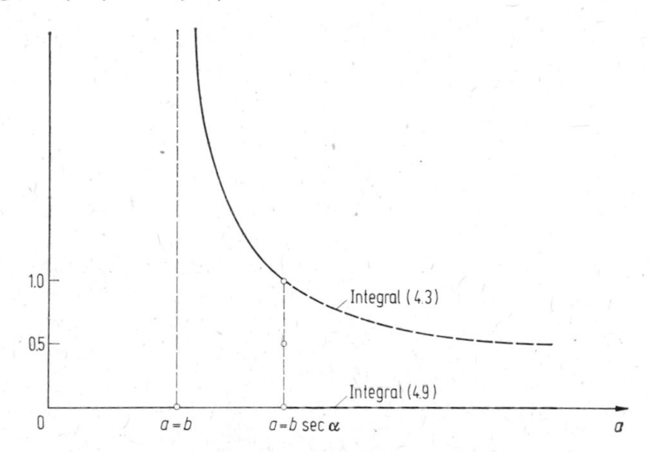

Fig. 8. Geometric illustration of the difference between Weber-Schafheitlin integrals of complete and incomplete Bessel functions.

The analogous integrals for incomplete Struve functions,

$$H_0(\alpha, at) = \frac{2}{\pi} \int_0^\alpha \sin(at\cos\theta)\, d\theta$$

can be evaluated by exactly the same steps. Therefore we give just the final results:

$$\int_0^\infty H_0(\alpha, at) \sin bt\, dt = \frac{1}{\sqrt{a^2 - b^2}} \Omega(a, b)\, \Omega(b/a, \cos\alpha); \tag{4.11}$$

$$\int_0^\infty H_0(\alpha, at) \cos bt\, dt = \begin{cases} \dfrac{1}{\pi} \dfrac{1}{\sqrt{a^2 - b^2}} \ln \dfrac{\sin\theta_0 + \sin\alpha}{|\sin\theta_0 - \sin\alpha|}, & a > b; \\ -\dfrac{2}{\pi} \dfrac{1}{\sqrt{b^2 - a^2}} \operatorname{arctg} \dfrac{a \sin\alpha}{\sqrt{b^2 - a^2}}, & a < b. \end{cases} \tag{4.12}$$

From (4.11) for $\alpha = \pi/2$ there follows the well known result for Struve functions (c.f. [11, p. 792]):

$$\int_0^\infty H_0(at) \sin bt\, dt = \frac{1}{\sqrt{a^2 - b^2}} \begin{Bmatrix} 1, & a > b; \\ 0, & a < b, \end{Bmatrix} \tag{4.13}$$

since $\Omega(b/a, 0) \equiv 1$.

In conclusion we note that analogous improper integrals may be evaluated for the Bessel and Struve functions of the first order, $J_1(\alpha, at)$ and $H_1(\alpha, at)$. To show this we write them as:

$$J_1(a, z) = \frac{2z}{\pi} \int_0^{\alpha} \sin^2\theta \cos(z\cos\theta)\, d\theta$$

$$= \frac{2}{\pi}\int_0^{\alpha} \cos\theta \sin(z\cos\theta)\, d\theta - \frac{2}{\pi}\sin\alpha \sin(z\cos\alpha); \tag{4.14}$$

$$H_1(\alpha, z) = \frac{2z}{\pi}\int_0^{\alpha} \sin^2\theta \sin(z\cos\theta)\, d\theta$$

$$= -\frac{2}{\pi}\int_0^{\alpha} \cos\theta\cos(z\cos\theta)\, d\theta + \frac{2}{\pi}\sin\alpha\cos(z\cos\alpha), \tag{4.15}$$

and then to use the transformations presented above.

5. Improper Integrals of Hankel Type Involving Incomplete Cylindrical Functions

Many definite integrals of incomplete cylindrical functions may be evaluated if properties of the integrals

$$\frac{1}{2\pi i}\int \varphi(z)\, E_\nu^{\pm}(\alpha, az)\, dz; \quad \frac{1}{2\pi i}\int \varphi(z)\varepsilon_\nu(w, az)\, dz;$$

$$\frac{1}{2\pi i}\int \varphi(z)\, E_\nu^{\pm}(\alpha, az)\, E_\mu^{\pm}(\alpha, bz)\, dz$$

taken over suitable contours are known, where $\varphi(z)$ is an algebraic function. We shall limit ourselves here to the simplest such integrals, our aim being chiefly to illustrate methods of computation and conditions required for convergence.

We consider an integral along the closed contour L of the form

$$\frac{1}{2\pi i}\int_L \frac{z^{\mu-1}E_\nu^{+}(\alpha, az)}{(z^2 - r^2)^{m+1}}\, dz, \tag{5.1}$$

where for simplicity we assume $\mathrm{Im}(r) > 0$. We may take L to be the contour consisting of semicircles of radii R and ϱ, centered at the origin, and sections of the real axis as shown in Fig. 9. We shall now find conditions under which the contributions to (5.1) from the integrals over the two semicircles vanish as $\varrho \to 0$ and $R \to \infty$. For small $|z|$,

$$E_\nu^{\pm} \sim z^\nu. \tag{5.2}$$

Therefore the contribution to (5.1) from the inner semicircle will vanish in the limit as $\varrho \to 0$, if $Re(\nu + \mu) > 0$.

For large values of $z = Re^{i\Phi}$, $E_\nu^\pm(\alpha, z)$ behaves (c.f. § 10 of Chapter II) as

$$E_\nu^\pm(\alpha, z) \approx \sqrt{\frac{2}{\pi z}}\, e^{\pm i(z - \nu\pi/2 - \pi/4)} \pm 2i\, \frac{z^{\nu-1}}{A_\nu} \sin^{2\nu-1} \alpha e^{\pm iz\cos\alpha} \tag{5.3}$$

and the contribution to (5.1) from the large semicircle will be of magnitude

$$\begin{aligned}
&\frac{1}{2\pi i} \int\limits_{(c)} \frac{z^{\mu-1} E_\nu^\pm(\alpha, az)}{(z^2 - r^2)^{m+2}}\, dz \\
&\sim \frac{\text{const}}{R^{2m+5/2-\mu}} \int\limits_0^\pi e^{i\varphi(\mu - 5/2 - 2m) + iaR\cos\varphi - aR\sin\varphi}\, d\varphi \\
&+ \frac{\text{const}}{R^{2m+3-\mu-\nu}} \int\limits_0^\pi e^{i\varphi(\mu+\nu-3-2m) + (iaR\cos\varphi - aR\sin\varphi)\cos\alpha}\, d\varphi,
\end{aligned} \tag{5.4}$$

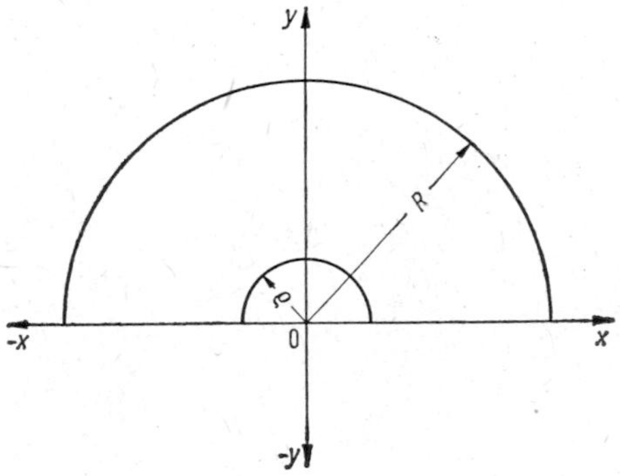

Fig. 9. The contour L.

provided the requirements $a > 0$ and $\cos\alpha > 0$ are satisfied. In this case the integrals on the right side are determined chiefly by contributions from the vicinity of points in the interval of integration for which $\sin\varphi \sim \varphi$ or $\sin\varphi \sim \pi - \varphi$. Computations show that the magnitudes of these integrals for large R are of the order R^{-1} (c.f. also § 9 of Chapter II). Therefore Eq. (5.4) for $aR \gg 1$ takes the form

$$\frac{1}{2\pi i} \int\limits_{(c)} \frac{z^{\mu-1} E_\nu^+(\alpha, az)}{(z^2 - r^2)^{m+1}}\, dz \sim \frac{\text{const}}{R^{2m+7/2-\mu}} + \frac{\text{const}}{R^{2m+4-\mu-\nu}}. \tag{5.5}$$

From this it follows that the contribution to the integral (5.1) from the large semicircle will become negligibly small as $aR \to \infty$, provided the conditions $\operatorname{Re}(\mu) < 2m + 7/2$ and $\operatorname{Re}(\mu + \nu) < 2m + 4$ are met. If $a = 0$ or $\cos\alpha = 0$, the extra conditions become more stringent: $\operatorname{Re}\mu < 2m + 5/2$ and $\operatorname{Re}(\mu + \nu) < 2m + 3$.

Thus conditions for which the contributions to (5.1) from the large and small semicircles vanish as $\varrho \to 0$ and $R \to \infty$ are:

$$\left.\begin{aligned} 0 < \operatorname{Re}(\mu + \nu) < 2m + 3; &\quad \cos\alpha = 0;\\ 0 < \operatorname{Re}(\mu + \nu) < 2m + 4; &\quad \cos\alpha > 0;\\ \operatorname{Re}\mu < 2m + 7/2; &\quad a > 0. \end{aligned}\right\} \tag{5.6}$$

If $\operatorname{Im}(r) > 0$ and conditions (5.6) are satisfied, then the integrand of (5.1) has just one singularity inside the contour L, a simple pole at $z = r$. Therefore, by the calculus of residues we obtain

$$\frac{1}{2\pi i}\int\limits_L \frac{z^{\mu-1} E_\nu^+(\alpha, az)}{(z^2 - r^2)^{m+1}}\,dz = \frac{1}{2\pi i}\int\limits_{-\infty}^{\infty} \frac{x^{\mu-1} E_\nu^+(\alpha, ax)}{(x^2 - r^2)^{m+1}}\,dx$$

$$= \frac{1}{2\pi i}\int^{(r+)} \frac{z^{\mu-1} E_\nu^+(\alpha, az)}{(z^2 - r^2)^{m+1}}\,dz.$$

Now noting that on the negative real axis, the argument of z is $i\pi$, we get

$$\frac{1}{2\pi i}\int\limits_{-\infty}^{\infty} \frac{x^{\mu-1} E_\nu^+(\alpha, ax)}{(x^2 - r^2)^{m+1}}\,dx$$

$$= \frac{1}{2\pi i}\int\limits_0^{\infty} \frac{x^{\mu-1}[E_\nu^+(\alpha, ax) - e^{i\mu\pi} E_\nu^+(\alpha, e^{i\pi} ax)]}{(x^2 - r^2)^{m+1}}\,dx.$$

Taking account also of the relations (3.3) and (1.6) of Chapter II

$$\begin{aligned} E_\nu^+(\alpha, ax) &= J_\nu(\alpha, ax) + iH_\nu(\alpha, ax);\\ E_\nu^+(\alpha, e^{i\pi}ax) &= e^{i\pi\nu}[J_\nu(\alpha, ax) - iH_\nu(\alpha, ax)], \end{aligned} \tag{5.7}$$

and also of the identity

$$\frac{1}{(z^2 - r^2)^{m+1}} = \frac{1}{m!}\left(\frac{d}{dr^2}\right)^m \left(\frac{1}{z^2 - r^2}\right), \tag{5.8}$$

we find after a series of transformations

$$\begin{aligned} &\int\limits_0^{\infty} \frac{x^{\mu-1}\{-J_\nu(\alpha, ax)\sin[(\nu+\mu)\pi/2] + H_\nu(\alpha, ax)\cos[(\nu+\mu)\pi/2]\}}{(x^2 - r^2)^{m+1}}\,dx\\ &\qquad = \frac{\pi e^{-i(\nu+\mu)\pi/2}}{2m!}\left(\frac{d}{dr^2}\right)^m [r^{\mu-2} E_\nu^+(\alpha, ar)]. \end{aligned} \tag{5.9}$$

If r is purely imaginary, $r = ik$ $(k > 0)$, then the last equation takes the form:

$$\int_0^\infty \frac{x^{\mu-i}\left\{-\sin\left[(\nu+\mu)\frac{\pi}{2}\right] J_\nu(\alpha, ax) + \cos\left[(\nu+\mu)\frac{\pi}{2}\right] H_\nu(\alpha, ax)\right\}}{(x^2+k^2)^{m+1}}\, dx$$
$$= \frac{\pi}{m!}(-1)^{m+1}\left(\frac{d}{dk^2}\right)^m [k^{\mu-2} F_\nu^-(\alpha, ak)], \tag{5.10}$$

where

$$F_\nu^-(\alpha, ak) = \frac{1}{2} e^{-i\nu\pi/2} E_\nu^+(\alpha, iak) = \frac{(ak)^\nu}{A_\nu} \int_0^\alpha e^{-ak\cos\theta} \sin^{2\nu}\theta\, d\theta. \tag{5.11}$$

From these equations it is evident that (5.1) can be expressed by modified functions and has the same structure as the Hankel integral involving cylindrical functions (c.f. Ref. [10, p. 465]).

From (5.10) we can obtain several interesting special cases. For $m = 0$, (5.10) takes the form

$$\int_0^\infty \frac{x^{\mu-1}\left\{-\sin\left[(\nu+\mu)\frac{\pi}{2}\right] J_\nu(\alpha, ax) + \cos\left[(\nu+\mu)\frac{\pi}{2}\right] H_\nu(\alpha, ax)\right\}}{x^2+k^2}\, dx \tag{5.12}$$
$$= -\pi k^{\mu-2} F_\nu^-(\alpha, ak).$$

Now taking $\nu + \mu$ equal to one and three, according to conditions (5.6), we find for the Hankel integrals involving incomplete Bessel functions,

$$\int_0^\infty \frac{J_\nu(\alpha, ax)}{x^2+k^2} \frac{dx}{x^\nu} = \frac{\pi}{k^{\nu+1}} F_\nu^+(\alpha, ak). \tag{5.13}$$

$$\int_0^\infty \frac{J_\nu(a, ax)}{x^2+k^2} \frac{dx}{x^{\nu-2}} = -\frac{\pi}{k^{\nu-1}} F_\nu^-(\alpha, ak). \tag{5.14}$$

Moreover, taking $\mu + \nu = 2$, we find the corresponding integral for incomplete Struve functions,

$$\int_0^\infty \frac{H_\nu(\alpha, ax)}{x^2+k^2} \frac{dx}{x^{\nu-1}} = \frac{\pi}{k^\nu} F_\nu^-(\alpha, ak). \tag{5.15}$$

For $\alpha = \pi/2$ these equations reduce to the well known expressions for the Hankel integrals of Bessel and Struve functions.

Concluding this section we remark that the evaluation of particular cases of the Hankel integrals, for example (5.13) and (5.14), may be carried out directly by substituting the appropriate integral representation for the incomplete cylindrical function involved and interchanging the order of integration. As an illustration we shall obtain Eq. (5.13) in this way. We have

$$\int_0^\infty \frac{1}{x^2+k^2} \frac{J_\nu(\alpha, ax)}{x^\nu} dx = \frac{2a^\nu}{A_\nu} \int_0^\alpha \sin^{2\nu}\theta \, d\theta \int_0^\infty \frac{\cos(ax\cos\theta)}{x^2+k^2} dx$$
$$= \frac{a^\nu}{A_\nu} \int_0^\alpha \sin^{2\nu}\theta \, d\theta \int_{-\infty}^\infty \frac{e^{iax\cos\theta}}{x^2+k^2} dx. \tag{5.16}$$

The last integral, according to the calculus of residues, is equal to $(\pi/R) \exp(-ak\cos\theta)$, and thus (5.13) follows immediately.

We see from the examples considered above that the asymptotic behavior of incomplete cylindrical functions determined earlier can be used to evaluate various contour integrals of these functions analogous to the contour integrals of cylindrical functions discussed in Ref. [10, pp. 465—475]. It is only necessary to emphasize that the conditions for convergence of these integrals involving incomplete cylindrical functions are more restrictive than are the conditions for convergence of the same integrals with the corresponding cylindrical functions. This is because the asymptotic behavior of incomplete cylindrical functions involves an interaction between two parts, as, for example, in Eq. (5.3). Thus, for evaluation of a contour integral of the type (5.1) involving the Hankel function $\mathscr{H}_\nu^{(1)}(az)$ the conditions for convergence are: $\operatorname{Re}(\mu+\nu) > 0$ to guarantee convergence at zero, and $\operatorname{Re}(\mu) < 2m + 7/2$ $(\alpha > 0)$ to guarantee convergence at infinity. The same integral of the incomplete cylindrical function $E_\nu^+(\alpha, az)$, as we saw above, requires that the additional conditions $\operatorname{Re}(\mu+\nu) < 2m+4$ for $\alpha < \pi/2$ and $\operatorname{Re}(\mu+\nu) < 2m+3$ for $\alpha = \pi/2$ be met to guarantee convergence of the integral at infinity.

6. Definite Integrals Containing Incomplete Cylindrical Functions

In addition to the improper integrals considered earlier, many integrals with finite limits involving incomplete cylindrical functions can be expressed in closed form either in terms of incomplete cylindrical functions or by other known functions. To illustrate methods of evaluating such integrals we limit ourselves to incomplete cylindrical functions of the Poisson form.

We consider first integrals in the forms

$$\int_0^z z^{\nu+1} E_\nu^+(w, z)\, dz; \tag{6.1}$$

$$\int_0^z z^{-\nu+1} E_\nu^+(w, z)\, dz. \tag{6.2}$$

To evaluate these we make use of the recursion relations (2.22) and (2.23) of Chapter II:

$$z\frac{\partial E_\nu^+}{\partial z} + \nu E_\nu^+ = zE_{\nu-1}^+ - 2\frac{z^\nu}{A_\nu}\cos w \sin^{2\nu-1} w e^{iz\cos w}; \tag{6.3}$$

$$z\frac{\partial E_\nu^+}{\partial z} - \nu E_\nu^+ = -zE_{\nu+1}^+ + i2\frac{z^{\nu+1}}{A_{\nu+1}}\sin^{2\nu+1} w e^{iz\cos w}, \tag{6.4}$$

where $A_\nu = 2^\nu \Gamma(\nu + 1/2)\,\Gamma(1/2)$. Multiplying (6.3) by $z^{\nu-1}$, we have

$$\frac{\partial}{\partial z}(z^\nu E_\nu^+) = z^\nu E_{\nu-1}^+ - 2\frac{z^{2\nu-1}}{A_\nu}\cos w \sin^{2\nu-1} w e^{iz\cos w}. \tag{6.5}$$

Integrating the last equation between the limits zero and z, we get

$$\int_0^z z^\nu E_{\nu-1}^+(w, z)\, dz = z^\nu E_\nu^+(w, z) + 2\frac{\cos w \sin^{2\nu-1} w}{A_\nu}\int_0^z z^{2\nu-1} e^{iz\cos w}\, dz.$$

Or, replacing ν with $\nu + 1$, we find

$$\int_0^z z^{\nu+1} E_\nu^+(w, z)\, dz = z^{\nu+1} E_{\nu+1}^+(w, z) + 2\frac{\cos w \sin^{2\nu+1} w}{A_{\nu+1}}\int_0^z z^{2\nu+1} e^{iz\cos w}\, dz. \tag{6.6}$$

The integral on the right side of this equation can be evaluated by elementary functions if the index ν takes on integer or half-odd values. In particular, when $\nu = 0$ (6.6) takes the form:

$$\int_0^z zE_0^+(w, z)\, dz = zE_1^+(w, z) + 2\,\frac{\tan w}{\pi}(e^{iz\cos w} - 1) - \frac{2}{\pi} iz \sin w e^{iz\cos w}. \tag{6.7}$$

For arbitrary ν this integral can be expressed in terms of the incomplete gamma function (c.f. § 9, Chapter II):

$$\begin{aligned}\int_0^z z^{\nu+1} E_\nu^+(w, z)\, dz &= z^{\nu+1} E_{\nu+1}^+(w, z)\\ &+ 2\frac{e^{-i\pi(\nu+1)}}{A_{\nu+1}}\tan^{2\nu+1} w \cdot \gamma(2\nu + 2, iz/\cos w).\end{aligned} \tag{6.8}$$

Similarly, multiplying (6.4) by $z^{-\nu}$ and taking account of

$$z^{-\nu}\frac{\partial E_\nu^+}{\partial z} - \nu z^{-\nu-1} E_\nu^+ = \frac{\partial}{\partial z}(z^{-\nu} E_\nu^+),$$

we find

$$\begin{aligned}\int_0^z z^{-\nu+1} E_\nu^+(w, z)\, dz &= -z^{-\nu+1} E_{\nu-1}^+(w, z)\\ &+ 2\frac{\sin^{2\nu-1} w}{A_\nu \cos w}[e^{iz\cos w} - 1] + \frac{2}{A_{\nu-1}}\int_0^w \sin^{2\nu-1} w\, dw.\end{aligned} \tag{6.9}$$

If, in formulae (6.6) and (6.9), we take $w = \pi/2$ and regard z as real, then equating the real and imaginary parts separately gives well known formulae for integrals of Bessel and Struve functions, which in our case take the form:

$$\begin{aligned}
&\int_0^z z^{\nu+1} J_\nu(z)\, dz = z^{\nu+1} J_{\nu+1}(z);\\
&\int_0^z z^{-\nu+1} J_\nu(z)\, dz = -z^{-\nu+1} J_{\nu-1}(z) + \frac{1}{2^{\nu-1}\,\Gamma(\lambda)};\\
&\int_0^z z^{-\nu+1} H_\nu(z)\, dz = -z^{-\nu+1} H_{\nu-1}(z) + \frac{2z}{A_\nu};\\
&\int_0^z z^{\nu+1} H_\nu(z)\, dz = z^{\nu+1} H_{\nu+1}(z).
\end{aligned} \tag{6.10}$$

Corresponding integrals involving incomplete cylindrical functions of the Bessel form may be evaluated by similar means; in particular, for an integral of type (6.1) with $\varepsilon_\nu(w, z)$,

$$\begin{aligned}
\int_0^z z^{\nu+1} \varepsilon_\nu(w, z)\, dz &= z^{\nu+1} \varepsilon_{\nu+1}(w, z)\\
&- \frac{z^{\nu+2}}{\pi i(\nu+2)} + \frac{e^{-i(\nu+1)w}}{\pi i} \int_0^z z^{\nu+1} e^{z \sinh w}\, dz.
\end{aligned} \tag{6.11}$$

Here also for integer ν the right hand side is easily evaluated. In the general case it may be expressed in terms of the incomplete γ-function.

We turn now to the evaluation of certain integrals with respect to the variable w and as illustrations we limit ourselves again to incomplete cylindrical functions of the Poisson form. We consider the integral

$$u_{k,\nu}(w, z) = \int_0^w \sin^k \theta E_\nu^+(\theta, z) \cos\theta\, d\theta. \tag{6.12}$$

By the definition of $E_\nu^+(\theta, z)$ we have

$$u_{k,\nu}(w, z) = 2\frac{z^\nu}{A_\nu} \int_0^w \sin^k\theta \cos\theta\, d\theta \int_0^\theta e^{iz\cos t} \sin^{2\nu} t\, dt.$$

Integrating by parts, we obtain

$$u_{k,\nu}(w, z) = \frac{1}{k+1} \sin^{k+1} w E_\nu^+(w, z) - 2\frac{z^\nu}{(k+1)A_\nu} \int_0^w e^{iz\cos t} \sin^{2\nu+1+k} t\, dt. \tag{6.13}$$

Here the second term is not difficult to express by means of incomplete cylindrical functions of order $\nu + (1+k)/2$, whence, after some easy transformations, we find

$$u_{k,\nu}(w, z) \tag{6.14}$$

$$= \frac{1}{k+1}\left[\sin^{k+1} w E_\nu^+(w, z) - \left(\frac{2}{z}\right)^{(k+1)/2} \frac{\Gamma\left(\nu + 1 + \frac{k}{2}\right)}{\Gamma\left(\nu + \frac{1}{2}\right)} E^+_{\nu+(k+1)/2}(w, z)\right].$$

We point out some particular cases of this equation. If we set $w = \pi$, we get for $k + 1 > 0$

$$u_{k,\nu}(\pi, z) = \int_0^\pi \sin^k \theta \cos \theta E_\nu^+(\theta, z)\, d\theta$$

$$= -2\left(\frac{2}{z}\right)^{(k+1)/2} \frac{1}{k+1} \frac{\Gamma\left(\nu + 1 + \frac{k}{2}\right)}{\Gamma\left(\nu + \frac{1}{2}\right)} J_{\nu+(k+1)/2}(z). \tag{6.15}$$

Analogously we obtain an expression for the function $u_{k,\nu}(w, z)$ for $w = \pi/2$. For this we use the relation $E_\nu^+(\pi/2, z) = J_\nu(z) + iH_\nu(z)$ and equate real and imaginary parts of (6.14) separately, taking z to be real. We find

$$u^c_{k,\nu}(\pi/2, z) = \frac{z^\nu}{A_\nu} \int_0^{\pi/2} \sin^k \theta \cos \theta\, d\theta \int_0^\theta \cos(z \cos t) \sin^{2\nu} t\, dt \tag{6.16}$$

$$= \frac{1}{k+1}\left[J_\nu(z) - \left(\frac{2}{z}\right)^{(k+1)/2} \frac{\Gamma\left(\nu + 1 + \frac{k}{2}\right)}{\Gamma\left(\nu + \frac{1}{2}\right)} J_{\nu+(k+1)/2}(z)\right];$$

$$u^s_{k,\nu} = \frac{z^\nu}{A_\nu} \int_0^{\pi/2} \sin^k \theta \cos \theta\, d\theta \int_0^\theta \sin(z \cos t) \sin^{2\nu} t\, dt$$

$$= \frac{1}{k+1}\left[H_\nu(z) - \left(\frac{2}{z}\right)^{(k+1)/2} \frac{\Gamma\left(\nu + 1 + \frac{k}{2}\right)}{\Gamma\left(\nu + \frac{1}{2}\right)} H_{\nu+(k+1)/2}(z)\right]. \tag{6.17}$$

These results may be applied to integrals of the form

$$\int_0^w f(\theta) \cos \theta E_\nu^+(\theta, z)\, d\theta,$$

where $f(\theta)$ is a polynomial in $\sin \theta$,

$$f(\theta) = \sum_{k=0}^n a_k \sin^k \theta. \tag{6.18}$$

It is not difficult to reduce the following integrals to this form:

$$\int_0^w \sin 2k\theta E_\nu^+(\theta, z)\, d\theta; \tag{6.19}$$

$$\int_0^w \cos(2k-1)\,\theta E_\nu^+(\theta, z)\, d\theta, \tag{6.20}$$

where k is an integer. Actually, the well known formulae for $\sin(2k\theta)$ and $\cos[(2k-1)\theta]$ [11, p. 42] may be written as

$$\sin 2k\theta = \frac{\cos\theta}{2} \sum_{l=1}^{k} \frac{(-1)^{l+1}}{(2l-1)!} \frac{(k+l-1)!}{(k-l)!} 2^{2l} \sin^{2l-1}\theta;$$

$$\cos(2k-1)\,\theta = \cos\theta \sum_{l=0}^{k} \frac{(-1)^l\, 2^{2l}}{(2l)!} \frac{(k+l-1)!}{(k-l-1)!} \sin^{2l}\theta.$$

Substituting these expressions into (6.19) and (6.20) we get

$$\int_0^w \sin 2k\theta E_\nu^+(\theta, z)\, d\theta = \frac{1}{2} \sum_{l=1}^{k} \frac{(-1)^{l+1}\, 2^{2l}}{(2l-1)!} \frac{(k+l-1)!}{(k-l)!} u_{2l-1,\nu}(w, z); \tag{6.21}$$

$$\int_0^w \cos(2k-1)\,\theta E_\nu^+(\theta, z)\, d\theta = \sum_{l=0}^{k} \frac{(-1)^l\, 2^{2l}\,(k+l-1)!}{(2l-1)!\,(k-l-1)!} u_{2l,\nu}(w, z), \tag{6.22}$$

where the function $u_{m,\nu}(w, z)$ is as defined above. These expressions may also be written in other forms, if the function $u_{2l-1,\nu}(w, z)$ is written according to expression (6.14) and account is taken of the fact that

$$\sum_{l=0}^{k} \frac{(-1)^l\, 2^{2l}\,(k+l-1)!}{(2l)!\,(k-l)!} \sin^{2l} w = \frac{\cos 2kw}{k}. \tag{6.23}$$

Equation (6.21) then looks like

$$\int_0^w \sin 2k\theta E_\nu^+(\theta, z)\, d\theta = -\frac{\cos 2kw}{2k} E_\nu^+(w, z) + \frac{1}{2} \sum_{l=0}^{k} \frac{(-1)^l\, 2^{2l}\,(k+l-1)!}{(2l)!\,(k-l)!} \left(\frac{2}{z}\right)^l \frac{\Gamma\left(\nu+l+\frac{1}{2}\right)}{\Gamma\left(\nu+\frac{1}{2}\right)} E_{\nu+l}(w, z). \tag{6.24}$$

A similar expression can be obtained for integrals of the form (6.20).

From the results just obtained it is evident that for $w = \pi/2$ and $w = \pi$ these integrals may be expressed, as above, as finite sums of Bessel and Struve functions.

We consider now the integral

$$q_\nu(w, y, z) = \int_0^w e^{iy\cos\theta} E_\nu^+(\theta, z) \sin\theta \, d\theta. \tag{6.25}$$

Integrating by parts, we find

$$q_\nu(w, y, z) = i \frac{e^{iz\cos w} E_\nu^+(w, z)}{y} - 2i \frac{z^\nu}{y A_\nu} \int_0^w e^{i(z+y)\cos\theta} \sin^{2\nu}\theta \, d\theta$$

or

$$\begin{aligned} q_\nu(w, y, z) &= \int_0^w e^{iy\cos\theta} \sin\theta E_\nu^+(\theta, z) \, d\theta \\ &= \frac{i}{y}\left[E_\nu^+(w, z)\, e^{iy\cos w} - \left(\frac{z}{y+z}\right)^\nu E_\nu^+(w, y+z)\right]. \end{aligned} \tag{6.26}$$

Thus integrals like (6.25) are quite simple to evaluate by means of incomplete cylindrical functions. If (6.26) is solved for $E_\nu^+(w, z+y)$ the result may be treated as an addition formula in integral form for incomplete cylindrical functions.

We point out some particular cases. If $y = -z$, then, noting that

$$\lim_{x\to 0} \frac{E_\nu^+(w, x)}{x^\nu} = 2 \frac{1}{A_\nu} \int_0^w \sin^{2\nu} t \, dt,$$

we find

$$q_\nu(w, -z, z) = -\frac{i}{z}\left[E_\nu^+(w, z)\, e^{-iz\cos w} - 2\frac{z^\nu}{A_\nu}\int_0^w \sin^{2\nu} t \, dt\right]. \tag{6.27}$$

Also, for $y = z$ we have

$$q_\nu(w, z, z) = \frac{i}{z}\left[E_\nu^+(w, z)\, e^{iz\cos w} - \frac{1}{2^\nu} E_\nu^+(w, 2z)\right]. \tag{6.28}$$

For $w = \pi/2$ or $w = \pi$ these integrals are easily expressed by means of Bessel and Struve functions.

The methods presented in this section are applicable to analogous integrals containing other incomplete cylindrical functions of Poisson as well as Bessel or Sonine-Schlaefli forms.

Chapter VII

Application of Incomplete Cylindrical Functions to Problems of Wave Propagation and Diffraction

In this and succeeding chapters we shall consider some problems whose solutions are connected with the evaluation of incomplete cylindrical functions. The asymptotic behavior of these functions determined earlier permits us to study analytically those physical situations — often of great interest — in which certain parameters are either very large or very small. Often, however, there arises the need for quantitative analysis in intermediate cases, when asymptotic analysis is inadequate. In such cases, obviously, it is necessary to have tables of incomplete cylindrical functions. In the general case these functions depend on three parameters w, z, and ν, each of which may be complex. It is evident, therefore, that comprehensive tabulation of these functions in their general form is a forbiddingly difficult problem.

In practice, a large fraction of the incomplete cylindrical functions which actually appear are of order zero or one. Therefore short tables of the incomplete functions

$$J_n^{\mathbf{!}}(\alpha, x),\ H_n^{\mathbf{!}}(\alpha, x),\ F_n^{\pm}(\alpha, x) \ \text{ for } n = 0{,}1$$

are given in the appendix for real α and x. As we shall see, tables of these functions alone are not always sufficient. However, several investigators have produced tables for certain functions, by means of which other incomplete cylindrical functions can be evaluated.

It is therefore appropriate to begin our study of the application of the incomplete cylindrical functions to problems of physics and engineering with an exposition of the relations between these finctions and other special integrals tabulated in the literature, brief characteristics of which may also be found in guides to mathematical tables such as references [19, 20 and 27].

Such relations allow on the one hand the possibility of finding numerical values of incomplete cylindrical functions in each specific case. On the other hand, the manifestation of such relations demonstrates the wide variety of theoretical and applied problems which lead to incomplete cylindrical functions.

1. Connections between Incomplete Cylindrical Functions and some Tabulated Special Integrals

In [9], in connection with the solution of an engineering problem of optimization of motion on intersecting trajectories, incomplete integrals of the form

$$\left.\begin{aligned} u &= \int_0^\sigma \cos(r \sin t - pt)\, dt; \\ v &= \int_0^\sigma \sin(r \sin t - pt)\, dt, \end{aligned}\right\} \tag{1.1}$$

are introduced. These differ from the incomplete Anger function $A_\nu(\sigma, z)$ and the incomplete Weber function $B_\nu(\sigma, z)$ only by a constant multiplier, i.e.

$$\left.\begin{aligned} A_p(\sigma, r) &= \pi u; \\ B_p(\sigma, r) &= -\pi v. \end{aligned}\right\} \tag{1.2}$$

In [9] the properties of these functions are studied; they are expanded in Fourier series, and tables are compiled for σ between 0 and π and $0 \leq r = p \leq 0.5$.

Further, in a series of works devoted to the study of oscillating wings in supersonic flow, integrals of the form

$$\begin{aligned} Jc(\lambda, z) &= \int_0^z J_0(\lambda t) \cos t\, dt; \quad & Nc(\lambda, z) &= \int_0^z N_0(\lambda t) \cos t\, dt; \\ Js(\lambda, z) &= \int_0^z J_0(\lambda t) \sin t\, dt; \quad & Ns(\lambda, z) &= \int_0^z N_0(\lambda t) \sin t\, dt, \end{aligned} \tag{1.3}$$

arise: (c.f. [4] and its bibliography). In these works various properties of the functions (1.3) are studied and fairly extensive tables are given for real values of the parameters λ and z in the intervals $0.1 \leq \lambda \leq 1$, $0 \leq z \leq 5$. The linear combinations of these functions

$$\left.\begin{aligned} Je_0\left(\frac{i}{\lambda}, \frac{z}{\lambda}\right) &= \lambda [Jc(\lambda, z) - iJs(\lambda, z)]; \\ Ne_0\left(\frac{i}{\lambda}, \frac{z}{\lambda}\right) &= \lambda [Nc(\lambda, z) - iNs(\lambda, z)] \end{aligned}\right\} \tag{1.4}$$

are particular cases of the more general incomplete integrals of Lipschitz-Hankel, in terms of which the incomplete cylindrical functions $E_\nu^\pm(w, z)$ are expressed.

In particular, setting $\nu = 0$ in Eq. (4.29) of Chapter II, we have

$$E_0^+(w, z) = \frac{2w}{\pi} J_0(z) + i \tan w \left[N_0(z)\, Je_0\left(\frac{1}{\cos w}, z \cos w\right) - J_0(z)\, Ne_0\left(\frac{1}{\cos w}, z \cos w\right)\right]. \tag{1.5}$$

Using (1.4) and the fact that $\cos w = 1/\lambda \geq 1$, i.e.

$$\left.\begin{aligned} w &= \arccos \frac{1}{\lambda} = -i\beta; \\ \beta &= \ln\left(\frac{1}{\lambda} + \sqrt{\frac{1}{\lambda^2} - 1}\right); \\ \tan w &= -i\sqrt{1 - \lambda^2}, \end{aligned}\right\} \tag{1.6}$$

we find after simplification

$$\begin{aligned} E_0^+(-i\beta, x) \equiv \mathscr{H}_0(\beta, x) = &- \frac{2i\beta}{\pi} J_0(x) \\ &+ \sqrt{1 - \lambda^2}\left[N_0(x)\, Jc\left(\lambda, \frac{x}{\lambda}\right) - J_0(x)\, Nc\left(\lambda, \frac{x}{\lambda}\right)\right] \\ &+ i\sqrt{1 - \lambda^2}\left[N_0(x)\, Js\left(\lambda, \frac{x}{\lambda}\right) - J_0(x)\, Ns\left(\lambda, \frac{x}{\lambda}\right)\right], \end{aligned} \tag{1.7}$$

where $\mathscr{H}_0(\beta, x)$ is the incomplete Hankel function of real argument.

Similarly we obtain for $E_0^-(-i\beta, x)$

$$\begin{aligned} E_0^-(-i\beta, x) = &- \frac{2i\beta}{\pi} J_0(x) \\ &+ i\sqrt{1 - \lambda^2}\left[N_0(x)\, Js\left(\lambda, \frac{x}{\lambda}\right) - J_0(x)\, Ns\left(\lambda, \frac{x}{\lambda}\right)\right] \\ &- \sqrt{1 - \lambda^2}\left[N_0(x)\, Jc\left(\lambda, \frac{x}{\lambda}\right) - J_0(x)\, Nc\left(\lambda, \frac{x}{\lambda}\right)\right]. \end{aligned} \tag{1.8}$$

Equations (1.7) and (1.8) permit one to find numerical values from the aforementioned tabulation of the functions (1.3) for the incomplete cylindrical functions $E_0^\pm(w, x)$ for purely imaginary parameter $w = -i\beta$, or, what is the same thing, values of the incomplete Hankel function of zero order.

At the same time these equations show that incomplete cylindrical functions find application in the study of moving bodies in supersonic flow.

In passing we note that in connection with a similar problem, integrals of the form

$$\left.\begin{aligned} &\int_0^1 t^n e^{-i\omega t} J_0\left(\frac{\omega}{M} t\right) dt; \\ &\omega = 2k \frac{M^2}{M^2 - 1} \end{aligned}\right\} \tag{1.9}$$

have been tabulated (c.f. [20, p. 79]) for $n = 0, 1, 2, \ldots, 11$; $1,2 \leq M \leq 5$, $0 \leq k \leq 2$.

These integrals can also be indentified with incomplete cylindrical functions. In particular, for $n = 0$ and $n = 1$ they can be expressed in terms of incomplete Lipschitz-Hankel integrals involving Bessel functions.

Other tabulated special integrals, which permit the evaluation of incomplete cylindrical functions of order zero, may be found in [28], where six-figure tables of various generalized integral functions are given. For our purpose the following are of interest:

$$S(a, x) = \int_0^x \frac{\sin u}{u} \, dt; \tag{1.10}$$

$$C(a, x) = \int_0^x \frac{1 - \cos u}{u} \, dt; \tag{1.11}$$

$$Es(a, x) = \int_0^x \frac{e^{-u} \sin u}{u} \, dt; \tag{1.12}$$

$$Ec(a, x) = \int_0^x \frac{1 - e^{-u} \cos u}{u} \, dt; \tag{1.13}$$

$$E(a, x) = \int_0^x \frac{1 - e^{-u}}{u} \, dt, \tag{1.14}$$

where in each case $u = \sqrt{a^2 + t^2}$. Some additional information on these tables is found in [19]. Integrals (1.10) and (1.11) are called respectively, the generalized sine and cosine integral. The remaining forms are called generalized exponential integrals.

We now investigate connections between these functions and incomplete cylindrical functions. The first such relation for the generalized sine and cosine integrals was pointed out in [8 and 29].

We first consider (1.10) $S(a, x) = \int_0^x \frac{\sin \sqrt{a^2 + t^2}}{\sqrt{a^2 + t^2}} \, dt$, and introduce a new variable of integration by $t = ia \cdot \sin(\theta)$.
We find

$$S(a, x) = i \int_0^w \sin(a \cos \theta) \, d\theta, \tag{1.15}$$

where

$$\left. \begin{aligned} w \arcsin \frac{x}{ia} &= -i\beta; \\ \beta &= \ln\left[\frac{x}{a} + \sqrt{\left(\frac{x}{a}\right)^2 + 1}\,\right]. \end{aligned} \right\} \tag{1.16}$$

Recalling the definition of the incomplete cylindrical function $E_0^{\pm}(w, z)$, we have

$$\begin{aligned} S(a, x) &= \frac{\pi}{4}\left[E_0^{+}(-i\beta, a) - E_0^{-}(-i\beta, a)\right] \\ &= \frac{\pi}{4}\left[\mathscr{H}_0(\beta, a) - E_0^{-}(-i\beta, a)\right]. \end{aligned} \tag{1.17}$$

Analogously, from (1.11) we find

$$\begin{aligned} C(a, x) &= \beta - \frac{\pi i}{4}\left[E_0^{+}(-i\beta, a) + E_0^{-}(-i\beta, a)\right] \\ &= \beta - \frac{\pi i}{4}\left[\mathscr{H}_0(\beta, a) + E_0^{-}(-i\beta, a)\right]. \end{aligned} \tag{1.18}$$

Solving these equations for $\mathscr{H}_0$ and E_0^{-}, we get

$$\mathscr{H}_0(\beta, a) = \frac{2}{\pi}\left[S(a, x) + iC(a, x) - i\beta\right]; \tag{1.19}$$

$$E_0^{-}(-i\beta, a) = \frac{2}{\pi}\left[iC(a, x) - S(a, x) - i\beta\right], \tag{1.20}$$

where β is defined by the second of Eqs. (1.16). These formulae express the incomplete Hankel function of zero order for real arguments in terms of other tabulated functions. Because they are simpler, they may be more satisfactory for numerical evaluation than (1.17) and (1.8).

We turn now to consideration of (1.12) and, by means of the substitution $t = a \cdot \sinh(u)$, transform it into

$$\begin{aligned} Es(a, x) &= \int_0^x \frac{\sin\sqrt{a^2 + t^2}}{\sqrt{a^2 + t^2}} \exp\left(-\sqrt{a^2 + t^2}\right) dt \\ &= \frac{1}{2i}\int_0^\beta e^{-(1-i)a\cosh u}\, du - \frac{1}{2i}\int_0^\beta e^{-(1+i)a\cosh u}\, du \end{aligned}$$

or, recalling the definition of the incomplete MacDonald function $K_\nu(w,z)$ (c.f. § 1, Chapter II) we have

$$Es(a, x) = \frac{i}{2}\left[K_0(\beta, \sqrt{2}\, a e^{i\pi/4}) - K_0(\beta, \sqrt{2}\, a e^{-i\pi/4})\right]. \tag{1.21}$$

Similarly we find from definition (1.13)

$$\begin{aligned} Ec(a, x) &= \int_0^x \frac{1 - \cos\sqrt{a^2 + t^2}\exp\left(-\sqrt{a^2 + t^2}\right)}{\sqrt{a^2 + t^2}}\, dt \\ &= \beta - \frac{1}{2}\left[K_0(\beta, \sqrt{2}\, a e^{i\pi/4}) + K_0(\beta, \sqrt{2}\, a e^{-i\pi/4})\right]. \end{aligned} \tag{1.22}$$

From these equations we obtain

$$K_0(\beta, \sqrt{2}\, a e^{\pm i\pi/4}) = \beta - Ec(a, x) \mp iEs(a, x), \tag{1.23}$$

where, as before, β is defined by the second of Eqs. (1.16). Equation (1.23) allows one to evaluate the incomplete MacDonald function of zero order, $K_0(w, z)$, for real $w = \beta$ and complex values of z having arguments $\pm\pi/4$. In addition tables of the integral (1.14) permit, the evaluation of this same function for real $w = \beta$ and $z = x$. In fact, carrying out a similar transformation in the integral $E(a, x)$, we find

$$K_0(\beta, a) = -E(a, x) + \beta. \tag{1.24}$$

In the next section we shall show that $K_1(\beta, x)$ may also be evaluated in terms of tabulated special integrals.

Up to now we have been considering tabulated integrals which arc related to the incomplete cylindrical functions of Poisson form. We also include incomplete Weber and Lipschitz-Hankel integrals in the class of incomplete cylindrical functions, since these integrals are elements in the construction of incomplete cylindrical functions of the Poisson form, as well as of those of the Bessel and Sonine-Schlaefli forms. These integrals arise in various forms in many applied and theoretical problems. The incomplete Lipschitz-Hankel integral of the modified Bessel function

$$Ie_0(a, x) = \int_0^z e^{-at} I_0(t)\, dt \tag{1.25}$$

and the incomplete Weber integral of the same function

$$\tilde{Q}_0(x, z) = \frac{1}{2x} e^x \int_0^z t I_0(t) \exp\left(-\frac{t^2}{4x}\right) dt \tag{1.26}$$

occur especially often for real values of the parameters α, x, and z. The first of these integrals plays an important role in the study of oscillating wings in supersonic flow [4, 30, 31]. It also arises in the study of resonant absorption in media with finite dimensions [32—34]. The incomplete Weber integral (1.26) occurs in the study of thc propagation of electromagnetic waves along transmission lines by the solution of the telegrapher's equation [21]; in the study of transition processes and propagation of electromagnetic signals in ionized gases [35, 36]; in expressions for unsteady heat transfer between liquid and porous bodies; in the theory of ion exchange [7]; in investigation of the motion of carriers in semiconductors [37]; in computing the density perturbation behind rapidly moving bodies in rarified gases [38]; and in computing the characteristics of some nuclear cascades [39]. This same integral often arises in the solution of many problems in probability theory and mathematical statistics; in computation of the probability of a point falling within an arbitrary circle in the case of two dimensional circular normal distribution [40]; in study of the distribution of the envelope of a sinusoidal signal in the presence of Gaussian noise [41, 42]; and in investigation of the limiting distributions for one dimensional

Markov chains for two components [43]. A series of other applications of (1.26) may be found in the bibliographies of [6, 22 and 44]. Because of the importance of this integral for quantitative computations, it is widely tabulated. The most complete tables are found in [45]. The properties of (1.3), (1.25) and (1.26) have been the subject of a series of investigations [4, 6, 44, 46].

In § 5 of Chapter IV we established the following relation between the last two integrals:

$$\tilde{Q}_0\left(\frac{cz}{2}, z\right) = \frac{1}{2} e^{cz}\left[1 + \frac{1-c^2}{2c} Ie_0(a, z) - I_0(z)\, e^{-az}\right], \tag{1.27}$$

where

$$a = (1 + c^2)/2c. \tag{1.28}$$

Equation (1.27) makes it possible to evaluate either of these from tables of the other.

If we replace c with $1/c$ (1.27) and (1.28) we obtain

$$\tilde{Q}_0\left(\frac{z}{2c}, z\right) = \frac{1}{2} e^{z/c}\left[1 - \frac{1-c^2}{2c} Ie_0(a, z) - I_0(z)\, e^{-az}\right]. \tag{1.29}$$

Now combining (1.27) and (1.29), we find

$$e^{-cz}\tilde{Q}_0\left(\frac{cz}{2}, z\right) + e^{-z/c}\tilde{Q}_0\left(\frac{z}{2c}, z\right) = 1 - I_0(z)\, e^{-az}, \tag{1.30}$$

from which we see that the incomplete Weber integral $Q_0(cz/2, z)$ for real c is completely defined it we know its values in the interval $-1 < c < 1$.

For these same integrals the following expressions were obtained in § 6 of Chapter IV:

$$\tilde{Q}_0\left(\frac{cz}{2}, z\right) = \left[Y_1\left(\frac{z}{c}, z\right) + Y_2\left(\frac{z}{c}, z\right)\right] \exp\left[\frac{z}{2}\left(c - \frac{1}{c}\right)\right]; \tag{1.31}$$

$$Ie_0(a, z) = \frac{1}{\sqrt{a^2-1}}\left\{1 - e^{-az}\left[I_0(z) + 2Y_1\left(\frac{z}{c}, z\right) + 2Y_2\left(\frac{z}{c}, z\right)\right]\right\}, \tag{1.32}$$

where as before the connection between c and a is defined by (1.28) and $Y_1(z/c, z)$ and $Y_2(z/c, z)$ are Lommel functions of two purely imaginary arguments. The available extensive tables of these functions permit the evaluation of these integrals in a sufficiently wide region of their parameters. However, these tables do not permit evaluation of the incomplete Lipschitz-Hankel integral $Ie_0(a, z)$ for $-1 < a < 1$, because they include only real values of c, for which the absolute value $|a|$, according to expression (1.28), is always larger than one. For the evaluation of $Ie_0(a, z)$ for values $-1 < a < 1$, the tables of the incomplete cylindrical functions $F_0^-(\alpha, z)$ and $F_0^-(\alpha, z)$ appended below may be used, since, according to Eq. (6.16) of Chapter IV, we have

$$Ie_0(a, z) = \frac{\pi z}{\sin\alpha} \{F_0^-(\alpha, z)\, I_1(z) - F_1^-(\alpha, z)\, I_0(z)\} + z e^{-z\cos\alpha} I_0(z), \quad (1.33)$$

where $a = \cos\alpha$.

Consequently, the incomplete cylindrical functions of orders zero and one of the Poisson as well as the Bessel form may be calculated for certain values of their arguments with the aid of tables already available in the literature. Moreover, some incomplete Lipschitz-Hankel and Weber integrals, also related to the class of incomplete cylindrical functions, are either themselves tabulated or may be evaluated with the aid of tables for cylindrical functions of two arguments. The tables appended below enlarge the set of tabulated values of all these functions.

For the sake of completeness and to help in choosing tables for practical applications of the results of this and other sections, a collection of computational formulae and tables is given at the end of the book.

2. Absorption of Radiation in the Earth's Atmosphere

Changes in the composition of the atmosphere of the earth under the influence of external radiation have great significance in various applied problems. One of the fundamental characteristic effects of radiation is the quantitative absorption of energy expended on ionization and disassociation of molecules of the atmosphere. The problem of determining the coefficient of radiation absorption for the earth's atmosphere, taking account of its inhomogenity, is of interest here as an example of the application of incomplete cylindrical functions to the analysis of important physical problems. We remark in advance that [47 and 48] contain tables which permit the evaluation of the incomplete MacDonald function $K_1(\beta, z)$.

We give here a brief outline of the essential features of the problem, following [47]. Let 0 be the center of the earth, a its radius and let $0S$

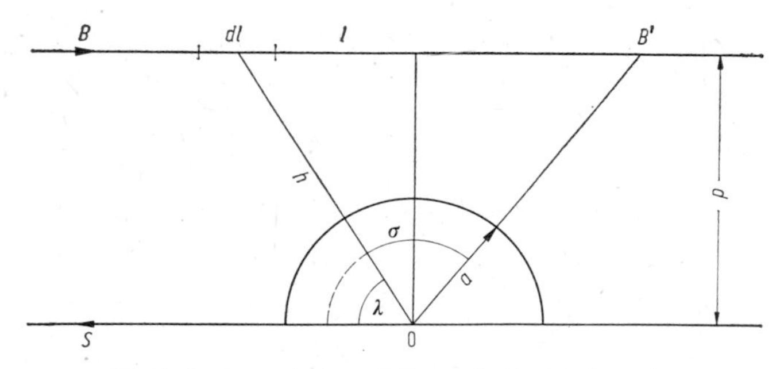

Fig. 10. Incidence of solar radiation on the Earth's atmosphere.

be in the direction of the sun (c.f. Fig. 10). Assume that there impinges a one-dimensional bundle of solar radiation in the direction BB', at a distance p from the center of the earth. The amount of radiation dS absorbed in an element of length dl is given by

$$dS = -AS\varrho\, dl\,, \tag{2.1}$$

where A is the absorption coefficient, S is the quantity of radiation at element dl, and ϱ is the atmospheric density, which usually changes exponentially with the height h above the surface of the earth

$$\varrho = \varrho_0 \exp\,(-h/H)\,, \tag{2.2}$$

l/H being the damping decrement. Introducing angles λ and σ as shown in Fig. 10, we find

$$\left.\begin{aligned} h &= p \operatorname{cosec} \lambda - a\,; \\ l &= -p \cot \lambda\,, \end{aligned}\right\} \tag{2.3}$$

since $dl = p \operatorname{cosec}^2 (\lambda)\, d\lambda$. Substituting these relations in (2.1), we obtain

$$dS = -AS\varrho_0 p \operatorname{cosec}^2 \lambda \exp\left[\frac{a - p \operatorname{cosec} \lambda}{H}\right] d\lambda\,. \tag{2.4}$$

Integrating this equation between the limits $\lambda = 0\,(l = -\infty)$ and $\lambda = \sigma$, we find

$$S = S_0 \exp\left[-Ap\varrho_0 \int\limits_0^\sigma \exp\left(\frac{a - p \operatorname{cosec} \lambda}{H}\right) \frac{d\lambda}{\sin^2 \lambda}\right]. \tag{2.5}$$

Consequently, the amount of radiation absorbed per unit volume, according to (2.1), is determined by

$$\begin{aligned} \frac{dS}{dl} &= I = I_0 \exp\left(1 + \frac{h_0 - h}{H}\right) \\ &\times \exp\left[-\frac{p}{H} \int\limits_0^\sigma \exp\left(\frac{a + h_0 - p \operatorname{cosec} \lambda}{H}\right) \frac{d\lambda}{\sin^2 \lambda}\right], \end{aligned} \tag{2.6}$$

where for convenience we have put $h_0 = H \ln\,(A\varrho_0 H)\,;\; I_0 = S_\infty/H \exp 1$. Now, introducing the notations

$$\left.\begin{aligned} z &= \frac{h - h_0}{H}\,; \\ x &= \frac{a + h_0}{H} + z\,; \\ p &= x \sin \sigma\,, \end{aligned}\right\} \tag{2.7}$$

we find

$$I = I_0 \exp\,[1 - z - f(x, \sigma) \exp\,(-z)]\,, \tag{2.8}$$

where the function $f(x, \sigma)$ has the form

$$f(x, \sigma) = x \sin \sigma e^x \int_0^\sigma \frac{1}{\sin^2 \lambda} \exp\left(-x \frac{\sin \sigma}{\sin \lambda}\right) d\lambda. \tag{2.9}$$

In [47] a detailed study of the properties of $f(x, \sigma)$ is given, including series and asymptotic expansions, and also a short table for values of σ between 0 and $\pi/2$. For $\sigma > \pi/2$ the formula obtained can be written as

$$f(x, \sigma - \pi) = f(x, \sigma) + 2e^{x(1-\sin\sigma)} f(x \sin \sigma, \pi/2), \tag{2.10}$$

taking $f(y, \pi/2) = -ye^y K_1(y)$ where $K_1(y)$ is a MacDonald function. More detailed tables of $f(x, \sigma)$ are given in Ref. [48].

It is possible to show that $f(x, \sigma)$ is related to the class of incomplete cylindrical functions and can be expressed by means of incomplete MacDonald functions. Taking $u = 1/\sin \lambda$ in (2.9), we have

$$f(x, \sigma) = x \sin \sigma e^x \int_{\cosh\beta}^{\infty} e^{-xu\sin\sigma} \frac{u\,du}{\sqrt{u^2 - 1}}, \tag{2.11}$$

where $\cosh \beta = (1/\sin \sigma) > 1$, or

$$\beta = \cosh^{-1} (1/\sin \sigma) = \ln\left(\operatorname{ctg} \frac{\sigma}{2}\right). \tag{2.12}$$

Integrating (2.11) by parts gives

$$\begin{aligned} f(x, \sigma) &= x \sin \sigma e^x \left(-e^{-x} \sinh \beta + x \sin \sigma \int_{\cosh\beta}^{\infty} \sqrt{u^2 - 1}\, e^{-xu\sin\sigma}\, du\right) \\ &= -x \sin \sigma \sinh \beta + x \sin \sigma e^x \left(x \sin \sigma \int_1^{\infty} e^{-xu\sin\sigma} \sqrt{u^2 - 1}\, du\right. \\ &\quad \left. - x \sin \sigma \int_1^{\cosh\beta} e^{-xu\sin\sigma} \sqrt{u^2 - 1}\, du\right). \end{aligned} \tag{2.13}$$

But the first term in round brackets is the MacDonald function $K_1(x \sin \sigma)$, and the second is none other that the incomplete MacDonald function $K_1(\beta, x \sin \sigma)$. Consequently we have

$$\begin{aligned} f(x, \sigma) &= -x \sin \sigma \sinh \beta \\ &+ x \sin \sigma e^x [K_1(x \sin \sigma) - K_1(\beta, x \sin \sigma)]. \end{aligned} \tag{2.14}$$

Thus the study of absorption of solar radiation in the atmosphere leads one to the analysis of incomplete cylindrical functions. On the other hand, since $f(x, \sigma)$ is tabulated, it is useful to solve (2.14) for $K_1(\beta, z)$. We find

$$K_1(\beta, x \sin \sigma) = K_1(x \sin \sigma) - \cot \sigma e^{-x} - \frac{1}{x \sin \sigma} e^{-x} f(x, \sigma). \tag{2.15}$$

Here β is defined by (2.12), i.e. $\sinh \beta = \cot \sigma$.

This equation allows one to find numerical values for the incomplete MacDonald function $K_1(\beta, z)$ for a wide range of the parameters $\beta = \ln [\cot (\sigma/2)]$ and $z = x \sin \sigma$ from the tables of $f(x, \sigma)$ contained in Ref. [48].

3. Radiation from a Vertical Dipole on the Earth's Surface

In the theory of propagation of radio waves along the surface of the earth there is considerable interest in problems of propagation in inhomogeneous, layered, and in inhomogeneous-layered strata. The inhomogeneous structure of the earth's surface distorts the phase of the electromagnetic field. The complicated structure of the earth may also influence the electromagnetic field, and this allows the possibility of deducing something about the thickness of the layers and their conductivity. The problem of propagation of radio waves in inhomogeneous strata is therefore of great practical significance, and a series of theoretical and experimental studies has been devoted to it (c.f., e.g. [49, 50] and their bibliographies).

In connection with such studies the problem of finding the field set up by a vertical dipole located on the surface of the earth is of great importance. This problem has been considered in detail in many works, all having the basic intention of giving a precise mathematical solution.

Here we present one of the methods of solution, following the references cited except for a few changes for convenience of exposition.

Let a vertical electric dipole be situated on the surface of the earth at the point (0, 0, 0). It is required to find an analytic expression for the components of the electromagnetic field so generated. If we introduce the Hertz vector Π with only the single component π_z, directed along the axis of the dipole, different from zero, then Maxwell's equations in cylindrical coordinates r, z, and φ may be written as

$$\left(\frac{\partial^2}{\partial r^2} + \frac{1}{r}\frac{\partial}{\partial r} + \frac{1}{r^2}\frac{\partial^2}{\partial \varphi^2} + \frac{\partial^2}{\partial z^2} + k_0^2\right)\Pi = -4\pi a \delta(\boldsymbol{R}); \tag{3.1}$$

$$\left(\frac{\partial^2}{\partial r^2} + \frac{1}{r}\frac{\partial}{\partial r} + \frac{1}{r^2}\frac{\partial^2}{\partial \varphi^2} + \frac{\partial^2}{\partial z^2} + k^2\right)\Pi_1 = -\frac{4\pi a}{\varepsilon}\,\delta(\boldsymbol{R}), \tag{3.2}$$

where Π and Π_1 refer respectively to the atmosphere and to the earth, $k^2 = \varepsilon k_0^2$, k_0 is the wave number in the atmosphere, ε is the complex permeability, $R = \sqrt{r^2 + z^2}$, and $\delta(\boldsymbol{R})$ is the Dirac delta function. Equations (3.1) and (3.2) differ from the corresponding equations of [49, p. 216], in that their right hand sides already express the presence of a point-source of dipole moment situated at the origin. The coefficient a which appears here is a constant characteristic of the dipole.

The problem is consequently that of to finding solutions of Maxwell's equations (3.1), (3.12), satisfying the following boundary conditions at the interface between the two media:

$$\Pi(r, \varphi, 0) = \varepsilon \Pi_1(r, \varphi, 0); \tag{3.3}$$

$$\frac{\partial}{\partial z}[\Pi(r, \varphi, z) - \Pi_1(r, \varphi, z)]/_{z=0} = 0. \tag{3.4}$$

According to classical theory, the general solution of each of Eqs. (3.1), (3.2) is a linear combination of the general solution of the homogeneous equation and an arbitrary particular solution of the inhomogeneous equation. Particular solutions of such equations may be found by well known methods (c.f. e.g. [5, Vol. 1, Ch. VII]) and in our case they take the form:

$$\left.\begin{aligned} a\int\frac{e^{ik_0|\boldsymbol{R}-\boldsymbol{R}_0|}}{|\boldsymbol{R}-\boldsymbol{R}_0|}\,\delta(\boldsymbol{R}_0)\,d\boldsymbol{R}_0 &= \frac{ae^{ik_0R}}{R};\\ R &= \sqrt{r^2+z^2}\end{aligned}\right\} \tag{3.5}$$

for Eq. (3.1), and

$$\frac{ae^{ik_0R}}{\varepsilon R} \tag{3.6}$$

for Eq. (3.2).

The general solution of the homogeneous equations may be found by the method of separation of variables as the superposition of particular solutions

$$J_0(\nu r)\exp\left(-z\sqrt{\nu^2-k_0^2}\right), \quad z>0; \tag{3.7}$$

$$J_0(\nu r)\exp\left(z\sqrt{\nu^2-k^2}\right), \quad z<0, \tag{3.8}$$

where $J_0(\nu r)$ is the Bessel function, and ν is the separation parameter. Thus the general solutions of (3.1) and (3.2) may be written

$$\Pi = \frac{a}{\sqrt{z^2+r^2}}\exp\left(ik_0\sqrt{z^2+r^2}\right) + \int_0^\infty f(\nu)\,J_0(\nu r)\exp\left(-z\sqrt{\nu^2-k_0^2}\right)d\nu,\ z>0 \tag{3.9}$$

$$\Pi_1 = \frac{a}{\varepsilon\sqrt{z^2+r^2}}\exp\left(ik\sqrt{z^2+r^2}\right) + \int_0^\infty \varphi(\nu)\,J_0(\nu r)\exp\left(z\sqrt{\nu^2-k^2}\right)d\nu,\ z<0,$$

where $f(\nu)$ and $\varphi(\nu)$ are at present, arbitrary functions, except that it is necessary to take $\operatorname{Re}\sqrt{\nu^2-k_0^2} \geq 0$ and $\operatorname{Re}\sqrt{\nu^2-k^2} \geq 0$ to insure convergence of their repective integrals. Substituting these solutions into the boundary conditions (3.3) and (3.4), we obtain

$$\begin{aligned} &\int_0^\infty [\varepsilon\varphi(\nu) - f(\nu)]\,J_0(\nu r)\,d\nu = \frac{a}{r}\left(e^{ik_0r} - e^{ikr}\right);\\ &\int_0^\infty \left[\sqrt{\nu^2-k^2}\,\varphi(\nu) + \sqrt{\nu^2-k_0^2}\,f(\nu)\right]J_0(\nu r)\,d\nu = 0.\end{aligned} \tag{3.10}$$

Multiplying these equations by $r J_0(\nu' r)$ and integrating from $r = 0$ to $r = \infty$, we find

$$\begin{aligned} \varepsilon\varphi(\nu) - f(\nu) &= \nu a\left(\frac{1}{\sqrt{\nu^2 - k_0^2}} - \frac{1}{\sqrt{\nu^2 - k^2}}\right); \\ \sqrt{\nu^2 - k^2}\,\varphi(\nu) &+ \sqrt{\nu^2 - k_0^2}\, f(\nu) = 0. \end{aligned} \tag{3.11}$$

To obtain this last result we used the orthogonality property

$$\int_0^\infty J_0(\nu r)\, J_0(\nu' r)\, r\, dr = \frac{\delta(\nu - \nu')}{\nu} \tag{3.12}$$

and the relation

$$\int_0^\infty e^{ikr} J_0(\nu r)\, dr = \frac{1}{\sqrt{\nu^2 - k^2}}, \tag{3.13}$$

obtained from (4.2) and (4.3) of Chapter VI.

We now solve the system (3.11) for $f(\nu)$ and $\varphi(\nu)$, and find

$$\left.\begin{aligned} f(\nu) &= \frac{a\nu}{\sqrt{\nu^2 - k_0^2}} \cdot \frac{\sqrt{\nu^2 - k_0^2} - \sqrt{\nu^2 - k^2}}{\varepsilon\sqrt{\nu^2 - k_0^2} + \sqrt{\nu^2 - k^2}}; \\ \varphi(\nu) &= -\frac{a\nu}{\sqrt{\nu^2 - k^2}} \cdot \frac{\sqrt{\nu^2 - k_0^2} - \sqrt{\nu^2 - k^2}}{\varepsilon\sqrt{\nu^2 - k_0^2} + \sqrt{\nu^2 - k^2}}. \end{aligned}\right\} \tag{3.14}$$

Substituting these into (3.9) and developing the particular solutions contained in them with respect to the same complete system of functions (3.7) or (3.8), i.e.

$$\frac{e^{ik\sqrt{z^2+r^2}}}{\sqrt{z^2 + r^2}} = \int_0^\infty J_0(\nu r)\, e^{\mp z\sqrt{\nu^2 - k^2}} \frac{\nu\, d\nu}{\sqrt{\nu^2 - k^2}}, \tag{3.15}$$

we find the following expression for the desired Hertz vector:

$$\left.\begin{aligned} \Pi &= a(1 + \varepsilon) \int_0^\infty \frac{J_0(\nu r)\, e^{-z\sqrt{\nu^2 - k_0^2}}}{\varepsilon\sqrt{\nu^2 - k_0^2} + \sqrt{\nu^2 - k^2}}\, \nu\, d\nu; \\ \Pi_1 &= \frac{a(1 + \varepsilon)}{\varepsilon} \int_0^\infty \frac{J_0(\nu r)\, e^{z\sqrt{\nu^2 - k^2}}}{\varepsilon\sqrt{\nu^2 - k_0^2} + \sqrt{\nu^2 - k^2}}\, \nu\, d\nu. \end{aligned}\right\} \tag{3.16}$$

If we set $a = \varepsilon/(1 + \varepsilon)$, these expressions reduce to the corresponding formulae of [49].

We show now that in the particular case of practical interest for which the radiator and receiver are located close to the surface of the earth, i.e. for $z = 0$, the Hertz vector may be expressed in terms of incomplete cylindrical functions. Obviously, it is sufficient to show this for $\Pi(r, 0)$, because conditions (3.3) and (3.4) hold for $z = 0$. To do this, we use the following identity which is found, for example, in [49, p. 219]:

$$\frac{1}{\varepsilon\sqrt{v^2-k_0^2}+\sqrt{v^2-k^2}} = \frac{\sqrt{\varepsilon}}{1-\varepsilon^2}\int\limits_{\sqrt{1+1/\varepsilon}}^{\sqrt{1+\varepsilon}} \frac{1}{\sqrt{v^2-\mu^2u^2}}\, d\left(\frac{1}{\sqrt{u^2-1}}\right), \tag{3.17}$$

there we have used the relation $k^2 = \varepsilon k_0^2$ and set $\mu^2 = \varepsilon k_0^2/(1+\varepsilon)$. Substituting (3.17) into the expression for $\Pi(r, 0)$ for the system (3.16), and interchanging the order of integration, we find

$$\Pi(r, 0) = \frac{a\sqrt{\varepsilon}}{1-\varepsilon}\int\limits_{\sqrt{1+1/\varepsilon}}^{\sqrt{1+\varepsilon}} d\left(\frac{1}{\sqrt{u^2-1}}\right)\int\limits_0^\infty \frac{J_0(vr)\, v\, dv}{\sqrt{v^2-\mu^2u^2}}. \tag{3.18}$$

The inner integral, according to (3.15) is equal for $z = 0$ to $\exp(i\mu ur)/r$ Therefore (3.18) can be written

$$\Pi(r, 0) = \frac{a}{r}\,\frac{\sqrt{\varepsilon}}{1-\varepsilon}\int\limits_{\sqrt{1+1/\varepsilon}}^{\sqrt{1+\varepsilon}} e^{i\mu ur}\,\frac{d}{du}\left(\frac{1}{\sqrt{u^2-1}}\right) du. \tag{3.19}$$

Integrating by parts we find

$$\Pi(r, 0) = \frac{a}{r(1-\varepsilon)}\left(e^{ik_0 r\sqrt{\varepsilon}} - e^{ik_0 r}\right) - \frac{iak_0\varepsilon}{(1-\varepsilon)\sqrt{1+\varepsilon}}\int\limits_{\sqrt{1+1/\varepsilon}}^{\sqrt{1+\varepsilon}} e^{i\mu ur}\frac{du}{\sqrt{u^2-1}}. \tag{3.20}$$

But, according to (1.26) and (1.27) of Chapter II, the integral on the right may be expressed in terms of the incomplete Hankel function $\mathscr{H}_0(w, z)$, and so the final expression for the Hertz vector Π is

$$\begin{aligned}\Pi(r, 0) &= \frac{a}{r(1-\varepsilon)}\left(e^{ik_0 r\sqrt{\varepsilon}} - e^{ik_0 r}\right)\\ &+ \frac{\pi a k_0\varepsilon}{2(1-\varepsilon)\sqrt{1+\varepsilon}}\left[\mathscr{H}_0(\beta_0, z) - \mathscr{H}_0(\beta, z)\right],\end{aligned} \tag{3.21}$$

where

$$\left.\begin{aligned}\cosh\beta_0 &= \sqrt{1+\varepsilon};\\ \cosh\beta &= \sqrt{1+1/\varepsilon};\\ z &= k_0 r\sqrt{\varepsilon}/\sqrt{1+\varepsilon}.\end{aligned}\right\} \tag{3.22}$$

The equations obtained show that the problem of finding a mathematical description for the field on a plane surface due to a vertical dipole located in the same plane leads to the study of incomplete cylindrical functions. The detailed study of the properties of these functions given in the second chapter can therefore be applied specific cases. In particular, simple approximate expressions for the Hertz vector may be obtained by means of the results of § 10, Chapter II in the interesting case where the wave length $|k_0 r\sqrt{1+\varepsilon}| \gg 1$, and when the dielectric constant of the earth is such that $|\varepsilon| > 1$.

As a preliminary to this analysis we note that the expansion parameter for the incomplete cylindrical functions $E_\nu^\pm(w, z)$ is $|z \sin w|$ while the corresponding one for the incomplete Hankel function $\mathscr{H}_\nu(\beta, z)$ is $z \sinh \beta$. Since equations (3.21) and (3.22) imply

$$\left.\begin{aligned} z \sinh \beta_0 &= \frac{k_0 r \varepsilon}{\sqrt{1+\varepsilon}}; \\ z \sinh \beta &= \frac{k_0 r}{\sqrt{1+\varepsilon}} \end{aligned}\right\} \tag{3.23}$$

and since in the case at hand the wave length is large with respect to unity, we may use the asymptotic expansions for the incomplete Hankel functions. According to (10.4) and (10.14) of Chapter II, the asymptotic development of the incomplete Hankel function has the form

$$\begin{aligned} &\mathscr{H}_0(\beta, z) \approx \mathscr{H}_0^{(1)}(z) \\ &- \frac{2}{\pi z \sinh \beta} e^{iz\cosh\beta} \left(1 - i \frac{\cosh \beta}{z \sinh^2 \beta} + \frac{1 + 3\coth^2 \beta}{z \sinh^2 \beta}\right). \end{aligned} \tag{3.24}$$

For our problem, we have because of (3.22)

$$\begin{aligned} &\mathscr{H}_0(\beta_0, z) \approx \mathscr{H}_0^{(1)}(z) \\ &- \frac{2\sqrt{1+\varepsilon}}{\pi k_0 r \varepsilon} e^{ik_0 r\sqrt{\varepsilon}} \left[1 - i \frac{1+\varepsilon}{\varepsilon\sqrt{\varepsilon}} \frac{1}{k_0 r} + \frac{(1+\varepsilon)(3+4\varepsilon)}{\varepsilon^2} \frac{1}{(k_0 r)^2}\right]; \end{aligned} \tag{3.25}$$

$$\begin{aligned} &\mathscr{H}_0(\beta, z) \approx \mathscr{H}_0^{(1)}(z) \\ &- \frac{2\sqrt{1+\varepsilon}}{\pi k_0 r} e^{ik_0 r} \left[1 - i \frac{1+\varepsilon}{k_0 r} + \frac{(1+\varepsilon)(4+3\varepsilon)}{(k_0 r)^2}\right]. \end{aligned} \tag{3.26}$$

For the sake of brevity, the asymptotic expansion for the Hankel function $\mathscr{H}_0^{(1)}(z)$ has not been written out.

Substituting these developments in (3.21), we find the following approximate expression for the Hertz vector:

$$\begin{aligned} \Pi(r, 0) \approx i \frac{a\varepsilon(1+\varepsilon)}{(\varepsilon-1) k_0 r} \frac{e^{ik_0 r}}{r} &\left[1 + i \frac{4+3\varepsilon}{k_0 r}\right. \\ &\left.- \frac{1}{\varepsilon^{5/2}} e^{ik_0 r(\sqrt{\varepsilon}-1)} \left(1 - i \frac{3+4\varepsilon}{\varepsilon^{3/2}} \frac{1}{k_0 r}\right)\right]. \end{aligned} \tag{3.27}$$

The term containing $\exp[ik_0 r(\sqrt{\varepsilon}-1)]$, characterizes the radio wave propagated from radiator to receiver through the earth. This effect is in fact negligible if the radiator is located in the atmosphere (c.f. [49, p. 324]). In such a case (3.27) takes the still simpler form

$$\Pi(r, 0) \approx i \frac{a\varepsilon(1+\varepsilon)}{(\varepsilon-1) k_0 r} \frac{e^{ik_0 r}}{r} \left(1 + i \frac{4+3\varepsilon}{k_0 r}\right). \tag{3.28}$$

In [49, p. 226] there appears the approximation

$$\Pi(r, 0) \approx \frac{e^{ik_0 r}}{r} \frac{i\varepsilon}{k_0 r}\left(1 + \frac{1}{\varepsilon}\right). \tag{3.29}$$

In order to reconcile our result, (3.28), with (3.29), we set $a = \varepsilon/(1 + \varepsilon)$ as before. Then, for sufficiently large $|\varepsilon| \gg 1$ we find from (3.28):

$$\Pi(r, 0) \approx \frac{e^{ik_0 r}}{r} \frac{i\varepsilon}{k_0 r}\left(1 + \frac{1}{\varepsilon}\right)\left[1 + i\,\frac{4 + 3\varepsilon}{k_0 r}\right], \tag{3.30}$$

which differs from formula (3.29) only by the multiplier contained in the square brackets.

In addition to the expression obtained for $\Pi(r, 0)$ for large values of $k_0 r$ we give an approximation to this quantity for $k_0 r \ll 1$. It can be obtained immediately from (3.19) as

$$\Pi(r, 0) \approx \frac{a}{r}\left\{1 + ik_0 r\left[\frac{\sqrt{\varepsilon}}{1 + \sqrt{\varepsilon}} + \frac{\varepsilon}{(1 - \varepsilon)\sqrt{1 + \varepsilon}} \ln \frac{\sqrt{\varepsilon}\,(\sqrt{\varepsilon} + \sqrt{1 + \varepsilon})}{1 + \sqrt{1 + \varepsilon}}\right]\right\}. \tag{3.31}$$

Equations (3.28) and (3.31) permit us to approximate the Hertz vector in limiting cases when the wave length is very short or long. For its evaluation in intermediate regions it is necessary to use the exact formula (3.21) and tables for the incomplete Hankel function. If we neglect absorption, i.e. if we assume that the real part of the permeability is much larger than the imaginary part, then these evaluations may be effected with the aid of (1.7)—(1.19) and the corresponding tables for the functions contained there.

4. The Problem of Diffraction by a Wedge

The problem of wave diffraction by a wedge, described by the wave equation, has received wide treatment by various methods. In Ref. [51] a new method is presented, which is effective in the solution of all classical problems of diffraction by a wedge considered in the literature, as well as for a number of new problems. However, to apply this method it is necessary to treat a series of special integrals which cannot be expressed in terms of known and tabulated functions. We shall show below, that in many cases these integrals reduce to the incomplete cylindrical functions we have already discussed.

Assume some process described by the wave equation

$$\Delta^2 u - c^2 \frac{\partial^2 u}{\partial t^2} = 0, \tag{4.1}$$

takes place in the wedge-shaped region $r \geq 0$, $0 \leq \varphi \leq \alpha$ (we use the cylindrical coordinate system (r, φ, z), where ∇^2 is the Laplace operator and $1/c$ is the velocity of wave propagation. Let the Dirichlet conditions

$$u(r, 0) = u(r, \alpha) = 0 \tag{4.2}$$

be imposed on the bounding planes $\varphi = 0$ and $\varphi = \alpha$, where for definiteness we shall take $\pi \leq \alpha \leq 2\pi$. We further assume that for $t = t_0$ a source of disturbance is initiated at the point (r_0, φ), producing a field $u_0(r, \varphi, t)$ independent of z. This field eventually impinges on the boundaries of the region, on which conditions (4.2), must be satisfied. As a consequence there arises in the medium the complementary wave field $u(r, \varphi, t) = u - u_0$, characterising the reflection of waves u_0 from planes $\varphi = 0$ and $\varphi = \alpha$ and also including the effects of diffraction of the incoming waves by the wedge $r \geq 0$ and $\alpha \leq \varphi \leq 2\pi$ (Fig. 11), and satisfying the initial conditions

$$\left.\begin{aligned} (u - u_0)_{t=0} &= 0; \\ \frac{\partial}{\partial t}(u - u_0)_{t=0} &= 0. \end{aligned}\right\} \tag{4.3}$$

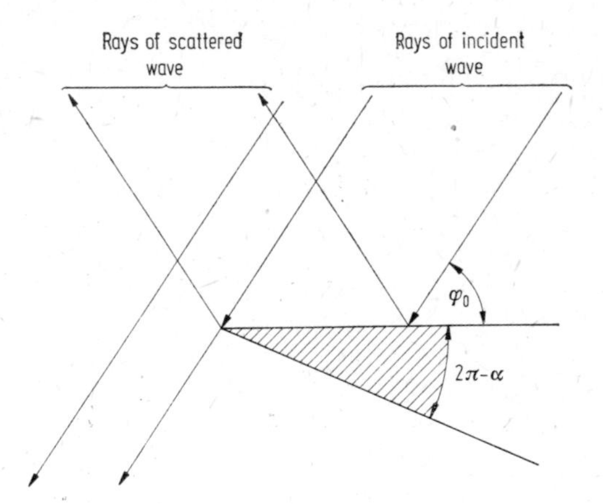

Fig. 11. Relations between a diffracting wedge and the incident and reflected rays.

It is required to find $u(v, \varphi, t)$ for a given field $u_0(r, \varphi, t)$. For convenience we assume the latter can be written as a combination of incoming and reflected waves, i.e.,

$$u_0(r, \varphi, t) \tag{4.4}$$

$$= \begin{cases} v[t + cr\cos(\varphi - \varphi_0)] - v[t + cr\cos(\varphi + \varphi_0)]; & 0 < \varphi < \pi - \varphi_0; \\ v[t + cr\cos(\varphi - \varphi_0)], & \pi - \varphi_0 < \varphi < \pi + \varphi_0; \\ 0, & \pi + \varphi_0 < \varphi < \alpha, \end{cases}$$

where the function $v(t)$ describes the form of the impulse of the incoming wave. In the case of a plane monochromatic wave $v(\tau)$ has the form

$$v(\tau) = \exp(i\omega\tau). \tag{4.5}$$

To solve the problem in this formulation we shall use the method of series, which is described in detail in [51]. We shall limit ourselves to the particular case in which the incoming wave is plane and monochromatic, has wave vector lying in the plane of the principal section of the wedge: $r \geq 0$; $\alpha < \varphi < 2\pi$, and makes an angle $\alpha < \varphi_0 < 2\pi - \alpha$ with the boundary $\varphi = 0$. Because according to (4.4) $u_0(r, \varphi, t)$ is an odd function of φ, the unknown solution $u(r, \varphi, t)$ will also be an odd function of φ, and has a series development on $0 < \varphi < \alpha$ with respect to the complete system of functions $\sin(m\pi\varphi/\alpha)$:

$$u = \sum_{m=1}^{\infty} f_m(r, t) \sin\frac{m\pi\varphi}{\alpha}. \tag{4.6}$$

Substituting (4.6) in (4.1) we find the following equation for the expansion coefficients f_m:

$$\frac{\partial^2 f_m}{\partial r^2} + \frac{1}{r}\frac{\partial f_m}{\partial r} - \frac{m^2\pi^2}{\alpha^2}\frac{f_m}{r^2} = c^2\frac{\partial^2 f}{\partial t^2}. \tag{4.7}$$

One method for solving this equation is to use the Laplace-Mellin transform in the form

$$f_m(r, t) = \frac{1}{2\pi i}\int_{\sigma-i\infty}^{\sigma+i\infty} F_\mu(s) J_\mu(-icrs) e^{st}\, ds, \tag{4.8}$$

where $\mu = m\pi/\alpha$; $J_\nu(z)$ is the Bessel function and $F_\mu(s)$ is at this point an unknown function. The next step in the method of series solution involves developing the $u_0(r, \varphi, t)$ in a similar series and substituting this into the initial and boundary conditions. We have

$$u_0(r, \varphi, t) = \sum_{m=1}^{\infty} f_m^0(r, t) \sin\left(\frac{m\pi}{\alpha}\varphi\right), \tag{4.9}$$

where the expansion coefficients $f_m^0(r, t)$ are defined by the well-known formula

$$f_m^0(r, t) = \frac{2}{\alpha}\int_0^\alpha u_0(r, \varphi, t) \sin\left(\frac{m\pi}{\alpha}\varphi\right) d\varphi. \tag{4.10}$$

Substituting the Mellin integral for $u_0(r, \varphi, t)$ for the dependence on t and taking account of (4.4) and (4.5), we find after interchanging the order of integration

$$f_m^0(r, t) = \frac{1}{2\pi i}\int_{\sigma-i\infty}^{\sigma+i\infty} F(s)\, \theta(-icsr) e^{st}\, ds, \tag{4.11}$$

where

$$\theta(-iy) = \frac{2}{\alpha}\left[\int_0^{\pi+\varphi_0} e^{y\cos(\varphi-\varphi_0)} \sin\left(\frac{m\pi}{\alpha}\varphi\right) d\varphi - \int_0^{\pi-\varphi_0} e^{y\cos(\varphi+\varphi_0)} \sin\left(\frac{m\pi}{\alpha}\varphi\right) d\varphi\right], \tag{4.12}$$

and $F(s)$ has the form

$$F(s) = \int_0^\infty v(\tau)\, e^{-s\tau}\, d\tau, \tag{4.13}$$

which according to (4.5) for a monochromatic plane wave is equal to

$$F(s) = \frac{1}{s - i\omega}. \tag{4.14}$$

By means of the substitutions $\varphi - \varphi_0 = \theta$ and $\varphi + \varphi_0 = -\theta$ in the first and second of Eq. (4.12), respectively, the function $\theta(-iy)$ may be written as

$$\begin{aligned}\theta(-iy) &= \frac{2}{\pi}\int_{-\pi}^{\pi} e^{y\cos\theta} \sin\left[\frac{m\pi}{\alpha}(\varphi+\varphi_0)\right] d\varphi \\ &= \frac{2}{\alpha}\sin\left(\frac{m\pi}{\alpha}\varphi_0\right)\int_{-\pi}^{\pi} e^{y\cos\theta - im\pi\theta/\alpha}\, d\theta.\end{aligned} \tag{4.15}$$

Introducing the new variable of integration $u = i\theta$ and recalling the definition of the incomplete Hankel function of the Bessel form (c.f. Eq. (1.7) of Chapter III) we find

$$\theta(-iy) = \frac{2\pi}{\alpha} e^{i\mu\pi/2} \sin(\mu\varphi_0)\, h_\mu(i\pi, -iy), \tag{4.16}$$

where $\mu = m\pi/\alpha > 0$ and $h_\mu(w, z)$ is the incomplete Hankel function.

Substituting (4.14) and (4.16) into (4.11), we obtain

$$f_m^0(r, t) = \frac{1}{\alpha i} e^{i\mu\pi/2} \sin(\mu\varphi_0) \int_{\sigma-i\infty}^{\sigma+i\infty} \frac{h_\mu(i\pi, -isrc)}{s - i\omega}\, e^{st}\, ds, \tag{4.17}$$

from which we see that the coefficients of the expansion of a plane wave in terms of the complete system of functions $\sin(m\pi\varphi/\alpha)$ can be expressed in terms of the incomplete Hankel function of the Bessel form $h_\mu(w, z)$.

For the steady state case, $f_m^0(r, t) \equiv f_m^0(r, 0)$, these coefficients are found to be

$$f_m^0(r, 0) = \frac{2\pi}{\alpha} e^{i\mu\pi/2} \sin(\mu\varphi_0)\, h_\mu(i\pi, kr), \tag{4.18}$$

where $k = \omega c$.

The function $h_\mu(i\pi, z)$ also can be expressed in terms of the incomplete cylindrical function $\Phi_\mu(w, z)$. According to (7.8) of Chapter III, we have

$$h_\mu(i\pi, z) = \Phi_\mu\left(\frac{3}{2}\pi i, z\right) - \Phi_\mu\left(-\frac{i\pi}{2}, z\right).$$

Moreover, taking $b = 3\pi i/2$ in the first of Eqs. (6.8) of Chapter III and $b = -i\pi/2$ in the second, we find

$$\Phi_\mu\left(\frac{3}{2}\pi i, z\right) = \mathscr{H}_\mu^{(1)}(z) - e^{-i\mu\pi}\Phi_{-\mu}\left(-\frac{i\pi}{2}, z\right);$$

$$\Phi_\mu\left(-\frac{i\pi}{2}, z\right) = -\mathscr{H}_\mu^{(2)}(z) - e^{i\mu\pi}\Phi_{-\mu}\left(-\frac{i\pi}{2}, z\right).$$

Using these expressions for $\Phi_\mu(3/2\pi i, z)$ and $\Phi_{-\mu}(-i\pi/2, z)$, we obtain finally

$$\begin{aligned} h_\mu(i\pi, z) &= \mathscr{H}_\mu^{(1)}(z) + e^{-2i\mu\pi}\mathscr{H}_\mu^{(2)}(z) \\ &\quad + (e^{-2i\mu\pi} - 1)\,\Phi_\mu\left(-\frac{i\pi}{2}, z\right). \end{aligned} \tag{4.19}$$

We remark that Ref. [51] introduces the auxilliary function $\varphi_\mu^-(z)$, which is connected with the incomplete cylindrical function $\Phi_\mu(-i\pi/2, z)$ by

$$\varphi_\mu^-(z) = -\frac{1}{2}\mathscr{H}_\mu^{(2)}(z) - \frac{1}{2}\Phi_\mu\left(-\frac{i\pi}{2}, z\right).$$

where

$$\varphi_\mu^-(z) = \frac{1}{2\pi i}\int_{-i\pi/2}^{\infty - i(\pi+\lambda)} e^{z\sinh u - \nu u}\, du.$$

Substituting this expression into (4.19), we find

$$h_\mu(i\pi, z) = 2J_\mu(z) + 4ie^{-i\pi\mu}\sin\mu\pi\varphi_\mu^-(z), \tag{4.20}$$

which, together with (4.18) agrees with the results of [51].

The asymptotic behavior of the function $\varphi_\mu^-(z)$ may be found without difficulty from the fundamental Eq. (6.38) of Chapter III. We have

$$\begin{aligned} \varphi_\mu^-(z) &\approx -\frac{1}{4}\mathscr{H}_\mu^{(2)}(z) + \frac{\mu}{2\pi z}e^{-i(z-\mu\pi/2)} \\ &\approx -\frac{1}{4}\sqrt{\frac{2}{\pi z}}\,e^{-i(z-\mu\pi/2-\pi/4)} + \frac{\mu}{2\pi z}e^{-i(z-\mu\pi/2)}. \end{aligned} \tag{4.21}$$

If now we substitute the representations (4.6) and (4.9) into initial conditions (4.3) and simultaneously make use of (4.8), (4.17), (4.20), and the asymptotic expansion (4.21), we find the following expression for the unknown function $F_\mu(s)$:

$$F_\mu(s) = \frac{4\pi}{\alpha}e^{i\mu\pi/2}\frac{\sin\mu\varphi_0}{s - i\omega}. \tag{4.22}$$

Consequently the expansion coefficients $f_m(r, t)$ in (4.6) take the form

$$f_m(r, t) = \frac{4\pi}{\alpha}e^{i\mu\pi/2}\sin\mu\varphi_0\,\frac{1}{2\pi i}\int_{\sigma - i\infty}^{\sigma + i\infty}\frac{J_\mu(-icrs)}{s - i\omega}e^{st}\,ds. \tag{4.23}$$

We now transform the integral on the right. Introducing the new variable of integration $z = -icrs$, we obtain

$$\frac{1}{2\pi i} \int_{\sigma-i\infty}^{\sigma+i\infty} \frac{J_\mu(-icrs)}{s - i\omega} e^{st}\, ds = \frac{1}{2\pi i} \int_{-i\sigma'-\infty}^{-i\sigma'+\infty} \frac{J_\mu(z)}{z - cr\omega} \exp\left(i \frac{tz}{cr}\right) dz, \qquad (4.24)$$

where σ' may be taken as an arbitrarily small positive quantity.

We consider (4.24) along the closed contour shown in Fig. 12, with the cut along the negative real axis. Noting that in [51], $t/cr \geq 1$, and using the asymptotic representation of $J_\mu(z)$

$$J_\mu(z) \approx \sqrt{\frac{\pi}{2z}} \cos\left(z - \frac{\mu\pi}{2} - \frac{\pi}{4}\right),$$

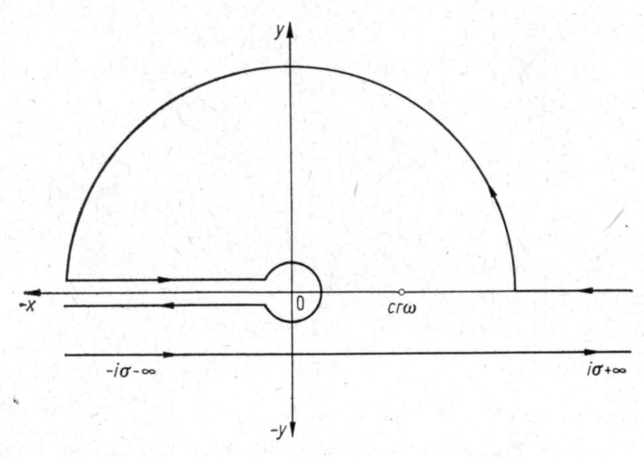

Fig. 12. Closed integration contour with cut along negative real axis.

for $|z| \gg 1$, we convince ourselves that the integral over the semicircle of large radius is negligible. Therefore the contributions of interest are from the residue $2\pi i J_\mu(cr\omega)\, e^{i\omega t}$ at the pole $z = cr\omega$, and from the integrals along the sides of the cut. If we use the facts that $z = x \exp(i\pi)$, along the upper edge and $z = x \exp(-i\pi)$, along the lower edge, and note that

$$J_\mu(x e^{\pm i\pi}) = e^{\pm i\mu\pi} J_\mu(x),$$

we find

$$\frac{1}{2\pi i} \int_{-i\sigma'-\infty}^{-i\sigma'+\infty} \frac{J_\mu(z)}{z - cr\omega} \exp\left(i \frac{tz}{cr}\right) dz = J_\mu(cr\omega)\, e^{i\omega t} + \frac{\sin\mu\pi}{\pi} \int_0^\infty \frac{J_\mu(x)}{x + cr\omega} \exp\left(-i \frac{tx}{cr}\right) dx. \qquad (4.25)$$

The last integral on the right hand side of this equation may be evaluated by means of incomplete cylindrical functions of the Bessel form. To see this, we use the easily established identity

$$\frac{1}{x + cr\omega} = \int_0^\infty e^{-\alpha(x+cr\omega)}\, d\alpha. \tag{4.26}$$

valid for $\operatorname{Re}(x + cr\omega) > 0$: Applying it in the last integral and interchanging the order of integration, we find

$$\begin{aligned} &\int_0^\infty \frac{J_\mu(x)}{x + cr\omega} \exp\left(-i\frac{xt}{cr}\right) dx \\ &= \int_0^\infty e^{-\alpha c\omega r}\, d\alpha \int_0^\infty e^{-x(\alpha + it/cr)} J_\mu(x)\, dx. \end{aligned} \tag{4.27}$$

But for the last integral

$$\int_0^\infty J_\mu(x) \exp\left[-x\left(\alpha + i\frac{t}{cr}\right)\right] dx = \frac{\left[\sqrt{\left(\alpha + i\frac{t}{cr}\right)^2 + 1} - \alpha + i\frac{t}{cr}\right]^\mu}{\sqrt{\left(\alpha + i\frac{t}{cr}\right)^2 + 1}} \tag{4.28}$$

(c.f., e.g. [10, p. 422] and also (1.2) of Chapter VI for $\alpha = \pi$, $b = 1$ and $\lambda = 0$). Substituting (4.28) in (4.27) and introducing a new variable of integration u by means of $\alpha + it/(cr) = e^{i\pi} \sinh u$, we find

$$u = \ln\left[\left(\frac{t}{icr} - \alpha\right) + \sqrt{\left(\frac{t}{icr} - \alpha\right)^2 + 1}\right].$$

After some simplifications we get

$$\int_0^\infty \frac{J_\mu(x)}{x + cr\omega} \exp\left(-i\frac{tx}{cr}\right) dx = e^{i\omega t} \int_{-\infty}^{-i\pi/2-\beta} e^{kr\sinh u + \mu u}\, du,$$

where $\cosh\beta = t/(cr) \geq 1$ and $\omega c = k$. But the integral on the right is precisely the incomplete cylindrical function $\Phi_{-\mu}(w, z)$, (c.f. Eq. (6.5) of Chapter III). In other words,

$$\int_0^\infty \frac{J_\mu(x)}{x + kr} \exp\left(-i\frac{tx}{cr}\right) dx = \pi i e^{i\omega t} \Phi_{-\mu}\left(-\beta - \frac{i\pi}{2}, kr\right). \tag{4.29}$$

Now substituting (4.29) in (4.23), and the result in (4.23), we obtain the expansion coefficients $f_m(r, t)$ in terms of Bessel functions and the incomplete cylindrical function $\Phi_{-\mu}$

$$\begin{aligned} f_m(r, t) = \frac{4\pi}{\alpha} \sin\mu\varphi_0 &\left[J_\mu(kr) + i\sin\mu\pi\Phi_{-\mu}\left(-\beta - \frac{i\pi}{2}, kr\right)\right] \\ &\times \exp\left[i\left(\omega t + \frac{\mu\pi}{2}\right)\right]. \end{aligned} \tag{4.30}$$

The first term in this equation, on substitution in Eq. (4.6), gives the solution of the steady-state problem of diffraction of a plane monochromatic wave. The second term describes the transient part of the solution. As much as we are considering the case $t/cr > 1$, i.e. $\cosh\beta > 1$, we may use the asymptotic expansion of $\Phi_{-\mu}(-\beta - i\pi/2, kr)$ for $kr\cdot\sinh\beta \gg 1$ to estimate the transient effect. [c.f. Eq. (6.31) of Chapter IV for $\alpha = -\beta$, $\nu = -\mu$]

$$\Phi_{-\mu}\left(-\beta - \frac{i\pi}{2}, kr\right) \approx \frac{1}{\pi kr\sinh\beta}\exp\left[i\left(kr\cosh\beta + \mu\frac{\pi}{2}\right) - \beta\mu\right]$$
$$\approx -\frac{\left[\frac{t}{cr} - \sqrt{\left(\frac{t}{cr}\right)^2 - 1}\right]^\mu}{\pi kr\sqrt{\left(\frac{t}{cr}\right)^2 - 1}}\cdot e^{-i(\omega t + \pi\mu/2)}. \tag{4.31}$$

We thus obtain the following approximate expression for the coefficients $f_m(r, t)$:

$$f_m(r, t) \approx \frac{4\pi}{\alpha}\sin\mu\varphi_0 J_\mu(kr)\exp\left[i\left(\omega t + \frac{\mu\pi}{2}\right)\right]$$
$$- \frac{4i}{\alpha kr}\sin\mu\varphi_0\sin\mu\pi\frac{\left[\frac{t}{cr} - \sqrt{\left(\frac{t}{cr}\right)^2 - 1}\right]^\mu}{\sqrt{\left(\frac{t}{cr}\right)^2 - 1}}. \tag{4.32}$$ [1]

If we substitute (4.32) into (4.6) we have

$$u \approx \frac{4\pi}{\alpha}e^{i\omega t}\sum_{m=0}^{\infty}\sin\mu\varphi_0\sin\mu\pi e^{i\mu\pi/2}J_\mu(kr)$$
$$- \frac{4i}{\alpha kr}\frac{1}{\sqrt{\left(\frac{t}{cr}\right)^2 - 1}}\sum_{m=0}^{\infty}\sin\mu\varphi_0\sin\mu\pi\sin\mu\varphi\left[\frac{t}{cr} - \sqrt{\left(\frac{t}{cr}\right)^2 - 1}\right]^\mu, \tag{4.33}$$

where $\mu = m\pi/\alpha$. The first term, as already noted, represents steady-state diffraction. The second term characterizes the process of establishing the stationary regime, and can be summed in closed form, so that (4.33) can be written:

$$u \approx \frac{4\pi}{\alpha}e^{i\omega t}\sum_{m=0}^{\infty}\sin\mu\varphi_0\sin\mu\varphi J_\mu(kr)\,e^{i\mu\pi/2}$$
$$- \frac{ip}{\alpha kr\sqrt{\left(\frac{t}{cr}\right)^2 - 1}}\left(\frac{\sin\psi_1}{1 - 2p\cos\psi_1 + p^2} + \frac{\sin\psi_2}{1 - 2p\cos\psi_2 + p_2}\right. \tag{4.34}$$
$$\left. - \frac{\sin\psi_3}{1 - 2p\cos\psi_3 + p^2} - \frac{\sin\psi_4}{1 - 2p\cos\psi_4 + p^2}\right),$$

[1] We note in passing that in [51] on page 37, there seems to be a misprint: There the coefficient of $1/kr$ is multiplied by $1/\omega t$.

where we have introduced the notations

$$\psi_1 = \frac{\pi}{\alpha}(\pi + \varphi_0 - \varphi); \qquad \psi_3 = \frac{\pi}{\alpha}(\pi + \varphi_0 + \varphi);$$

$$\psi_2 = (\pi - \varphi_0 + \varphi)\frac{\pi}{\alpha}; \qquad \psi_4 = (\pi - \varphi_0 - \varphi)\frac{\pi}{\alpha}; \tag{4.35}$$

$$p = \left[\left(\frac{t}{cr}\right) - \sqrt{\left(\frac{t}{cr}\right)^2 - 1}\right]^{\pi/\alpha}.$$

For large $t/cr \gg 1$, $p \sim [+2t/(cr)]^{-\pi/\alpha}$, we have

$$\begin{aligned} u \approx \frac{4\pi}{\alpha} e^{i\omega t} \sum_{m=0}^{\infty} \sin \mu\varphi_0 \sin \mu\pi J_\mu(kr)\, e^{i\mu\pi/2} \\ - \frac{8i}{akr}\left(\frac{cr}{2t}\right)^{1+\pi/\alpha} \sin\frac{\pi^2}{\alpha} \sin\frac{\pi\varphi}{\alpha} \sin\frac{\pi\varphi_0}{\alpha}. \end{aligned} \tag{4.36}$$

From these equations we see that the terms which characterize the establishment of the stationary regime of the diffraction process decay like $[t^{-(1+\pi/\alpha)}]$ as $t \to \infty$. The corresponding estimate obtained in [51] indicates decay like $t^{-(2+\pi/\alpha)}$.

5. Diffraction of Waves by a Screen of Given Form

In the preceeding section the problem of scattering of waves by the wedge $r \geq 0$, $0 \leq \varphi \leq \alpha$ was solved for the case when the wave vector of the incident wave lay in the plane of the principal section of the wedge. We consider below the scattering of a wave or particle by objects of finite dimensions (in particular screens of diverse forms). In this case the interaction of the incident wave or particle with the given object will be characterized by the effects of an interaction potential $V(\boldsymbol{r})$, which differs from zero only in the region of space occupied by the object.

Let a plane wave fall on a screen of given form. The interaction of this wave with the screen is described by the Schrödinger equation

$$(\nabla^2 + k^2)\,\Psi(\boldsymbol{r}) = V(\boldsymbol{r})\,\Psi(\boldsymbol{r}), \tag{5.1}$$

where k is the wave number.

If we consider (5.1) as an inhomogeneous differential equation with inhomogeneous term $V(\boldsymbol{r})\,\psi(\boldsymbol{r})$, its solution may be written formally as the following integral equation (c.f., e.g. [5, Vol. II, Ch. 12]):

$$\Psi(\boldsymbol{r}) = \Psi_1(\boldsymbol{r}) - \frac{1}{4\pi}\int \frac{e^{ik|\boldsymbol{r}-\boldsymbol{r}_0|}}{|\boldsymbol{r}-\boldsymbol{r}_0|} V(\boldsymbol{r}_0)\,\Psi(\boldsymbol{r}_0)\,d\boldsymbol{r}_0, \tag{5.2}$$

where $\Psi_1(\boldsymbol{r})$ is the solution of the corresponding homogeneous equation, which for our case is the incoming plane wave

$$\Psi_1(\boldsymbol{r}) = \exp(i\boldsymbol{k}\boldsymbol{r}). \tag{5.3}$$

The integration of the second term is taken over the region occupied by the scattering object.

In the following we shall be interested in the intensity of the wave at distances much larger than the linear dimensions of the scattering object. For these points we may take

$$|\boldsymbol{r} - \boldsymbol{r}_0| \approx r - r_0 \cos\theta\,, \tag{5.4}$$

where θ is the angle between the vector $\boldsymbol{r}$ of the observation point and the vector $\boldsymbol{r}_0$ of the source point in the scattering object. Substituting (5.4) and (5.3) in (5.2), we find

$$\Psi(\boldsymbol{r}) \approx \mathrm{e}^{ikr} - \frac{1}{4\pi}\frac{\mathrm{e}^{ikr}}{r}\int \mathrm{e}^{-i\boldsymbol{k}'\boldsymbol{r}_0} V(\boldsymbol{r}_0)\,\Psi(\boldsymbol{r}_0)\,d\boldsymbol{r}_0, \tag{5.5}$$

where $\boldsymbol{k}'$ is the wave vector of the scattered wave; $\boldsymbol{k}'\boldsymbol{r}_0 = kr_0\cos(\theta)$. This solution evidently satisfies the boundary conditions at infinity. The coefficient of $\exp(ikr)/r$ in the diverging wave is called the scattering amplitude:

$$f(\vartheta, \varphi) = -\frac{1}{4\pi}\int \mathrm{e}^{-i\boldsymbol{k}'\boldsymbol{r}_0}\, V(\boldsymbol{r}_0)\,\Psi(\boldsymbol{r}_0)\,d\boldsymbol{r}_0. \tag{5.6}$$

The square of the modulus of this amplitude determines the intensity of the scattered wave at the observation point (ϑ, φ).

In order to determine the quantity $f(\vartheta, \varphi)$ precisely, it is necessary to know the exact solution of (5.1) for the given potential $V(\boldsymbol{r})$. In general this problem is analytically intractable, so in most cases recourse is made to some approximate method of solution. If the interaction potential can be considered small, then in many practical cases it is justifiable, as a first approximation, to take for the function $\Psi(\boldsymbol{r}_0)$ in the integrand of (5.6) the undisturbed wave $\exp(ik \cdot \boldsymbol{r}_0)$, i.e.

$$f = -\frac{1}{4\pi}\int \mathrm{e}^{i(\boldsymbol{k}-\boldsymbol{k}')\boldsymbol{r}_0}\, V(\boldsymbol{r}_0)\,d\boldsymbol{r}_0. \tag{5.7}$$

Here $\boldsymbol{k}$ and $\boldsymbol{k}'$ ($|\boldsymbol{k}| = |\boldsymbol{k}'|$) are the wave of the vectors incident and scattered waves, respectively, and the integration is carried out over the region where the interaction potential $V(\boldsymbol{r}_0)$ is non-zero. In the theory of particle scattering this approximation is usually called the first Born approximation, while in optics it is also referred to as the Kirchoff approximation.

We shall consider in detail cases in which the screen is plane, and the wave vector $\boldsymbol{k}$ of the incident wave is perpendicular to the plane of the screen. If the z-axis is taken in the direction of $\boldsymbol{k}$, then the effects of the interaction of the incident wave with the screen may be characterized by the following function:

$$V = -4\pi c\,\Omega(D_0, D)\,\delta(z)\,, \tag{5.8}$$

where c is a constant, characterizing the intensity of the interaction; $\delta(z)$ is the Dirac delta function, and $\Omega(D_0, D)$ is a unit step function, equal to one for all points D of the screen, and zero elsewhere.

Substituting (5.8) into (5.7), we find

$$f(\vartheta, \varphi) = c \int e^{-i\varkappa\varrho} \, d\boldsymbol{S}_0 = c \int e^{-i\varkappa\varrho\cos(\varphi-\varphi_0)} \, d\boldsymbol{S}_0. \tag{5.9}$$

Here $\varkappa$ is the projection of the vector $(\boldsymbol{k}' - \boldsymbol{k})$ on to the plane of the screen and, since $\boldsymbol{k}$ is perpendicular to the plane of the screen, $|\varkappa| = |\boldsymbol{k}'| \sin(\vartheta) = |\boldsymbol{k}| \sin(\vartheta)$; ϑ and φ are the coordinates of the observed point. Here the integrations (with respect to ϱ and φ_0) are carried out over the whole surface of the screen.

Equation (5.9) determines the Fraunhofer diffraction of the plane screen to within a constant multiplier. If the scattering screen has the form of a circle of radius R, then this equation takes the form

$$f(\vartheta, \varphi) = c \int_0^R \int_0^{2\pi} e^{-i\varkappa\varrho\cos(\varphi_0-\varphi)} \varrho \, d\varrho \, d\varphi_0 = 2\pi c \int_0^R J_0(\varkappa\varrho) \, \varrho \, d\varrho. \tag{5.10}$$

Here $J_0(\varkappa\varrho)$ is the Bessel function of zero order. Taking into account Eq. (6.10) of Chapter VI

$$\int_0^R J_0(\varkappa R) \, R \, dR = \frac{R J_1(\varkappa R)}{\varkappa},$$

we get the well known formula for the scattering amplitude of a plane circular screen

$$f(\vartheta, \varphi) = 2\pi c R^2 \frac{J_1(\varkappa R)}{\varkappa R}. \tag{5.11}$$

We now show that if the screen is a segment of a circle of angle $\alpha < 2\pi$ and radius R, the scattering amplitude $f(\vartheta, \varphi)$ may be expressed in terms of incomplete cylindrical functions. For, in this case (5.9) can be written

$$\begin{aligned} f(\vartheta, \varphi) &= c \int_0^R \int_0^{\alpha} e^{i\varkappa\varrho\cos(\varphi_0-\varphi)} \varrho \, d\varrho \, d\varphi_0 \\ &= c \int_0^R \varrho \, d\varrho \int_{-\varphi}^{\alpha-\varphi} e^{-ik\varrho\cos\theta} \, d\theta. \end{aligned} \tag{5.12}$$

Or, according to the definition of the incomplete cylindrical function $E_\nu^-(w, z)$ of the Poisson form, c.f. § 1 of Chapter II, we have

$$f(\vartheta, \varphi) = \frac{\pi c}{2} \int_0^R [E_0^-(\alpha - \varphi, \varkappa\varrho) + E_0^-(\varphi, \varkappa\varrho)] \, \varrho \, d\varrho. \tag{5.13}$$

Using in addition (6.7) of Chapter VI, we find the following expression for the scattering amplitude:

$$f(\vartheta, \varphi; \alpha) = \frac{cR^2}{2} [F(\alpha - \varphi, \varkappa R) + F(\varphi, \varkappa R)], \tag{5.14}$$

where $\varkappa = k \sin(\vartheta)$ and for brevity we have set

$$F(\psi, \varrho) = \frac{E_1^-(\psi, \varrho)}{\varrho} + \frac{2i}{\pi\varrho} \sin\psi e^{-i\varrho\cos\psi} + \frac{2}{\pi\varrho^2} \tan\psi (e^{-i\varrho\cos\psi} - 1). \tag{5.15}$$

Equation (5.15) is suitable for numerical evaluation only when $0 \leq \alpha \leq \pi$. For $\pi \leq \alpha \leq 2\pi$ values of $f(\vartheta, \varphi', \alpha)$ may be found from

$$f(\vartheta, \varphi; \alpha) = 2\pi c R^2 \frac{J_1(\varkappa R)}{\varkappa R} + f(\vartheta, \varphi; \alpha - 2\pi), \tag{5.16}$$

which is not difficult to obtain from (5.15) and (3.13) of Chapter II:

$$E_1^-(2\pi - w, \varrho) = -E_1^-(w, \varrho) + 4J_1(\varrho). \tag{5.17}$$

It is also evident from (5.14) that $(\vartheta, \varphi, \alpha)$ is a symmetric function of φ about $\varphi = \alpha/2$, so that $f(\vartheta, 0; \alpha) \equiv f(\vartheta, \alpha; \alpha)$. Since $f(\vartheta, \varphi, 0) \equiv 0$, we readily recover the well known result (5.11) from (5.16) for $\alpha = 2\pi$.

Separating the real and imaginary parts of (5.15) and using the fact that $E_1^-(\psi, \varrho) = J_1(\psi, \varrho) - iH_1(\psi, \varrho)$, we have

$$\begin{aligned} F(\psi, \varrho) = {} & \frac{J_1(\psi, \varrho)}{\varrho} + \frac{2}{\pi\varrho} \sin\psi \sin(\varrho\cos\psi) + \frac{2}{\pi\varrho^2} \tan\psi\, [\cos(\varrho\cos\psi) - 1] \\ & - i\left[\frac{H_1(\psi, \varrho)}{\varrho} - \frac{2\sin\psi}{\pi\varrho} \cos(\varrho\cos\psi) + \frac{2\tan\psi}{\pi\varrho^2} \sin(\varrho\cos\psi)\right], \end{aligned} \tag{5.18}$$

from which it is clear that the scattering amplitude of a circular sector can be expressed by means of elementary functions and the incomplete Bessel and Struve functions $J_1(\psi, \varrho)$ and $H_1(\psi, \varrho)$, respectively.

The equations obtained here may be used to evaluate the diffraction field in other cases, for example, when the screen is a circular segment. This is because the diffraction by a triangular screen is easy to evaluate. To see this, let the screen be a right triangle with sides a and b. Then Eq. (5.9) gives

$$\begin{aligned} f(\vartheta, \varphi) &= c \int_0^a e^{i\varkappa_a x}\, dx \int_0^{b - bx/a} e^{i\varkappa_b y}\, dy \\ &= c\, \frac{ab}{(a\varkappa_a - b\varkappa_b)} \left(\frac{1 - e^{i\varkappa_b b}}{\varkappa_b b} - \frac{1 - e^{i\varkappa_a a}}{a\varkappa_a} \right), \end{aligned} \tag{5.19}$$

where $\varkappa_a = k \sin\vartheta \cos\varphi$; $\varkappa_b = k \sin\vartheta \sin\varphi$. These results may be used for all triangles, since each can be divided into two right triangles. Therefore Eqs. (5.14)—(5.18) for a circular sector, together with the last formula gives the possibility of evaluating scattering amplitudes for any circular segment. It is evident, that the required quantity $f(\vartheta, \varphi)$ is determined, according to (5.9), as the difference between the values of $f(\vartheta, \varphi)$ for the appropriate circular sector (α, R) and the values for two similar right triangles with sides $a = R \sin(\alpha/2)$ and $b = R \cos(\alpha/2)$.

We now determine an approximate expression for the scattering amplitude in the limiting case when $\varkappa R = kR \sin(\vartheta) \ll 1$ and $|\varkappa R \sin\psi| = |kR \sin\vartheta \sin\psi| \gg 1$. For $\varkappa R \ll 1$ we have

$$f(\vartheta, \varphi; \alpha) \approx \frac{cR^2}{2}\left\{\alpha - \frac{\varkappa^2 R^2}{8}[\alpha + \sin\alpha \cos(\alpha - 2\varphi)] - i\frac{4\varkappa R}{3}\sin\frac{\alpha}{2}\cos\left(\frac{\alpha}{2} - \varphi\right)\right\}, \tag{5.20}$$

where $\varkappa = k \sin(\vartheta)$.

The asymptotic expansion of $f(\vartheta, \varphi; \alpha)$ for large $\varkappa R \gg 1$ depends on the relation between the parameters φ and α for, according to (10.32)—(10.37) of Chapter II, the asymptotic behavior of $E_\nu(w, z)$ changes as w passes through $w = \pi$ and $w = 0$. By virtue of (5.16) it is sufficient to consider the case $0 \leq \alpha \leq \pi$. For these values of α, $f(\vartheta, \varphi; \alpha)$ as a function of φ may be written:

$$f(\vartheta, \varphi; \alpha) = \frac{\pi c R^2}{2}\begin{cases} F(\alpha - \varphi, \varkappa R) + F(\varphi, \varkappa R), \; 0 < \varphi \leq \alpha; \\ -F(\varphi - \alpha, \varkappa R) + F(\varphi, \varkappa R), \; \alpha \leq \varphi < \pi; \\ -F(\varphi - \alpha, \varkappa R) + F^*(\varphi, \varkappa R), \; \pi < \varphi < \pi + \alpha; \\ -F^*(\varphi - \alpha, \varkappa R) + F^*(\varphi, \varkappa R), \; \pi + \alpha \leq \varphi < 2\pi. \end{cases} \tag{5.21}$$

Here we have used the fact that $F(\psi, x)$ is an odd function of ψ, (c.f. (5.15)) and the asterisk on F indicates that in order to find the asymptotic behavior of F from (5.15), (10.35) of Chapter II must be used to determine the asymptotic behavior of $E_1(\psi, x)$, since the parameter ψ of this function is larger than π. Omitting further details, we give here an approximate expression for the scattering amplitude for large $\varkappa R \gg 1$, correct to order $1/\varkappa R^2$:

$$f(\vartheta, \varphi; \alpha) \approx \frac{\pi c R^2}{2}[\Pi(\alpha, -\varphi, \varkappa R) + \Pi(\varphi, \varkappa R)] + \begin{Bmatrix} 2\dfrac{\mathscr{H}_1^{(2)}(\varkappa R)}{\varkappa R}; & 0 < \varphi < \alpha; \\ 0; & \alpha < \varphi < \pi; \\ 2\dfrac{\mathscr{H}_1^{(1)}(\varkappa R)}{\varkappa R}; & \pi < \varphi < \pi + \alpha; \\ 0; & \pi + \alpha < \varphi < 2\pi, \end{Bmatrix} \tag{5.22}$$

where for brevity the asymptotic development of the Hankel functions is not written out and the notation

$$\Pi(\psi, \varrho) = \frac{2\cot\psi}{\pi\varrho^2}e^{-i\varrho\cos\psi} - \frac{2\tan\psi}{\pi\varrho^2}(e^{-i\varrho\cos\psi} - 1) \tag{5.23}$$

is introduced. On the boundaries $\varphi = \alpha; \pi$, and $\pi + \alpha$ the asymptotic behavior of $f(\vartheta, \varphi; \alpha)$ can be determined as the mean value of this function as one approaches the given boundary from the right and from the left.

We note here some other diffraction problems, in which incomplete cylindrical functions appear. In Ref. [2] the phenomenon of diffraction by small cracks in a curvilinear surface is studied in detail. In particular, the determination of the amplitude of the wave diffracted by a straight cylinder with a narrow crack perpendicular to its cross-section is connected with the evaluation of the integral

$$\int_0^\alpha e^{i\xi\cos\theta}\, d\theta, \tag{5.24}$$

which represents, except for a constant multiplier of $2/\pi$, the incomplete cylindrical function $E_0^+(\alpha, \xi)$:

$$\int_0^\alpha e^{i\xi\cos\theta}\, d\theta = \frac{\pi}{2} E_0^+(\alpha, \xi) = \frac{\pi}{2} J_0(\alpha, \xi) + i\frac{\pi}{2} H_0(\alpha, \xi). \tag{5.25}$$

There is interest in developments of this integral as Taylor and Fourier series. These are given in [2]. We have already given the Fourier expansion (c.f. Eqs. (7.14)—(7.17) of Chapter II). The two expansions in powers of ξ are written in our notation

$$\begin{aligned} J_0(\alpha, \xi) &= \alpha - \frac{sp}{2^2}\xi^2 + \frac{s\xi^4}{4^2}\left(\frac{c^3}{3^2} + \frac{p}{2^2}\right) + \cdots \\ &+ (-1)^{n-1}\frac{s\xi^{2n}}{(2n)^2}\left[\frac{c^{2n-1}}{(2n-1)!} + \frac{c^{2n-2}}{(2n-3)!\,(2n-2)^2} + \cdots \right. \\ &\left. + \frac{p}{2^2 \cdot 4^2 \cdots (2n-2)^2}\right] + \cdots; \end{aligned} \tag{5.26}$$

$$\begin{aligned} H_0(\alpha, \xi) &= s\left[\xi - \frac{\xi^2}{3^2}\left(\frac{c^2}{2^2} + 1\right) + \frac{\xi^4}{5^2}\left(\frac{c^4}{4!} + \frac{c^2}{2!\,3^2} + \frac{1}{3^2}\right)\right. \\ &+ \cdots + (-1)^n \frac{\xi^{2n+1}}{(2n+1)^2}\left(\frac{c^{2n}}{(2n)!} + \frac{c^{2n-2}}{(2n-2)!\,(2n-1)^2}\right. \\ &\left.\left. + \cdots + \frac{1}{3^2 \cdot 5^2 \cdots (2n-1)^2}\right) + \cdots\right], \end{aligned} \tag{5.27}$$

where for brevity we have set $c = \cos(\alpha)$, $s = \sin\alpha$ and $p = \cos\alpha + \alpha/\sin\alpha$. It is evident that (5.26) and (5.27) may be obtained from the series for $E_0^+(w, z)$ given in § 7 of Chapter II if explicit expressions are written for the coefficients $c_{m,0}$.

In the study of diffraction phenomena at a point of intersection of optical rays, i.e. at a focus, there arises the quantity B,

$$B = \iint e^{-ikr\cos\theta}\, d\Omega, \tag{5.28}$$

representing the incoming wave bundle as a superposition of plane waves [52, p. 411]. If the limits of the solid angle correspond to a right cone of vertex angle 2α, then the expression for B takes the form

$$B = 2\pi \int_0^\alpha e^{-ikr\cos\theta} \sin\theta \, d\theta . \tag{5.29}$$

In other words, (5.29) may be expressed in terms of the incomplete cylindrical function $E_\nu^-(w, z)$ with half odd index, which in turn can be expressed by elementary functions.

If the waves from infinity converge in a focal line, then corresponding to the expression for B we have

$$B(r, \varphi) = \int_{-\alpha}^{\alpha} e^{-ikr\cos(\varphi - \varphi_0)} \, d\varphi_0 . \tag{5.30}$$

In this case $B(r, \varphi)$ cannot be expressed by elementary functions. However, by means of obvious transformations it is not difficult to express $B(r, \varphi)$ by means of incomplete cylindrical functions:

$$B(r, \varphi) = \frac{\pi}{2} \left[E_0^-(\alpha - \varphi, kr) + E_0^-(\alpha + \varphi, kr)\right]. \tag{5.31}$$

To conclude this section we note that in such problem areas as scattering of electromagnetic waves by cylinders with longitudinal cracks, and wave production by oscillations of pistons in tubes of various cross sections (c.f., e.g. [5, pp. 363—503]) integrals arise which are related to incomplete cylindrical functions. These cases may often be treated like the ones above.

6. Some Problems in the Theory of Diffraction in Optical Apparatus

The necessity for more precise determination of the diffraction properties of optical systems arises in a number of problems associated with the application of modern optical instruments. Diffraction of the light waves associated with a bounded cone of rays by the lens, mirror and diaphragm composing an optical system contributes to distortion of the image. Because of diffraction, points are imaged as circles, whose size limits the resolving power of the instrument. Therefore the problem of precisely determining the intensity of waves passing through an optical instrument to a given point of the image plane takes on considerable importance.

This problem has been considered in many works [8, 53, 54]. In [53], for example, it is pointed out that the classical theory of Struve is inapplicable for the computation of the distribution of the intensity of light fluctuations in the field of a microscope. Also in [53] are given a series of new formulae characterizing the diffracted images of point and line sources in a microscope.

When polarization is applied in the image field of an optical system, the image of a coherent line source passing outside the focus may be described by the following integral:

$$L = \int_0^{2\alpha} \cos (k_0 c \sin \theta) \exp (i k_0 d \cos \theta) \cos \theta \, d\theta, \tag{6.1}$$

where d is the object-to-focus distance, k_0 is the wave number, c is the distance of the observation point from the center of the image, and 2α is the aperture angle of the diaphragm.

In [54] and elsewhere, a general method is given for computing the distribution of intensity of light perturbations arising from the passage of light through various optical systems. All of these computations are connected with analysis of the so called transmission coefficient $g(s)$, which is related to the coefficient of transparency $f(x, y)$ of the optical system in the follwoing way:

$$g = \frac{1}{2\pi} \int\int f\left(x + \frac{1}{2} s, y\right) f^*\left(x - \frac{1}{2} s, y\right) dx \, dy, \tag{6.2}$$

where f^* is the complex conjugate of f and the integration is carried out over the region where $f(x, y) \neq 0$. This function embodies as the aberration of the system, and also its focussing defect. In the absence of aberration $f(x, y)$ has the form

$$f(x, y) = \begin{Bmatrix} \exp [i k_0 w (x^2 + y^2)], & x^2 + y^2 \leq 1; \\ 0, & x^2 + y^2 > 1, \end{Bmatrix} \tag{6.3}$$

where w characterizes the defect of the focus. Substituting (6.3) in (6.2) gives, on carrying out the integration with respect to x

$$g = \frac{4}{z} \int_0^{\sqrt{1-(s/2)^2}} \sin\left(\sqrt{1 - y^2} - \frac{1}{2} |s|\right) dy, \tag{6.4}$$

where $z = 2 k_0 w |s|$ and the parameter $|s| \leq 2$. For values of $|s| \geq 2$ the function $g(s)$, according to (6.3), is identically zero. If we now introduce a new variable of integration by the equation $y = \sin \theta$, we find

$$\begin{aligned} g = \frac{4}{z} \cos (z \, |s|/2) \int_0^{\alpha} \sin (z \cos \theta) \cos \theta \, d\theta \\ - \frac{4}{z} \sin (z \, |s|/2) \int_0^{\alpha} \cos (z \cos \theta) \cos \theta \, d\theta, \end{aligned} \tag{6.5}$$

where $\alpha = \arccos (|s|/2)$.

In this regard Ref. [55] shows that the distribution of intensity of light passing through the optical system of an interferometer is determined by the following integral:

$$q = \int_0^\alpha \cos^2 (K \cos \theta)\, d\theta, \tag{6.6}$$

where α is the aperture, and K is the phase characteristic.

There is thus an entire set of problems, all connected with the study of the diffractive properties of optical apparatus, and leading to the evaluation of the integrals L, g, and q, which cannot be expressed in terms of known and tabulated functions. The evaluation of the integral L was discussed in [56], and its evaluation led to the numerical integration of an associated differential equation.

In a later work [8], it is shown that all these integrals can be expressed in terms of incomplete Bessel and Struve functions. In that same reference, as we remarked earlier, the basic properties of these functions are studied, and tables are supplied over a fairly wide region of their arguments.

For completeness we give here expressions for the above-mentioned integrals in terms of incomplete cylindrical functions.

Turning to (6.1) and writing it in the form

$$L = \frac{1}{2} \int_{-\alpha}^{\alpha} \exp\, [ik_0 (d \cos \theta + c \sin \theta)]\, \cos \theta\, d\theta,$$

with the notation

$$\left.\begin{aligned} d &= R \cos \gamma; \quad c = R \sin \gamma; \\ R &= \sqrt{c^2 + d^2}; \ \tan \gamma = \frac{c}{d}, \end{aligned}\right\} \tag{6.7}$$

we find after some obvious changes in the variables of integration

$$L = \frac{1}{2} \int_{-\alpha-\gamma}^{\alpha-\gamma} \cos (u + \gamma) \exp (ik_0 R \cos u)\, du \tag{6.8}$$

$$= \frac{\cos \gamma}{2iR} \frac{\partial}{\partial k_0} \int_{-\alpha-\gamma}^{\alpha-\gamma} \exp (ik_0 R \cos u)\, du + \frac{\sin \gamma}{2} \int_{-\alpha-\gamma}^{\alpha-\gamma} \exp (ik_0 R \cos u)\, d (\cos u).$$

The first term can be immediately expressed in terms of the incomplete cylindrical function $E_0^+ (\omega, z)$ and is equal to

$$\frac{\pi \cos \gamma}{2iR} \frac{d}{dk_0} [E_0^+ (\alpha - \gamma, k_0 R) + E_0^+ (\alpha + \gamma, k_0 R)].$$

If we use Eq. (2.24) of Chapter II and carry out the integration in the second term of (6.8), we find

$$\begin{aligned} L = {} & \frac{i\pi \cos \gamma}{4} [E_1^+ (\alpha - \gamma, k_0 R) + E_1^+ (\alpha + \gamma, k_0 R)] \\ & + \frac{1}{2} \left[\cos \gamma \sin (\alpha - \gamma) - \frac{i \sin \gamma}{k_0 R}\right] e^{ik_0 R \cos(\alpha - \gamma)} \\ & + \frac{1}{2} \left[\cos \gamma \sin (\alpha + \gamma) + \frac{i \sin \gamma}{k_0 R}\right] e^{ik_0 R \cos(\alpha + \gamma)}. \end{aligned} \tag{6.9}$$

Separating real and imaginary parts, and using

$$E_\nu^+(w, z) = J_\nu^i(w, z) + iH_\nu(w, z),$$

we obtain, after some simplifications,

$$\begin{aligned} L = \frac{c}{k_0 R^2} \Big\{ &- \frac{\pi k_0 R d}{4c} \left[H_1(\alpha - \gamma, k_0 R) + H_1(\alpha + \gamma, k_0 R) \right] \\ &+ \frac{d^2}{c^2} y \cos y \cos x + x \sin y \sin x + \sin y \cos x \Big\} \\ + i \frac{c}{k_0 R^2} \Big\{ &\frac{\pi k_0 R d}{4c} \left[J_1(\alpha - \gamma, k_0 R) + J_1(\alpha + \gamma, k_0 R) \right] \\ &+ \frac{d^2}{c^2} y \cos y \sin x - x \sin y \cos x + \sin y \sin x \Big\}, \end{aligned} \tag{6.10}$$

where we have introduced the notations

$$\left. \begin{aligned} x &= k_0 d \cos \alpha \\ y &= k_0 c \sin \alpha, \end{aligned} \right\} \tag{6.11}$$

and R and γ are defined by (6.7). This formula expresses the integral L in terms of elementary functions and the incomplete Bessel and Struve functions $J_1(w, z)$ and $H_1(w, z)$.

We give an approximate expression for the integral under consideration, valid when $k_0 R < 1$ $(x < 1, y < 1)$. Retaining only terms of order less than 2 in $k_0 R$, we have

$$L \approx \sin \alpha + \frac{i k_0 d}{2} \left(\alpha + \frac{\sin 2\alpha}{2} \right) - \frac{k_0^2 d^2 \sin \alpha}{2} \left(1 - \frac{c^2 - d^2}{3d^2} \sin^2 \alpha \right). \tag{6.12}$$

On the other hand, for large values of $k_0 R$, i.e.

$$|k_0 R \sin(\alpha - \gamma)| \gg 1, \qquad |k_0 R \sin(\alpha + \gamma)| \gg 1,$$

an approximate expression for L may be obtained with the aid of asymptotic expansions of the incomplete cylindrical functions. In this case from (10.4) and (10.14) of Chapter II we find

$$\begin{aligned} &\text{for} \quad k_0 R \sin(\alpha + \gamma) = k_0 (d \sin \alpha + c \cos \alpha) \gg 1; \\ &\qquad E_1^+(\alpha + \gamma, k_0 R) \approx \mathscr{H}_1^{(1)}(k_0 R) \\ &\qquad + \frac{2i \sin(\alpha + \gamma)}{\pi} e^{i k_0 R \cos(\alpha + \gamma)} \left(1 + i \frac{\cos(\alpha + \gamma)}{k_0 R \sin^2(\alpha + \gamma)} \right); \\ &\text{for} \quad k_0 R \sin(\alpha - \gamma) = k_0 (d \sin \alpha - c \cos \alpha) \gg 1; \\ &\qquad E_1^+(\alpha - \gamma, k_0 R) \approx \pm \mathscr{H}_1^{(1)}(k_0 R) \\ &\qquad + \frac{2i \sin(\alpha - \gamma)}{\pi} e^{i k_0 R \cos(\alpha - \gamma)} \left(1 + \frac{i \cos(\alpha - \gamma)}{k_0 R \sin^2(\alpha - \gamma)} \right). \end{aligned} \tag{6.13}$$

Here $\mathscr{H}_1^{(1)}(z)$ is the Hankel function of the first order and first kind. In these equations the positive sign in the second formula applies for

$(\alpha - \gamma) > 0$ and the negative sign for $(\alpha - \gamma) < 0$ in keeping with the evaluation of (10.37) of Chapter II. Substituting these formulae in (6.10) and using the asymptotic development for the Hankel function (c.f. (10.6) of Chapter II), we find after a series of simplifications that for $k_0 R \gg 1$:

$$
\begin{aligned}
L \approx\; & \Omega(\alpha, \gamma) \cos\gamma \sqrt{\frac{\pi}{2k_0 r}} \cos\left(k_0 R - \frac{\pi}{4}\right) \\
& + \frac{1}{2k_0 R}\left\{\frac{\cos(\alpha - 2\gamma)}{\sin(\alpha - \gamma)} \sin[k_0 R \cos(\alpha - \gamma)]\right. \\
& \left. + \frac{\cos(\alpha + 2\gamma)}{\sin(\alpha + \gamma)} \sin[k_0 R \cos(\alpha + \gamma)]\right\} \\
& + i\Omega(\alpha, \gamma) \cos\gamma \sqrt{\frac{\pi}{2k_0 R}} \sin\left(k_0 R - \frac{\pi}{4}\right) \\
& + i\frac{1}{2k_0 R}\left\{\frac{\cos(\alpha + 2\gamma)}{\sin(\alpha + \gamma)} \sin[k_0 R \cos(\alpha + \gamma)]\right. \\
& \left. - \frac{\cos(\alpha - 2\gamma)}{\sin(\alpha - \gamma)} \cos[k_0 R \cos(\alpha - \gamma)]\right\}.
\end{aligned}
\tag{6.14}
$$

Here $\Omega(\alpha, \gamma)$ is the unit step function, equal to one for $\alpha > \gamma > 0$ and zero for $\alpha < \gamma$.

Thus it follows from (6.14) that the character of the diffracted image for sufficiently large $k_0 R$ depends on the relation between the aperture and the quantity $\gamma = \arctan(c/d)$. Consequently, for $\alpha > \gamma > 0$ the fundamental contribution to L is given by a term proportional to $(1/k_0 R)^{1/2}$. If, however, $\alpha < \gamma$, this term in general vanishes and $L \sim (1/k_0 R)$.

We turn now to the consideration of (6.5), which we write as:

$$
\begin{aligned}
g = & -\frac{4}{z} \cos(z|s|/2) \frac{d}{dz} \int_0^\alpha \cos(z \cos\theta)\, d\theta \\
& - \frac{4}{z} \sin(z|s|/2) \frac{d}{dz} \int_0^\alpha \sin(z \cos\theta)\, d\theta,
\end{aligned}
\tag{6.15}
$$

where $\alpha = \arccos(|s|/2)$. Recalling the definitions of the incomplete Bessel and Struve functions (c.f., § 1 of Chapter II) we have

$$
g = -\frac{2\pi}{z} \cos(z|s|/2) \frac{dJ_0(\alpha, z)}{dz} - \frac{2\pi}{z} \sin(z|s|/2) \frac{dH_0(\alpha, z)}{dz}. \tag{6.16}
$$

Or, applying (2.25) of Chapter II

$$
\begin{aligned}
\frac{dJ_0(\alpha, z)}{dz} &= -J_1(\alpha, z) - \frac{2\sin\alpha}{\pi} \sin(z\cos\alpha); \\
\frac{dH_0(\alpha, z)}{dz} &= -H_1(\alpha, z) + \frac{2\sin\alpha}{\pi} \cos(z\cos\alpha)
\end{aligned}
$$

and noting that $\cos\alpha = |s|/2$, we find, after some elementary transformations that

$$g = \frac{2\pi}{z}\left[\cos\left(z\,|s|/2\right) J_1(\alpha, z) + \sin\left(z\,|s|/2\right) H_1(\alpha, z)\right], \tag{6.17}$$

where $z = 2k_0 w|s|$ and $|s| \leq 2$. For small values of $z = 2k_0 w|s| < 1$, we obtain by means of expansions (7.6)—(7.10) of Chapter II the following approximation:

$$\begin{aligned} g \approx 2\alpha - \sin 2\alpha - \frac{z^2}{8}\Big[&2\alpha(1+s^2) - s^2\sin 2\alpha \\ &- \frac{1}{2}\sin 4\alpha - \frac{16}{3}|s|\sin^3\alpha\Big]. \end{aligned} \tag{6.18}$$

On the other hand, for large values of the parameter $|2k_0 ws\sin\alpha| = |k_0 ws\sqrt{4-s^2}| \gg 1$, we can use the asymptotic development for the incomplete Bessel and Struve functions, c.f. (10.20) and (10.21) of Chapter II:

$$\begin{aligned} J_1(\alpha, z) &\approx J_1(z) - \frac{\sin\alpha}{\pi}\left[2\sin(z\cos\alpha) - \frac{\cos\alpha\cos(z\cos\alpha)}{z\sin^2\alpha}\right]; \\ H_1(\alpha, z) &\approx N_1(z) + \frac{\sin\alpha}{\pi}\left[2\cos(z\cos\alpha) + \frac{\cos\alpha\sin(z\cos\alpha)}{z\sin^2\alpha}\right]. \end{aligned} \tag{6.19}$$

Substituting these expressions in (6.17) and using the asymptotic expansions for the Bessel and Neumann functions, and also making use of the fact that $\cos\alpha = |s|/2$, we find

$$g \approx \frac{2\sqrt{2}}{\sqrt{\pi z}\,z}\left\{\sin\left[z\left(1 - \frac{|s|}{2}\right) - \frac{\pi}{4}\right] + \sqrt{\frac{2}{\pi z\,(4-s^2)}}\right\}. \tag{6.20}$$

Now, the third of the integrals introduced above (Eq. (6.6)) can immediately be expressed in terms of the incomplete Bessel function $J_0(a, 2K)$, i.e.,

$$q = \int_0^\alpha \cos^2(K\cos\theta)\,d\theta = \int_0^\alpha \frac{1+\cos(2K\cos\theta)}{2}\,d\theta = \frac{\alpha}{2} + \frac{\pi}{4}J_0(\alpha, 2K). \tag{6.21}$$

For $|2K| < 1$, we have

$$q \approx \alpha - \frac{K}{4}(2\alpha + \sin 2\alpha) \tag{6.22}$$

correct to terms of second order in $2K$. If $|2K\sin\alpha| \gg 1$, then from the asymptotic expansion of $J_0(\alpha, 2K)$, (see Eq. (10.23) of Chapter II) we find

$$q \approx \frac{\alpha}{2} + \frac{1}{4}\sqrt{\frac{\pi}{K}}\cos\left(2K - \frac{\pi}{4}\right) - \frac{\sin(2K\cos\alpha)}{4K\sin\alpha}. \tag{6.23}$$

Thus all three of the integrals considered above can be expressed in terms of elementary functions and incomplete Bessel and Struve functions of order zero and one. The tables appended below will permit one to find numerical values for formulae (6.10), (6.17), and (6.21) over a reasonably wide range of their parameters. For very small and very large values of the relevant parameters, estimates of these integrals may be found from (6.12)—(6.14), (6.18)—(6.20), (6.22), and (6.23).

We consider yet another problem, connected with the passage of light through a cylindrical conductor. Such a problem is of interest in the technology of scintillation counters, and also in the study of undistorted transmission of images at a distance by means of light conductors. In [57] conductors with circular bounding surfaces are considered, for the case where a point source radiating isotropically (or with a cosine distribution) is located in the center of one of the fundamental conductors. The light flux incident on a second fundimental conductor is made up of the sum of direct transmission and the results of one, two, etc. reflections from the bounding surface.

The collective characteristics of such a device are described by the transmission coefficient. In [57] it is shown that for an isotropic source and in the absence of absorption within the light-conductors, the computation of the transmission coefficient is connected with the evaluation of the integral

$$S = \int_{\xi}^{\infty} \frac{x e^{-\sigma x}}{(1 + x^2)^{3/2}} dx, \tag{6.24}$$

where $\xi = \sigma R/L$; $\sigma = -L \cdot \ln(r)/2R$, and $r < 1$ is the reflection coefficient of the bounding surface of the conductor, which is of radius R and length L.

This integral cannot be evaluated by means of known and tabulated functions. Therefore, the authors of the reference cited applied an approximate method to its computation, based on the interpolation formula $\sqrt{1 + x^2} \approx 0.5 + 0.9x$ in the region $1 < x < 4$.

We show that the integral S is related to the class of incomplete cylindrical functions. For this we observe that

$$\frac{x}{(1 + x^2)^{3/2}} = -\frac{d}{dx}\left(\frac{1}{\sqrt{1 + x^2}}\right).$$

After integrating by parts we find

$$S = \frac{r^3}{\sqrt{1 + (\sigma R/L)^2}} - \sigma \int_{\xi}^{\infty} \frac{e^{-\sigma x}}{\sqrt{1 + x^2}} dx. \tag{6.25}$$

Introducing a new variable of integration by means of the equation $x = \sinh(u - i\pi/2)$, we obtain

$$\int_{\xi}^{\infty} \frac{e^{-\sigma x}}{\sqrt{1+x^2}}\,dx = \int_{w}^{\infty + i\pi/2} e^{i\sigma \cosh u}\,du = -\frac{\pi i}{2}\mathscr{H}_0^{(1)}(\sigma) + \frac{\pi i}{2}\mathscr{H}_0(w,\sigma), \tag{6.26}$$

where $w = i\pi/2 + \sinh^{-1}\xi = i\pi/2 + \ln(\xi + \sqrt{1+\xi^2})$; $\mathscr{H}_0^{(1)}(\sigma)$ is the Hankel function of first order, and $\mathscr{H}_0(w, z)$ is the incomplete Hankel function. Thus the expression for S takes the form

$$S = \frac{r^3}{\sqrt{1+\xi^2}} + \frac{i\sigma\pi}{2}\left[\mathscr{H}_0^{(1)}(\sigma) - \mathscr{H}_0(w,\sigma)\right]. \tag{6.27}$$

In other words, S can be expressed as a combination of known functions and an incomplete Hankel function.

Equation (6.27) permits one to obtain approximations for S in the limiting cases

$$|\sigma \sinh w| = \sigma\sqrt{1+\xi^2} \ll 1 \quad \text{or} \quad \sigma\sqrt{1+\xi^2} \gg 1.$$

Using the expansion for the Hankel function for $\sigma < 1$, (see (1.7)–(1.12) of Chapter I)

$$\mathscr{H}_0^{(1)}(\sigma) \approx 1 + \frac{2i}{\pi}\left(\ln\frac{\sigma}{2} + c\right), \tag{6.28}$$

where $c = 0.57721\ldots$ is Euler's constant, and also expanding the incomplete Hankel function for $\sigma < 1$ and commensurate w,

$$\mathscr{H}_\nu(w,\sigma) \approx \frac{2}{\pi i} w = 1 + \frac{2}{\pi i}\ln\left(\xi + \sqrt{1+\xi^2}\right), \tag{6.29}$$

(see (7.5) of Chapter II), we find for $\sigma\sqrt{1+\xi^2} \gg 1$

$$S \approx \frac{r^3}{\sqrt{1+\xi^2}} + \sigma\left[\ln\frac{2}{\sigma(\xi + \sqrt{1+\xi^2})} - 0{,}57721\ldots\right]. \tag{6.30}$$

On the other hand, when $\sigma\sqrt{1+\xi^2} \gg 1$, we may use the asymptotic expansions for the incomplete Hankel function from (10.4) and (10.14) of Chapter II, and also the relation $\mathscr{H}_0(w, z) = E_0^+(-iw, z)$, to find

$$\mathscr{H}_0(w,\sigma) \approx \mathscr{H}_0^{(1)}(\sigma) - \frac{2e^{i\sigma\cosh w}}{\pi\sigma \sinh w}\left(1 - i\frac{\cosh w}{\sigma \sinh^2 w}\right). \tag{6.31}$$

Making the substitution $w = i\pi/2 + \sinh^{-1}\xi$, and using (6.27), we find for $\sigma\sqrt{1+\xi^2} \gg 1$,

$$S \approx \frac{r^3\xi}{1+\xi^2}\,\frac{1}{\sigma\sqrt{1+\xi^2}}, \tag{6.32}$$

where σ and ξ are determined from (6.24).

It is also of interest to write Eq. (6.27) in other forms. If we use the relation $\mathscr{H}_0(w, z) = E_0(-iw, z)$ and (4.12) of Chapter II, we find

$$\begin{aligned} S = \frac{r^3}{\sqrt{1+\xi^2}} &- \frac{\sigma\pi}{2} N_0(\sigma)\left[1 - \sqrt{1+\xi^2}\, Je_0(\xi, \sigma)\right] \\ &- \frac{\sigma\pi}{2} J_0(\sigma)\left[\frac{2}{\pi}\ln\left(\xi + \sqrt{1+\xi^2}\right) + \sqrt{1+\xi^2}\, Ne_0(\xi, \sigma)\right], \end{aligned} \tag{6.33}$$

where $Je_0(\xi, \sigma)$ and $Ne_0(\xi, \sigma)$ are the incomplete Lipschitz-Hankel integrals

$$\begin{aligned} Je_0(\xi, \sigma) &= \int_0^\sigma e^{-\xi u} J_0(u)\, du; \\ Ne_0(\xi, \sigma) &= \int_0^\sigma e^{-\xi u} N_0(u)\, du. \end{aligned} \tag{6.34}$$

Available tables of these integrals make it possible to evaluate S in regions of the variable $\sigma\sqrt{1+\xi^2}$ where the above asymptotic expansions are not suitable. The integrals (6.34) are of independent interest, because the incomplete cylindrical functions $E_0^{\pm}(i\beta + \pi/2, x)$ for real β and x can be expressed in terms of them.

In conclusion we note that S may be also expressed in terms of incomplete cylindrical functions of Bessel form. In fact, in taking $x = -\sinh u$ (6.25), we get

$$S = \frac{r^3}{\sqrt{1 + (\sigma R/L)^2}} - \pi i \sigma \Phi_0(-\sinh^{-1}\xi, \sigma), \tag{6.35}$$

where as before $\xi = \sigma R/L = -(\ln r)/2$; $\sigma = -(L \ln r)/2R$ and

$$\Phi_0(\alpha, \sigma) = \frac{1}{\pi i} \int_{-\infty}^{\alpha} e^{\sigma \sinh u}\, du$$

is the incomplete cylindrical function whose properties were discussed in § 6 of Chapter III.

Chapter VIII

Application of Incomplete Cylindrical Functions to some Problems of Solid State Theory and the Motion of Charged Particles in Electromagnetic Fields

In this chapter we consider some problems in solid-state physics and in the theory of motion of charged particles in electromagnetic fields, whose solutions lead to incomplete cylindrical functions.

The set of problems presented below by no means exhausts the extent of application of cylindrical functions in these branches of physics. We have considered only those problems which have been encountered more recently.

1. Computation of Radiation and Absorption Fields in the Theory of Multiphonon Processes

The transition probability for absorption or emission of light quanta is a basic characteristic interaction between various types of radiation and cristal lattices. If there exist quantum transitions for which the lattice energy changes by an amount many times larger than the energy of an oscillation quantum, then such transitions are called "multiphonon". The computation of transition probabilities is a difficult problem which leads to a search for various simplifying assumptions. If, for example, the crystal lattice contains defects (atoms or ions of impuritties, vacant cells, dislocations, etc.), then it is permissible to assume that electrons in the vicinity of a defect are coupled weakly to the electrons of the crystal. This allows one to consider the excitation process as adiabatic when the state of the strongly coupled electrons does not change, and to use perturbation techniques. We will not go into the details of these complex problems, which have been the subject of a large number of theoretical and empirical studies, c.f., for example, Ref. [58] and its bibliography. Here we limit ourselves to certain formulae for the computation of the spectra of radiation, whose analysis is facilitated with the aid of incomplete cylindrical functions.

In the adiabatic approximation and in the first approximation of perturbation theory, the evaluation of the form of the radiation spectrum, considering the actual line widths, leads to consideration of the following expression:

$$J(\Omega) = \frac{2}{\gamma\omega} \operatorname{Re} \int_0^\infty e^{-\gamma t/2\omega + \varphi_{-p}(it)}\, dt, \tag{1.1}$$

where the function $\varphi_{-p}(it)$ has the form

$$\varphi_{-p}(it) = ipt - \frac{\alpha}{2} \cot\frac{\beta}{2} + \frac{\alpha}{2\sinh\frac{\beta}{2}} \cos\left(t - i\frac{\beta}{2}\right). \tag{1.2}$$

if a small dispersive component is neglected. Here the parameters $\varrho\omega = \Omega - \Omega_0$, γ, w, p, α, and β are real quantities, the physical significance of which is explained in [58].

Substituting (1.2) into the integral (1.1), we get

$$J(\Omega) = \frac{2}{\gamma\omega} e^{-\alpha/2 \coth \beta/2} \cdot \operatorname{Re}\left[L_\mu\left(-\frac{\pi + \beta i}{2}, ix\right)\right], \tag{1.3}$$

where

$$\left.\begin{aligned} \mu &= p + i\frac{\gamma}{2\omega}; \\ x &= \frac{\alpha}{2\sinh\frac{\beta}{2}} \end{aligned}\right\} \tag{1.4}$$

and the function $L_\nu(w, z)$ is

$$L_\nu(w, z) = \int_0^\infty e^{i\nu u + iz\sin(w+u)} du. \tag{1.5}$$

We show now that the function $J(\Omega)$, characterizing the form of the absorption lines, may be expressed in terms of incomplete cylindrical functions. We consider in more detail the integral (1.5), whose evaluation we also encountered in the last section, assuming that ν has a positive imaginary part, i.e. $\nu = \alpha + i\beta$ and $\beta > 0$. This guarantees convergence of (1.5) for arbitrary complex values of w and z. Breaking up the interval of integration in (1.5) into segments of length 2π, we can represent $L_\nu(w, z)$ as:

$$L_\nu(w, z) = \sum_{k=0}^\infty \int_{2k\pi}^{2(k+1)\pi} e^{i\nu t + iz\sin(w+u)}\, du = \sum_{k=0}^\infty e^{i2k\pi\nu + i\pi\nu - i\nu w} \int_{-\pi+w}^{w+\pi} e^{i\nu t - iz\sin t}\, dt, \tag{1.6}$$

where the new variable of integration $t = u - 2k\pi - \pi + w$ has been introduced. Because $\operatorname{Im}(\nu) > 0$, the summation in (1.6) is easy to evaluate and after some simplifications we find

$$L_\nu(w, z) = \frac{ie^{-i\nu w}}{2 \sin \nu\pi} \int_{-\pi+w}^{\pi+w} e^{i\nu t - iz\sin t}\, dt$$
$$= \frac{ie^{-i\nu w}}{2 \sin \nu\pi} \left(\int_{-\pi}^{\pi} e^{i\nu t - iz\sin t}\, dt + \int_{\pi}^{\pi+w} e^{i\nu t - iz\sin t}\, dt + \int_{-\pi+w}^{-\pi} e^{i\nu t - iz\sin t}\, dt \right). \tag{1.7}$$

If now we introduce in the second integral the new variable of integration $t - \pi = \theta$, and in the third $\pi + t = \theta$, then we obtain

$$L_\nu(w, z) = i \frac{e^{-i\nu w}}{2 \sin \nu\pi} \left(\int_{-\pi}^{\pi} e^{i\nu t - iz\sin t}\, dt + 2i \sin \nu\pi \int_0^w e^{i\nu\theta + iz\sin\theta}\, d\theta \right). \tag{1.8}$$

But, by the definition of the Anger function, (see Eq. (1.14) of Chapter III), we have

$$\int_{-\pi}^{\pi} e^{i\nu t - iz\sin t}\, dt = 2 \int_0^{\pi} \cos(\nu t - z \sin t)\, dt = 2\pi A_\nu(z). \tag{1.9}$$

On the other hand, the second integral in (1.8) represents the incomplete cylindrical function $\varepsilon_\nu(iw, -z)$, (see § 1 of Chapter III) so that

$$L_\nu(w, z) = i \frac{e^{-i\nu w}\pi}{\sin \nu\pi} A_\nu(z) + \pi e^{-i\nu w\nu} \varepsilon_\nu(iw, -z). \tag{1.10}$$

Substituting (1.10) for $w = -\pi/2 - i\beta/2$ and $z = ix$ into Eq. (1.3), we find

$$J = \frac{2\pi}{\gamma\omega} e^{-\alpha/2 \operatorname{cth} \beta/2} \operatorname{Re} \left\{ e^{\mu(i\pi-\beta)/2} \left[\frac{iA_\mu(ix)}{\sin \mu\pi} + \varepsilon_\mu\left(i\frac{\pi}{2} - \frac{\beta}{2}, -ix\right) \right] \right\}. \tag{1.11}$$

This shows that determination of the absorption spectrum with account taken of the natural line widths is related to the analysis of incomplete cylindrical functions in the Bessel form, with complex index $\mu = p + i\gamma/2\omega$, $w = (i\pi - \beta)/2$ and purely imaginary argument $ix = i\alpha/[2 \sinh(\beta/2)]$.

From (1.11) it is not difficult to obtain expressions for $J(\Omega)$ in limiting cases from expansions of the incomplete cylindrical functions in Bessel form. We recall that the parameter in the expansions of this function was $|z \cosh w|$, which in the present case takes the values:

$$|z \cosh w| = \left| \frac{-i\alpha}{2 \sinh(\beta/2)} \cosh \frac{i\pi - \beta}{2} \right| = \frac{\alpha}{2} \tag{1.12}$$

for $\varepsilon_\mu(i\pi - \beta/2; -ix)$ and

$$|z \cosh w| = \left| \frac{i\alpha}{2 \sinh(\beta/2)} \cosh i\pi \right| = \frac{\alpha}{2 \sinh(\beta/2)} \tag{1.13}$$

for the Anger function $A_\mu(ix)$.

In connection with this we consider the following cases:

Let the parameter $\beta \gg 1$ (this corresponds physically to low temperatures, since $\beta = \pi\omega/kT$, (c.f. Ref. [58])) and let $\alpha/2 \ll 1$ (small departures from the equilibrium state). Under these conditions, both of the expansion parameters (1.12) and (1.13) have values less than unity and an approximate expression for $J(\Omega)$ may be obtained by means (7.2) and (7.4) of Chapter III for the incomplete cylindrical function $\varepsilon_\nu(w, z)$. Limiting ourselves to the first two terms, we have

$$\varepsilon_\mu(w, -ix) \approx \frac{1}{i\pi\mu}(1 - e^{-\mu w}) - \frac{x}{i\pi(1-\mu^2)}[e^{-\mu w}(\cosh w + \mu \sinh w) - 1]; \tag{1.14}$$

$$A_\mu(ix) \approx \frac{\sin \mu\pi}{\pi}\left(\frac{1}{\mu} + \frac{ix}{1-\mu^2}\right). \tag{1.15}$$

Substituting (1.14) and (1.15) into (1.11) and taking into a account the values of the parameters μ, x, and w in (1.13), we find after some simplifications

$$J(\Omega) \approx \frac{e^{-\alpha/2 \coth(\beta/2)}}{\omega^2}\left\{\frac{1}{p^2 + a^2} + \frac{\alpha(1 + p^2 + a^2)\coth\frac{\beta}{2} - 2p}{[(1+p)^2 + a^2][(1-p)^2 + a^2]}\right\}, \tag{1.16}$$

where $wp = \Omega - \Omega_0$ and $\alpha = \gamma/(2\omega)$. For $\alpha \to 0$ the last equation yields the usual resonance formula

$$J \approx \frac{1}{(\Omega - \Omega_0)^2 + \frac{\gamma^2}{4}}. \tag{1.17}$$

Conversely, when $\beta \ll 1$ (high temperatures) so that

$$x = \frac{\alpha}{[2\sinh(\beta/2)]} \gg 1,$$

an approximate expression for J may be obtained from the appropriate asymptotic expansions for the $A_\mu(z)$ and $\varepsilon_\nu(w, z)$. The expansion for the Anger function $A_\mu(ix) = A_\mu(\pi, ix)$ for $x \gg 1$ has the form

$$A_\mu(ix) \sim J_\mu(ix) = i^\mu I_\mu(x) \simeq \frac{i^\mu e^x}{\sqrt{2\pi x}}, \tag{1.18}$$

(c.f. the first equation of the system (7.10) of Chapter III for $\beta = \pi$). The asymptotic expansion for $\varepsilon_\nu(-i\pi/2 - \beta/2, -ix)$ follows from the evaluation of the third equation of the system (7.2) of the same chapter, taking $z = -ix = x \exp(-i\pi/2)$. For the leading term in this expansion we obtain

$$\varepsilon_\mu(i\pi/2 - \beta/2, -ix) \approx \frac{i}{\pi x}\,\frac{e^{x\cosh\beta/2 - \mu(i\pi/2 - \beta/2)}}{\sinh \beta/2}. \tag{1.19}$$

It is evident that because of the multiplier $\exp(i\pi/2 - \beta/2)$ in Eq. (1.11), the term $\varepsilon_\nu(i\pi/2 - \beta/2 - ix)$, although large itself, is negligible for J, because only the real part of the expression in curly brackets is required. Therefore on basis of (1.18) we find from (1.11) the following approximate values for $J(\Omega)$ when $\beta \ll 1$ $(x \gg 1)$:

$$J \approx \frac{2\pi}{\gamma w} e^{-\alpha/2 \coth \beta/2 + x} \frac{1}{\sqrt{2\pi x}} \operatorname{Re}\left(e^{\mu(i\pi - \beta/2)} \frac{i}{\sin \mu\pi}\right). \tag{1.20}$$

Substituting in (1.20) values $\mu = p + i\gamma/2\omega$, $x = \alpha/2 \sinh(\beta/2)$, we find

$$J \approx \frac{1}{\gamma\omega} \sqrt{\frac{\pi \sinh(\beta/2)}{\alpha}} \frac{\cos\left(2p\pi - \frac{\beta\gamma}{4\omega}\right) - e^{-\pi\gamma/\omega} \cos\frac{\beta\gamma}{4\omega}}{\cosh^2\frac{\pi\gamma}{2\omega} - \cos^2 p\pi} e^{-1/2\,[p\beta + \alpha \tanh(\beta/4)]}. \tag{1.21}$$

The formula just obtain clearly illustrates the resonant structure of the absorption lines for integer values of the parameter $p = (\Omega - \Omega_0)/\omega = m$. In these cases, the intensity at the resonant points is approximately equal to

$$J_{\text{res}} \approx \frac{1}{\gamma\omega} \sqrt{\frac{\pi \sinh(\beta/2)}{\alpha}} \cdot \frac{\cos\frac{\beta\gamma}{4\omega}\left(1 - e^{-\pi\gamma/\omega}\right)}{\sinh^2\frac{\pi\gamma}{2\omega}} e^{-1/2\,[m\beta + \alpha \tanh(\beta/4)]} \tag{1.22}$$

or, taking into account that $\beta \ll 1$ and $\gamma/\omega \ll 1$, we find

$$J_{\text{res}} \approx 4 \sqrt{\frac{\beta}{2\pi\alpha}} \frac{e^{-\beta/2\,(m + \alpha/4)}}{\gamma^2}, \quad m = 0, 1, 2, \ldots \tag{1.23}$$

In a similar way one can obtain other limiting cases from Eq. (1.11). For large values of $\mu = p + i(\gamma/2\omega)$, it obviously is necessary to use asymptotic developments of incomplete cylindrical functions for large values of the index (c.f. § 9 of Chapter III).

2. Scattering of Light by Atoms with Interference of Excited States

In a number of experimental studies of the scattering of light by atoms, resonant changes in the intensity of the scattered light depending on a magnetic field have been ascertained. In particular, the resonant scattering of light from cadmium vapor in a weak magnetic field is studied in [59]. This reference describes and interprets theoretically an experiment involving the interference of two excited states arising from modulation of the energy interval between them. The exciting light is a plane polarized wave of constant intensity propagating along the magnetic field. The magnetic field which causes splitting of the excited states is modulated so that the difference of the frequency of these levels changes according to the law

$$\omega_{21}(t) = \omega_H + \omega_1 \sin \Omega t. \tag{2.1}$$

The theory of this phenomenon shows that the intensity of light scattered from an atom having two neighboring excited states is given by the relation [59]

$$I(t) = K[\varrho_{11}|d^1_{\lambda 0}|^2 + \varrho_{22}|d^\lambda_{02}|^2 + 2\mathrm{Re}\,(\varrho_{11}d^\lambda_{10}d^\lambda_{20})], \tag{2.2}$$

where K is a multiplier relating the intensity of dipole radiation to the square of the dipole moment of the source; d^λ is an operator projecting the dipole moment on to a plane perpendicular to the direction of the observer, and ϱ is the density matrix of the atom whose variation with respect to time is determined by the system:

$$\left.\begin{aligned} \frac{d}{dt}\varrho_{11} &= -\gamma\varrho_{11} + F_{11}; \\ \frac{d}{dt}\varrho_{22} &= -\gamma\varrho_{22} + F_{22}; \end{aligned}\right\} \tag{2.3}$$

$$\frac{d}{dt}\varrho_{21} + i\omega_{21}\varrho_{21} = -\gamma\varrho_{21} + F_{21}. \tag{2.4}$$

Here γ is the width of the level, and the function F_{mn}, in the case where the incident wave propagates along the magnetic field and is polarized in the XOZ-plane, has the form

$$F_{mn} = Nh^{-2}d^x_{m0}d^x_{0n} < \varphi_{\omega 0} > I, \tag{2.5}$$

where I is proportional to the intensity of the incident light, $\varphi\omega_0$ is the spectral density for the incident light at the frequency ω_0, and N is the number of scattering atoms.

Because we are considering the case for which the intensity of the incident light does not change with time, the F_{mn} are all constants, and the solution of the system of Eqs. (2.3), (2.4) for the initial conditions $\varrho_{21}(0) = \varrho_1(0) = \varrho_{22}(0) = 0$ takes the form

$$\varrho_{11}(t) = \frac{F_{11}}{\gamma}(1 - \mathrm{e}^{-\gamma t}); \ \varrho_{22} = \frac{F_{22}}{\gamma}(1 - \mathrm{e}^{-\gamma t}); \tag{2.6}$$

$$\varrho_{21}(t) = F_{21}\int_0^t \exp\left\{-\gamma(t-t') - i\int_{t'}^t \omega_{21}(\tau)\,d\tau\right\}dt'. \tag{2.7}$$

Substituting the values of $\omega_{21}(t)$ from formula (2.1) in (2.7), we have

$$\varrho_{21}(t) = F_{21}\int_0^t \mathrm{e}^{-\gamma(t-t') - i\omega_H(t-t') + i(\omega_1/\Omega)(\cos\Omega t - \cos\Omega t')}\,dt'. \tag{2.8}$$

The expression thus obtained is not difficult to represent by means of incomplete cylindrical functions of Bessel form. To this end we introduce the new variable of integration $u = t(\Omega t' + \pi/2)$, and find after some simplifications

$$\varrho_{21}(t) = \frac{F_{21}}{i\Omega}\exp\left[-i\mu(\Omega t + \pi/2) + ix\cos\Omega t\right]\int_{i\pi/2}^{i(\pi/2+\Omega t)} \mathrm{e}^{-x\sinh u + \mu u}\,du. \tag{2.9}$$

Or, by the definition of the incomplete cylindrical function $\varepsilon_\nu(\omega, z)$, we have

$$\varrho_{21}(t) = \frac{\pi F_{21}}{\Omega} e^{-i\mu(\Omega t+\pi/2)+ix\cos\Omega t} [\varepsilon_{-\mu}(i\pi/2 + i\Omega t, -x) - \varepsilon_{-\mu}(i\pi/2, -x)], \tag{2.10}$$

where $\mu = (\omega_H - i\gamma)$ and $x = \omega_1/\Omega$. From this we see that the solution of Eq. (2.4) can be expressed in terms of the incomplete cylindrical function $\varepsilon_\nu(\omega, z)$; consequently its analysis in the general case can be effected by means of the properties of these functions.

In Ref. [59], the solution of the system (2.3), (2.4) was considered in detail for the case $\gamma t \gg 1$, for which it is possible to neglect transition phenomena. In this case (2.6) gives

$$\left.\begin{aligned} \varrho_{11} &= \frac{F_{11}}{\gamma}; \\ \varrho_{22} &= \frac{F_{22}}{\gamma}. \end{aligned}\right\} \tag{2.11}$$

Further, taking $t - t' = T$ in (2.7), we get

$$\varrho_{21}(t) = F_{21} \int_0^t \exp\left\{-(\gamma + i\omega_H)\tau + i\frac{\omega_1}{\Omega}[\cos\Omega t - \cos\Omega(t-\tau)]\right\} d\tau \tag{2.12}$$

and for $\gamma t \gg 1$ the integration may be taken over the interval $[0, \infty]$, so that we can express ϱ_{21} in our notation with a sufficient degree of precision as

$$\varrho_{21}(t) = \frac{F_{21} e^{ix\cos\Omega t}}{\Omega} \int_0^\infty \exp[-i\mu\xi - ix\cos(\Omega t - \xi)]\, d\xi. \tag{2.13}$$

The density $\varrho_{21}(t)$ may also be expressed by incomplete cylindrical functions. For, clearly, the integral in (2.13) is similar to that of (1.5), and consequently

$$\varrho_{21}(t) = \frac{F_{21}}{\Omega} e^{ix\cos\Omega t} L_{-\mu}(-i\pi/2 - i\Omega t, x)$$

Or, taking into account (1.10), we find

$$\begin{aligned} \varrho_{21}(t) &= \frac{\pi F_{21}}{\Omega} \exp[ix\cos\Omega t - i\mu(\pi/2 + \Omega t)] \\ &\times \left[\frac{A_{-\mu}(x)}{i\sin\mu\pi} + \varepsilon_{-\mu}(i\pi/2 + i\Omega t, -x)\right], \end{aligned} \tag{2.14}$$

where $A_{-\mu}(x)$ is the Anger function.

Substituting (2.11) and (2.14) into (2.2) and using (2.5), we find

$$I(t) = \frac{2cI}{\gamma}[1 - \Pi(t)], \tag{2.15}$$

where $c = NK\,|d_{01}|^4 < \varphi_{w0} > (2\hbar)^{-2}$ and $\Pi(t)$ denotes

$$\Pi(t) = \frac{\pi\gamma}{\Omega}\,\mathrm{Re}\Big\{\exp\,[2i\psi + ix\cos\Omega t - i\mu(\pi/2 + \Omega t)] \times \Big[\frac{A_{-\mu}(x)}{i\sin\pi\mu} + \varepsilon_{-\mu}(i\pi/2 + i\Omega t, -x)\Big]\Big\}. \tag{2.16}$$

Here ψ is the angle between the axis and the direction of the observation point.

The formulae obtained permit one to find simple approximate expressions for the intensity of the scattered light in the limiting cases of weak modulation, for which $x = \omega_1/\Omega \ll 1$, and strong modulation, for which $x \gg 1$. If $x \ll 1$, then, using developments (1.14) and (1.15) of the preceeding section, we obtain

$$\begin{aligned}\Pi(t) &= \frac{\gamma}{\Omega}\,\mathrm{Re}\Big\{\frac{e^{2i\psi}}{i\mu}\Big[1 + \frac{x}{1-\mu^2}(i\cos\Omega t + \mu\sin\Omega t)\Big]\Big\} \\ &= \beta\,\frac{\alpha\sin 2\psi + \beta\cos 2\psi}{\alpha^2+\beta^2} + \beta x\,\frac{(1-\alpha^2+\beta^2)\sin 2\psi - 2\alpha\beta\cos 2\psi}{[(1-\alpha)^2+\beta^2]\,[(1+\alpha)^2+\beta^2]}\sin\Omega t \\ &\quad - \beta x\,\frac{\beta(\beta^2-3\alpha^2+1)\sin 2\psi + \alpha(\alpha^2-3\beta^2-1)\cos 2\psi}{(\alpha^2+\beta^2)\,[(1-\alpha)^2+\beta^2]\,[(1+\alpha)^2+\beta^2]}\cos\Omega t,\end{aligned} \tag{2.17}$$

where we have introduced the notation $\beta = \gamma/\Omega$, $\alpha = \omega_H/\Omega$, $x = \omega_1/\Omega$.

For the analysis of more general cases it will prove convenient to transform Eq. (2.16). First of all, we recall that this formula was derived under the assumption $\gamma t \gg 1$. Usually one is interested in cases where the frequency of the modulation Ω is much larger than the width of the excited level γ, i.e. $\Omega \gg \gamma$, see [59]. Therefore in subsequent operations we may also consider $\Omega t \gg 1$, so that the quantity $\pi/2 + \Omega t$ arising in (2.16) may be taken as

$$\pi/2 + \Omega t = 2m\pi + \Delta, \tag{2.18}$$

where $m \gg 1$ and $|\Delta| \leq \pi$. In this case we can use the transformations given in § 7 of Chapter III for the derivation of formula (7.16) to evaluate $\varepsilon_\mu(i\pi/2 + i\Omega t, x)$. Then we have

$$\varepsilon_{-\mu}(2im\pi + i\Delta, -x) = \varepsilon_{-\mu}(i\Delta, -x) + \frac{\sin m\mu\pi}{\sin\mu\pi}\,\frac{1}{\pi i}\int_{-\pi i}^{\pi i} e^{-x\sinh u + \mu u}\,du.$$

Or, recalling the definition of the Anger function, we find

$$\varepsilon_{-\mu}(2im\pi + i\Delta, -x) = \varepsilon_{-\mu}(i\Delta, -x) + \frac{2e^{i\mu m\pi}\sin\mu m\pi}{\sin\mu\pi}\,A_{-\mu}(x). \tag{2.19}$$

Substituting this expressing in (2.16) and also taking $\pi/2 + \Omega t = 2m\pi + \Delta$, we obtain after some simplification

$$\begin{aligned}\Pi(t) = \frac{\pi\gamma}{\Omega} \operatorname{Re}\Big\{\exp\,(2i\psi + ix\sin\Delta - i\mu\Delta)\\ \times\Big[\frac{A_{-\mu}(x)}{i\sin\mu\pi} + e^{-2i\mu m\pi}\,\varepsilon_{-\mu}(i\Delta, -x)\Big]\Big\}.\end{aligned} \tag{2.20}$$

But for $\gamma t \sim 2m\pi\gamma/\Omega \gg 1$ the second term in the square brackets may be neglected, since $|\exp(2i\mu m\pi) \sim \exp(-2m\pi\gamma/\Omega) \ll 1$, and the function $\varepsilon_{-\mu}(i\Delta, -x)$ is bounded, i.e.

$$\begin{aligned}|\varepsilon_{-\mu}(i\Delta, -x)| &= \frac{1}{\pi}\left|\int\limits_0^\Delta e^{-ix\sin u + i\omega_H/\Omega + (\gamma/\Omega)u}\,du\right|\\ &< \frac{1}{\pi}\int\limits_0^\Delta e^{(\gamma/\Omega)u}du = \frac{1}{\pi}\frac{\Omega}{\gamma}(e^{\gamma\Delta/\Omega} - 1) \leq 1\end{aligned}$$

for $\gamma/\Omega \ll 1$ and $|\Delta| \leq \pi$. A sufficiently accurate expression for $\Pi(t)$ is thus

$$\Pi(t) = \frac{\pi\gamma}{\Omega}\operatorname{Re}\left[e^{i\varphi}\frac{A_{-\mu}(x)}{i\sin\mu\pi}\right], \tag{2.21}$$

where

$$\begin{aligned}\varphi &= 2\psi + x\cos\Omega t - \alpha(\Omega t + \pi/2 - 2m\pi) + i\beta\left(\Omega t + \frac{\pi}{2} - 2m\pi\right)\\ &= 2\psi + x\cos\Omega t - \alpha\left(\Omega t + \frac{\pi}{2} - 2m\pi\right),\end{aligned} \tag{2.22}$$

since for our m, $|\Omega t + \pi/2 - 2m\pi| < \pi$, and $\beta = \gamma/\Omega \ll 1$.

From the formula just obtained it is clear that the dependence of $\Pi(t)$ on time is contained only in the phase φ. This phase is periodic in t, with period $T = 2\pi/\Omega$, since by the structure of the given equations m must be replaced by $m + 1$ when Ωt is increased by 2π, i.e.

$$\begin{aligned}\varphi\left(t + \frac{2\pi}{\Omega}\right) &= 2\psi + x\cos\,(\Omega t + 2\pi)\\ &- \alpha\left[\Omega t + 2\pi + \frac{\pi}{2} - 2(m+1)\pi\right] \equiv \varphi(t).\end{aligned}$$

Thus under the assumptions $\gamma t \gg 1$, $\Omega \gg \gamma$, the function $\Pi(t)$ is itself periodic with period $T = 2\pi/\Omega$. For large modulation, $x = \omega_1/\Omega \gg 1$, and according to the first of Eqs. (7.10) of Chapter III, we have for $\beta = \pi$

$$\begin{aligned}A_{-\mu}(x) \sim J_{-\mu}(x) &\approx \sqrt{\frac{2}{\pi x}}\cdot\cos\left(x + \mu\frac{\pi}{2} - \frac{\pi}{4}\right)\\ &= \sqrt{\frac{2}{\pi x}}\cos\left(x + \alpha\frac{\pi}{2} - \frac{\pi}{4}\right).\end{aligned} \tag{2.23}$$

Here $\mu = \alpha - i\beta$ and the quantity $i\beta\pi/2$ is small, as in (2.22), i.e. $\beta = \gamma/\Omega \ll 1$.

Substituting (2.23) into (2.21), we get

$$\Pi(t) = \sqrt{\frac{2\pi}{x}}\,\beta \cos\left(x + \alpha\frac{\pi}{2} - \frac{\pi}{4}\right) \frac{\sin \pi\alpha \sin \varphi \cosh \beta\pi + \cos \alpha\pi \cos\varphi \cos \beta\pi}{\cosh^2 \beta\pi - \cos^2 \alpha\pi}, \tag{2.24}$$

where $\alpha = a_H/\Omega$, $\beta = \gamma/\Omega$ and $x = \omega_1/\Omega \gg 1$.

Thus, using the properties of incomplete cylindrical functions, we have obtained sufficiently simple formulae — (2.17), (2.21), and (2.24) — for determining the intensity of scattered light. In particular, the latter formula clearly illustrates the resonance phenomenon when ω_H is equal to an integral multiple of the modulating frequency Ω, i.e. when $\omega_H/\Omega = k$, $k = 0, 1, 2, \ldots$ For these values of α, the denominator in (2.24) takes the form $\cosh^2(\beta\pi) - 1 = \sinh^2(\beta\pi)$, which is very small for $\beta = \gamma/\Omega \ll 1$, and so the function $\Pi(t)$ at resonance turns out to be

$$\Pi(t) \approx \sqrt{\frac{2\pi}{x}}\,\beta \cos\left(x - \frac{\pi}{4} + k\frac{\pi}{2}\right) \cos k\pi \frac{\cos\varphi}{\sinh \beta\pi}. \tag{2.25}$$

3. Motion of Charged Particles in Constant Electric and Time-Varying Magnetic Fields[1]

We consider here the motion of charged particles in crossed electric and magnetic fields. The electric field is assumed constant and the magnetic field either increases monotonically, or changes periodically with time.

Let a charged particle with charge Ze and mass M move in an electromagnetic field $(\boldsymbol{E}, \boldsymbol{H})$. We assume that the electric field $\boldsymbol{E}$ is perpendicular to the magnetic field $\boldsymbol{H}$, that the electric field is $\boldsymbol{E} = \boldsymbol{E}_0 = \text{const.}$, and that $\boldsymbol{H} = \boldsymbol{H}_1 \sin(\Omega t + \alpha)$, where α is the initial phase.

The equations of motion of a charged particle in the field $(\boldsymbol{E}, \boldsymbol{H})$ in the nonrelativistic case have the form

$$M\frac{d\boldsymbol{v}}{dt} = Ze\left[\boldsymbol{E} + \frac{1}{c}[\boldsymbol{v}\,\boldsymbol{H}]\right], \tag{3.1}$$

where $\boldsymbol{v}$ is the velocity vector of the particle and c is the speed of light. If we introduce a rectangular Cartesian coordinate system with its x-axis in the direction of $\boldsymbol{E}$ and its z-axis in the direction of $\boldsymbol{H}$, we have

$$\left.\begin{aligned} &E_x = E_0;\; E_y = E_z = 0;\\ &H_z = H_1 \sin(\Omega t + \alpha);\; H_y = H_x = 0. \end{aligned}\right\} \tag{3.2}$$

[1] The results of this section were obtained in association with N. I. Leont'ev and N. V. Khaykhyan.

The equations of motion may then be written in the form

$$\left.\begin{aligned} \ddot{x} &= \frac{ZeE_0}{M} + \frac{ZeH_1}{Mc}\,\dot{y}\sin(\Omega t + \alpha);\\ \ddot{y} &= -\frac{ZeH_1}{Mc}\,\dot{x}\sin(\Omega t + \alpha);\\ \ddot{z} &= 0. \end{aligned}\right\} \tag{3.3}$$

The z-component is independently determined by the third equation of this system, and so in the following we shall study only the components of the motion in the (x, y) plane. For concreteness we shall study the solution of the system (3.3) which satisfies the initial conditions

$$\begin{aligned} x(0) &= \dot{x}(0) = 0;\\ y(0) &= \dot{y}(0) = 0, \end{aligned} \tag{3.4}$$

so that the initial phase in Eqs. (3.2), (3.3) characterizes the magnetic field strength at the instant $t = 0$.

Introducing the non-dimensional quantities

$$\begin{aligned} \varrho &= \frac{\omega_1}{\Omega}; & \Omega t &= \theta;\\ X &= \frac{x}{d}; & d &= \frac{ZeE_0}{M\omega_1^2};\\ Y &= \frac{y}{d}; & \omega_1 &= \frac{ZeH_1}{Mc} \end{aligned} \tag{3.5}$$

and the notations

$$\begin{aligned} U &= \frac{dX}{d\theta};\\ V &= \frac{dY}{d\theta};\\ \xi &= \cos(\theta + \alpha);\\ \xi_0 &= \cos\alpha, \end{aligned} \tag{3.6}$$

we write the system (3.3) in the form

$$\frac{dU}{d\xi} + \varrho V = -\frac{\varrho^2}{\sqrt{1-\xi^2}}; \quad \frac{dV}{d\xi} - \varrho U = 0. \tag{3.7}$$

Thus the problem has been reduced to finding solutions $U(z)$, $V(\xi)$, satisfying the initial conditions $U(\xi_0) = 0$ and $V(\xi_0) = 0$. It is not difficult to verify that these solutions are

$$\left.\begin{aligned} U &= -\varrho^2 \int_{\xi_0}^{\xi} \frac{\cos\varrho(\xi - t)}{\sqrt{1-t^2}}\,dt;\\ V &= -\varrho^2 \int_{\xi_0}^{\xi} \frac{\sin\varrho(\xi - t)}{\sqrt{1-t^2}}\,dt. \end{aligned}\right\} \tag{3.8}$$

Or, restoring the parameter θ, we obtain

$$\left.\begin{aligned} U &= \varrho^2 \int_{\alpha}^{\alpha+\theta} \cos \{\varrho [\cos (\theta + \alpha) - \cos \varphi]\} \, d\varphi; \\ V &= \varrho^2 \int_{\alpha}^{\alpha+\theta} \sin \{\varrho [\cos (\theta + \alpha) - \cos \varphi]\} \, d\varphi. \end{aligned}\right\} \tag{3.9}$$

We now show that the velocity components U and V can be expressed in terms of the incomplete Bessel function $J_0(w, \varrho)$ and the incomplete Struve function $H_0(w, \varrho)$. In fact, for the component V we have

$$\begin{aligned} V = {} & \varrho^2 \sin [\varrho \cos (\theta + \alpha)] \int_{\alpha}^{\alpha+\theta} \cos (\varrho \cos \varphi) \, d\varphi \\ & - \varrho^2 \cos [\varrho \cos (\theta + \alpha)] \int_{\alpha}^{\alpha+\theta} \sin (\varrho \cos \varphi) \, d\varphi \end{aligned}$$

or, by the definitions of the incomplete cylindrical functions, (see § 1 of Chapter II)

$$\begin{aligned} V = \frac{\pi\varrho^2}{2} \{ & [H_0(\alpha, \varrho) - H_0(\alpha + \theta, \varrho)] \cos [\varrho \cos (\theta + \alpha)] \\ & - [J_0(\alpha, \varrho) - J_0(\alpha + \theta, \varrho)] \sin [\varrho \cos (\theta + \alpha)]\}. \end{aligned} \tag{3.10}$$

Similarly we find for the component U:

$$\begin{aligned} U = \frac{\pi\varrho^2}{2} \{ & [H_0(\alpha + \theta, \varrho) - H_0(\alpha, \varrho)] \sin [\varrho \cos (\theta + \alpha)] \\ & + [J_0(\alpha + \theta, \varrho) - J_0(\alpha, \varrho)] \cos [\varrho \cos (\theta + \alpha)]\}. \end{aligned} \tag{3.11}$$

From (3.10) and (3.11) it is not difficult to obtain the following expression for the non-dimensional energy of the particle:

$$\begin{aligned} \frac{U^2 + V^2}{2} = \frac{\pi^2\varrho^4}{8} \{ & [J_0(\alpha + \theta, \varrho) - J_0(\alpha, \varrho)]^2 \\ & + [H_0(\alpha + \theta, \varrho) - H_0(\alpha, \varrho)]^2\}. \end{aligned} \tag{3.12}$$

In order to determine the position of the moving particle we substitute (3.11) and (3.12) into the system of (3.6) and carry out the indicated integrations. For the coordinate X we have

$$\begin{aligned} X = \int_0^{\theta} U \, d\varphi = \frac{\pi\varrho^2}{2} \Bigg\{ & \int_0^{\theta} [J_0(\alpha + \varphi, \varrho) - J_0(\alpha, \varrho)] \cos [\varrho \cos (\varphi + \alpha)] \, d\varphi \\ & + \int_0^{\theta} [H_0(\alpha + \varphi, \varrho) - H_0(\alpha, \varrho)] \sin [\varrho \cos (\alpha + \varphi)] \, d\varphi \Bigg\}. \end{aligned} \tag{3.13}$$

If we use the fact that

$$\frac{d}{d\psi} J_0(\psi + \alpha, \varrho) = \frac{2}{\pi} \cos [\varrho \cos (\psi + \alpha)],$$

then an integration by parts gives

$$\int_0^\theta J_0(\alpha + \varphi, \varrho) \cos [\varrho \cos (\alpha + \varphi)]\, d\varphi$$

$$= J_0(\alpha + \varphi, \varrho) \int_0^\varphi \cos [\varrho \cos (\psi + \alpha)]\, d\psi \,\Big|_{\varphi=0}^{\varphi=\theta}$$

$$- \int_0^\theta \cos [\varrho \cos (\psi + \alpha)]\, [J_0(\psi + \alpha, \varrho) - J_0(\alpha, \varrho)]\, d\psi,$$

from which follows

$$\begin{aligned}&\int_0^\theta J_0(\alpha + \varphi, \varrho) \cos [\varrho \cos (\alpha + \varphi)]\, d\varphi \\ &= \frac{\pi}{4} [J_0(\alpha + \theta, \varrho) + J_0(\alpha, \varrho)]\, [J_0(\alpha + \theta, \varrho) - J_0(\alpha, \varrho)].\end{aligned} \tag{3.14}$$

Similarly we obtain

$$\begin{aligned}&\int_0^\theta H_0(\alpha + \varphi, \varrho) \sin [\varrho \cos (\alpha + \varphi)]\, d\varphi \\ &= \frac{\pi}{4} [H_0(\alpha + \theta, \varrho) + H_0(\alpha, \varrho)]\, [H_0(\alpha + \theta, \varrho) - H_0(\alpha, \varrho)].\end{aligned} \tag{3.15}$$

Substituting (3.14) and (3.15) into (3.13), we find after some simplifications the following expression for X:

$$X = \frac{\pi^2 \varrho^2}{8} \{[J_0(\alpha + \theta, \varrho) - J_0(\alpha, \varrho)]^2 + [H_0(\alpha + \theta, \varrho) - H_0(\alpha, \varrho)]^2\}. \tag{3.16}$$

For the Y, we have by expression (3.6)

$$\begin{aligned}Y = \int_0^\theta V\, d\varphi &= \frac{-\pi\varrho^2}{2} \int_0^\theta [H_0(\alpha + \varphi, \varrho) - H_0(\alpha, \varrho)] \cos [\varrho \cos (\alpha + \varphi)]\, d\varphi \\ &+ \frac{\pi\varrho^2}{2} \int_0^\theta [J_0(\alpha + \varphi, \varrho) - J_0(\alpha, \varrho)] \sin [\varrho \cos (\alpha + \varphi)]\, d\varphi,\end{aligned} \tag{3.17}$$

and integration by parts gives

$$\int_0^\theta J_0(\alpha + \varphi, \varrho) \sin [\varrho \cos (\alpha + \varphi)]\, d\varphi$$

$$= J_0(\alpha + \varphi, \varrho) \int_0^\varphi \sin [\varrho \cos (\alpha + \psi)]\, d\psi \,\Big|_{\varphi=0}^{\varphi=\theta}$$

$$- \int_0^\theta \cos [\varrho \cos (\alpha + \varphi)]\, [H_0(\alpha + \varphi, \varrho) - H_0(a, \varrho)]\, d\varphi$$

$$= \frac{\pi}{2} [J_0(\alpha + \theta, \varrho)\, H_0(\alpha + \theta, \varrho)]$$

$$- J_0(\alpha, \varrho)\, H_0(\alpha, \varrho)] - \int_0^\theta H_0(\alpha + \varphi, \varrho) \cos [\varrho \cos (\alpha + \varphi)]\, d\varphi.$$

We find therefore from (3.17)

$$Y = \frac{\pi^2 \varrho^2}{4} \{ [J_0(\alpha + \theta, \varrho) - J_0(\alpha, \varrho)] [H_0(\alpha + \theta, \varrho) + H_0(\alpha, \varrho)] - 2 [D(\alpha + \theta, \varrho) - D(\alpha, \varrho)] \}, \tag{3.18}$$

where we have introduced the notation

$$D(\psi, \varrho) = \frac{2}{\pi} \int_0^{\psi} H_0(t, \varrho) \cos [\varrho \cos t] \, dt = \int_0^{\psi} \frac{d J_0(t, \varrho)}{dt} H_0(t, \varrho) \, dt. \tag{3.19}$$

Eqs. (3.10)—(3.12), (3.16) and (3.18) completely describe the motion of a charged particle in crossed electric and magnetic fields, where the electric field is constant, and the magnetic field is a periodic function of time. By means of tables of the incomplete Bessel and Struve functions, numerical values of the particle's velocity components, position vector, and kinetic energy field can be readily determined for any specific case.

In addition, an analysis of the resulting motion in the limiting cases $\varrho = \omega_1/\Omega \ll 1$ and $\varrho \gg 1$ may be carried out with the help of the appropriate expansions of the incomplete Bessel and Struve functions, as given in sections 7 and 10 of Chapter II. This analysis can be effected without difficulty for all θ and α. In connection with this we note that if $\alpha + \theta > \pi$ it is necessary to use the duplication formula given in § 3 of the same chapter,

$$\begin{aligned} J_0(\psi, \varrho) &= \frac{1}{2} [E_0^+(\psi, \varrho) + E_0^-(\psi, \varrho)] \\ &= 2k J_0(\varrho) + \frac{1}{2} [E_0^+(\psi - k\pi, \varrho) + E_0^-(\psi - k\pi, \varrho)] \\ &= 2k J_0(\varrho) + J_0(\psi - k\pi, \varrho); \end{aligned} \tag{3.20}$$

$$\begin{aligned} H_0(\psi, \varrho) &= \frac{1}{2} [E_0^+(\psi, \varrho) - E_0^-(\psi, \varrho)] \\ &= (-1)^k [E_0^+(\psi - k\pi, \varrho) - E_0^-(\psi - k\pi, \varrho)] \\ &= (-1)^k H_0(\psi - k\pi, \varrho), \end{aligned} \tag{3.21}$$

where $0 < \psi - k\pi < \pi$; $k = 1, 2, 3, \ldots$

From these formulae it is clear that $J_0(\psi, \varrho)$ increases with increasing ψ. The incomplete Struve function, however, is bounded and has period 2π. Thus the function $D(\psi, \varrho)$ introduced above is also bounded. Therefore as time goes an, X increases, and Y oscillates periodically with increasing amplitude. For sufficiently large times $\theta = \Omega t \approx k\pi$

($k \gg 1$), the coordinates X and Y may be written approximately as

$$\left.\begin{aligned} X &\approx \frac{[\pi\varrho K J_0(\varrho)]^2}{2}; \\ Y &\approx \frac{(\pi\varrho)^2}{2} k J_0(\varrho)\,[(-1)^k H_0(\alpha+\theta-k\pi,\varrho) + H_0(\alpha,\varrho)]. \end{aligned}\right\} \tag{3.22}$$

Here, as before $J_0(\varrho)$ is the Bessel function and k is determined by the condition $0 < \Omega t + \alpha - k\pi < \pi$.

These equations also permit one to estimate the dependence of X and Y on the parameter $\varrho = \omega_1/\Omega$ for sufficiently large times $\Omega t \gg 1$.

So far we have considered the problem of the motion of a charged particle moving with in a constant electric field and a periodic magnetic field. If the magnetic field $\boldsymbol{H}$ is increasing with time — for example, according to the law

$$H = H_1 \sinh \gamma t, \tag{3.23}$$

then the resulting motion may also be expressed in terms of incomplete cylindrical functions of purely imaginary argument $w = i\beta$, i.e. by incomplete Hankel functions.

The equations of motion in this case are

$$\frac{dU}{d\xi} = \frac{1}{\sqrt{1-\xi^2}} + \varrho V; \quad \frac{dV}{d\xi} = -\varrho U, \tag{3.24}$$

where

$$\varrho = \frac{\omega_0}{\gamma}; \quad \xi = \cosh \gamma t = \cosh \theta. \tag{3.25}$$

The solution of the system (3.24) which satisfies the initial conditions

$$V(1) = U(1) = 0, \tag{3.26}$$

is readily found, and has the form

$$\left.\begin{aligned} U &= \varrho^2 \int_1^{\xi} \frac{\cos \varrho(\xi - t)}{\sqrt{t^2-1}}\, dt; \\ V &= -\varrho^2 \int_1^{\xi} \frac{\sin \varrho(\xi - t)}{\sqrt{t^2-1}}\, dt, \end{aligned}\right\} \tag{3.27}$$

or, re-introducing the parameter θ ($\xi = \cosh\theta$, $t = \cosh\varphi$), ee have

$$\begin{aligned} U &= \varrho^2 \int_0^{\theta} \cos[\varrho(\cosh\theta - \cosh\varphi)]\, d\varphi \\ &= \varrho^2 \cos(\varrho\cosh\theta) \int_0^{\theta} \cos(\varrho\cosh\varphi)\, d\varphi \\ &+ \varrho^2 \sin(\varrho\cosh\theta) \int_0^{\theta} \sin(\varrho\cosh\varphi)\, d\varphi; \end{aligned} \tag{3.28}$$

$$V = -\varrho^2 \int_0^\theta \sin\left[\varrho\left(\cosh\theta - \cosh\varphi\right)\right] d\varphi$$

$$= \varrho^2 \cos\left(\varrho\cosh\theta\right) \int_0^\theta \sin\left(\varrho\cosh\varphi\right) d\varphi \tag{3.29}$$

$$- \varrho^2 \sin\left(\varrho\cosh\theta\right) \int_0^\theta \cos\left(\varrho\cosh\varphi\right) d\varphi .$$

Recalling definition (1.27) of Chapter II of the incomplete Hankel function $\mathscr{H}_0(\beta, x)$, we obtain finally

$$\left.\begin{aligned} U &= \frac{\pi}{2}\varrho^2 \{-\mathrm{Re}\,[\mathscr{H}_0(0, \varrho)] \cos\left(\varrho\cosh\theta\right) \\ &\quad + \mathrm{Im}\,[\mathscr{H}_0(\theta, \varrho)] \sin\left(\varrho\cosh\theta\right)\}; \\ V &= \frac{\pi\varrho^2}{2}\{\mathrm{Re}\,[\mathscr{H}_0(\theta, \varrho)] \sin\left(\varrho\cosh\theta\right) \\ &\quad + \mathrm{Im}\,[\mathscr{H}_0(\theta, \varrho)] \cos\left(\varrho\cosh\theta\right)\}, \end{aligned}\right\} \tag{3.30}$$

where the symbols Re and Im denote real and imaginary part as usual.

The relation between the incomplete Hankel functions and certain tabulated functions given in the first section of the previous chapter allows a numerical description of the charged particle's motion. Moreover, for $\gamma t \gg 1$ or $\varrho \cosh \gamma t \gg 1$ the motion may be approximated with the aid of the asymptotic developments given in § 10 of Chapter II.

We have seen that the incomplete Bessel, Struve, and Hankel functions of zero order aid in the description of a charged particle's motion in the presence of constant electric and periodic magnetic fields. Conversely, these functions may be expressed in terms of the trajectory of the charged particle, — for example, by the quantities $\dot{X}$ and X or by the velocity components $\dot{X}$ and $\dot{Y}$. The quantities X, $\dot{X}$ and $\dot{Y}$ are elements of a physical process which can be observed, and thus automatically and continuously provide numerical values and graphical representations of the incomplete cylindrical functions under consideration.

4. Motion of Charged Particles in Changing Electric and Magnetic Fields

In the preceeding section we considered the simplified problem of determining the motion of a charged particle in a constant electric field and a timevarying magnetic field. It is often necessary to study the motion of charged particles in an electromagnetic field ($\boldsymbol{E}, \boldsymbol{H}$) when both $\boldsymbol{E}$ and $\boldsymbol{H}$ are functions of time. If the motion takes place in a medium, then it is also necessary to consider the possibility of drag on the particle, to account for its collisions with other particles of the medium. The average loss of impulse caused by such collisions is characterized by the quantity Mv/τ_{ef}, where τ_{ef} is the mean collision time, which in the following we shall assume to be independent of the velocity. The equations of motion of a charged particle subject to drag can be written as (c.f., e.g. Ref. [60])

$$M\left(\frac{d\boldsymbol{v}}{dt} + \gamma \boldsymbol{v}\right) = Ze\left(\boldsymbol{E} + \frac{[\boldsymbol{vH}]}{c}\right), \tag{4.1}$$

where $\gamma = 1/\tau_{ef}$ is the mean collision frequency. We shall assume that the fields $\boldsymbol{E}$ and $\boldsymbol{H}$ are homogeneous in space and that their dependence on time is determined by:

$$\boldsymbol{H} = \boldsymbol{H}_0\left[1 + \frac{H_1}{H_0}\cos(\Omega t + \varphi)\right]; \tag{4.2}$$

$$\boldsymbol{E} = \boldsymbol{E}_0 + \boldsymbol{E}_1 \cos(\omega t + \psi), \tag{4.3}$$

where the phases φ and ψ characterize their respective fields for $t = 0$. Generally speaking, because the fields $\boldsymbol{H}$ and $\boldsymbol{E}$ are changing periodically with time, we should take account of the induced fields in (4.1). But for velocities small in comparison with the velocity of light, the induced magnetic field may be neglected. As for the induced electric field, it can be neglected because of the linearity of Eqs. (4.1). Therefore, in the following we shall study the changing velocity of the charged particle due only to the external fields defined in expressions (4.2), (4.3).

We introduce a rectangular Cartesian system of coordinates with z-axis parallel to $\boldsymbol{H}_0$. In this system Eqs. (4.1) take the form

$$\left.\begin{aligned} \frac{dv_x}{d\theta} + \beta v_x &= a_x + b_x \cos(\Delta\theta + \psi) + [\alpha + \beta\cos(\theta + \varphi)]\, v_y; \\ \frac{dv_y}{d\theta} + \beta v_y &= a_y + b_y \cos(\Delta\theta + \psi) - [\alpha + \beta\cos(\theta + \varphi)]\, v_x: \end{aligned}\right\} \tag{4.4}$$

$$\frac{dv_z}{d\theta} + \beta v_z = a_z + b_z \cos(\Delta\theta + \psi), \tag{4.5}$$

where we have introduced the notations:

$$\left.\begin{aligned} &\theta = \Omega t; \quad \alpha = \frac{\omega_H}{\Omega}; \quad \beta = \frac{\gamma}{\Omega}; \quad \varrho = \frac{\omega_1}{\Omega}; \\ &\Delta = \frac{\omega}{\Omega}; \quad \omega_H = Ze\frac{H_0}{Mc}; \quad \omega_1 = Ze\frac{H_1}{Mc}; \\ &\boldsymbol{a} = \frac{Ze}{M\Omega}\boldsymbol{E}_0; \quad \boldsymbol{b} = \frac{Ze}{M\Omega}\boldsymbol{E}_1. \end{aligned}\right\} \tag{4.6}$$

We limit ourselves to investigation of the velocity components only, but for completeness of the picture we shall take arbitrary initial velocities, i.e.

$$\boldsymbol{v}_{t=0} = \boldsymbol{v}_0 = (v_{0x}, v_{0y}, v_{0z}). \tag{4.7}$$

Equation (4.5) for the component v_z can be integrated independently, and gives

$$v_z = v_{0z}\,e^{-\beta\theta} + \frac{a_z}{\beta}(1 - e^{-\beta\theta})$$

$$+ b_z \frac{\beta\cos(\Delta\theta + \psi) + \Delta\sin(\Delta\theta + \psi) - (\beta\cos\psi + \Delta\sin\psi)\,e^{-\beta\theta}}{\beta^2 + \Delta^2}. \quad (4.8)$$

In the stationary regime, i.e. for sufficiently large $\beta\theta = \gamma t \gg 1$, this formula takes the form

$$v_z = \frac{a_z}{\beta} + b_z \frac{\beta\cos(\Delta\theta + \psi) + \Delta\sin(\Delta\theta + \psi)}{\beta^2 + \Delta^2}. \quad (4.9)$$

In this case, as we should expect, v_z does not depend on its initial value v_{0z}.

We now consider the determination of the velocity components in the plane perpendicular to the magnetic field. Equations (4.4) form a linear system of inhomogeneous equations with variable coefficients. A fundamental system of solutions for the associated homogeneous equations can be readily verified as

$$e^{-\beta\theta}\begin{Bmatrix}\sin[\alpha\theta + \varrho\sin(\theta + \varphi) - \varrho\sin\varphi]\\ \cos[\alpha\theta + \varrho\sin(\theta + \varphi) - \varrho\sin\varphi]\end{Bmatrix} \quad (4.10)$$

for component v_x and

$$e^{-\beta\theta}\begin{Bmatrix}\cos[\alpha\theta + \varrho\sin(\theta + \varphi) - \varrho\sin\varphi]\\ -\sin[\alpha\theta + \varrho\sin(\theta + \varphi) - \varrho\sin\varphi]\end{Bmatrix}. \quad (4.11)$$

for component v_y

The standard method of variation of parameters permits one to find a particular solution of the inhomogeneous system (4.4), so that the desired solution can be written

$$\left.\begin{aligned}
v_x &= e^{-\beta\theta}\{v_{0y}\sin\Phi + v_{0x}\cos\Phi\}\\
&\quad + a_\perp e^{-\beta\theta}\int_0^\theta e^{\beta u}\cos[\alpha(u - \theta) + \psi_0 + \varrho\sin(u + \varphi)\\
&\quad - \varrho\sin(\theta + \varphi)]\,du + b_\perp e^{-\beta\theta}\int_0^\theta e^{\beta u}\cos(\Delta u + \psi)\\
&\quad \times\cos[\alpha(u - \theta) + \psi_1 + \varrho\sin(u + \varphi) - \varrho\sin(\theta + \varphi)]\,du;\\
v_y &= e^{-\beta\theta}\{v_{0y}\cos\Phi - v_{0x}\sin\Phi\}\\
&\quad + a_\perp e^{-\beta\theta}\int_0^\theta e^{\beta u}\sin[\alpha(u - \theta) + \psi_0 + \beta\sin(u + \varphi)\\
&\quad - \varrho\sin(\theta + \varphi)]\,du + b_\perp e^{-\beta\theta}\int_0^\theta e^{\beta u}\cos(\Delta u + \psi)\\
&\quad \times\sin[\alpha(u - \theta) + \psi_1 + \varrho\sin(u + \varphi) - \varrho\sin(\theta + \varphi)]\,du,
\end{aligned}\right\}\quad (4.12)$$

where $\Phi = \alpha\theta + \varrho \sin(\theta + \varphi) - \varrho \sin\varphi$, and $a_\perp$ and $b_\perp$ are, respectively, the absolute values of the projections of the vectors $\boldsymbol{a}$ and $\boldsymbol{b}$ on the (x, y) plane, i.e.

$$\left.\begin{aligned} a_\perp &= \frac{Ze}{M\Omega}\sqrt{E_{0x}^2 + E_{0y}^2}\,; \quad \tan\psi_0 = \frac{E_{0y}}{E_{0x}}\,; \\ b_\perp &= \frac{Ze}{M\Omega}\sqrt{E_{1x}^2 + E_{1y}^2}\,; \quad \tan\psi_1 = \frac{E_{1y}}{E_{1x}}\,. \end{aligned}\right\} \tag{4.13}$$

Without loss of generality we may take one of the angles ψ, for instance ψ_0, equal to zero. This corresponds to choosing the direction of the axis z along the projection of the vector $\boldsymbol{E}_0$ on the (x, y) plane.

We now show that the velocity components v_x and v_y may be expressed in terms of incomplete cylindrical functions of the Bessel form. To this end we write Eqs. (4.12) as:

$$\left.\begin{aligned} v_x &= e^{-\beta\theta}(v_{0y}\sin\Phi + v_{0x}\cos\Phi) + a_\perp \operatorname{Re}[Q(\theta; 0, 0, 0)] \\ &\quad + \frac{1}{2} b_\perp \operatorname{Re}[Q(\theta; \Delta, \psi, \psi_1) + Q(\theta; -\Delta, -\psi, \psi_1)]; \\ v_y &= e^{-\beta\theta}(v_{0y}\cos\Phi - v_{0x}\sin\Phi) + a_\perp \operatorname{Im}[Q(\theta; 0, 0, 0)] \\ &\quad + \frac{1}{2} b_\perp \operatorname{Im}[Q(\theta; \Delta, \psi, \psi_1) + Q(\theta, -\Delta, -\psi, \psi_1)], \end{aligned}\right\} \tag{4.14}$$

where the auxilliary function Q has the form

$$\begin{aligned} Q(\theta; \Delta, \psi, \psi_1) &= e^{-\beta\theta + i(\psi+\psi_1) - i\alpha\theta - i\varrho\sin(\theta+\varphi)} \\ &\quad \times \int_0^\theta e^{i(\alpha+\Delta-i\beta) - i\varrho\sin(u+\varphi)}\, du. \end{aligned} \tag{4.15}$$

By the definition of the incomplete cylindrical function:

$$\varepsilon_\nu(w, z) = \frac{1}{\pi i}\int_0^w e^{z\sinh u - \nu u}\, du, \tag{4.15a}$$

Q can be written as:

$$\begin{aligned} Q(\theta; \Delta, \psi, \psi_1) &= \pi[\varepsilon_{-\mu}(i\theta + i\varphi, \varrho) - \varepsilon_{-\mu}(i\varphi, \varrho)] \\ &\quad \times \exp[i\mu\varphi + i(\psi + \psi_1) - \beta\theta - i\varrho\sin(\theta + \varphi) - i\alpha\theta)], \end{aligned} \tag{4.16}$$

where $\mu = \alpha + \Delta - i\beta$.

Equations (4.14) and (4.16) describe the motion of a charged particle, accounting for the effects of collisions, in an electromagnetic field of fairly general form (c.f. expressions (4.2) and (4.3)). The results of the preceeding section can be obtained readily from these solutions by setting $\gamma = 0$ $(\beta = 0)$; $E_1 = 0$ $(b = 0)$; $H_0 = 0$ $(\omega_H = 0)$; and $\varphi = \alpha - \pi/2$; $\boldsymbol{v}_0 = 0$.

The structure of (4.16) calls to mind relations derived earlier — for example expression (2.10). Consequently its analysis can be carried out by the methods set forth in the first two sections of this chapter.

In the case of stationary motion, when transition effects are negligible, i.e. when $\beta\theta = \gamma t \gg 1$, Eqs. (4.14) simplify somewhat. In fact, the additive terms which do not contain Q may be neglected, i.e.

$$\begin{aligned} v_x &= a_\perp \operatorname{Re}[Q(\theta, 0, 0, 0)] \\ &+ \frac{b_\perp}{2} \operatorname{Re}[Q(\theta;\ \Delta, \psi, \psi_1) + Q(\theta;\ -\Delta, -\psi, \psi_1)]; \\ v_y &= a_\perp \operatorname{Im}[Q(\theta;\ 0, 0, 0)] \\ &+ \frac{1}{2} b_\perp \operatorname{Im}[Q(\theta;\ \Delta, \psi, \psi_1) + Q(\theta; -\Delta, -\psi, \psi_1)]. \end{aligned} \tag{4.17}$$

After the change of integration $\xi = \theta - u$, the expression (4.15) for Q, in the stationary case, takes the form

$$\begin{aligned} Q(\theta; \Delta, \psi, \psi_1) &= \exp[i(\psi + \psi_1) - i\varrho \sin(\theta + \varphi) \\ &+ i\Delta\theta] \int_0^\infty e^{-\beta\xi - i(\alpha+\Delta)\xi - i\varrho\sin(-\theta-\varphi+\xi)} d\xi. \end{aligned} \tag{4.18}$$

Here the upper limit of integration should be θ, but for a stationary process $\beta\theta = \gamma t \gg 1$, and the integration can be carried out to infinity. This function can now be easily expressed in terms of the integral $L_\nu(\omega, z)$ considered earlier. We find

$$\begin{aligned} Q(\theta; \Delta, \psi, \psi_1) &= \exp[i(\psi + \psi_1) - i\varrho \sin(\theta + \varphi) + i\Delta\theta] \\ &\times L_{-\mu}(-\theta - \varphi, -\varrho)]. \end{aligned} \tag{4.19}$$

(4.19) permits further analysis of the stationary motion similar to that presented in sections 1 and 2 of this chapter.

In conclusion we mention some similar problems which arise in the study of cyclotron resonance in solids or plasmas situated in changing electromagnetic fields. For example, to obtain the results of Ref. [61], it is sufficient to take $E_0 = 0$ $(a = 0)$; $\varphi = \pi$; and $\psi = 0$ in Eqs. (4.17)—(4.19). One then obtains

$$\begin{aligned} v_x &= \frac{1}{2} b_\perp \operatorname{Re}[Q(\theta; \Delta, 0, \psi_1) + Q(\theta; -\Delta, 0, \psi_1)]; \\ v_y &= \frac{1}{2} b_\perp \operatorname{Im}[Q(\theta; \Delta, 0, \psi_1) + Q(\theta; -\Delta, 0, \psi_1)]. \end{aligned} \tag{4.20}$$

By (4.18), $Q(\theta; \Delta, 0, \psi)$ takes in this case the form

$$Q(\theta; \Delta, 0, \psi_1) = \exp(i\psi_1 + i\varrho\sin\theta + i\Delta\theta) \times \int_0^\infty e^{-\beta\xi - i(\alpha+\Delta)\xi + i\varrho\sin(\xi-\theta)}\, d\xi = L_{-\mu}(-\theta, \varrho)\exp(i\psi_1 + i\varrho\sin\theta + i\Delta\theta), \tag{4.21}$$

where $\mu = \alpha + \Delta - i\beta$.

Applying the analysis given above to (4.20) and (4.21), we can determine expressions for the components of the conductivity tensor of the medium in the corresponding limiting cases.

Chapter IX

Applications of Incomplete Cylindrical Functions to some Problems of Atomic and Nuclear Physics

1. Solution of a Particular Form of the Schrödinger Equation

It is well known that exact solutions of the Schrödinger equation can be obtained only in rare cases, for special forms of the interaction potential (c.f., e.g. [59] Vol. II, pp. 602—636). In general, the Schrödinger equation must be solved by some approximation scheme. We have already dealt with solutions of this type in studying diffraction by a screen of given form in § 5 of Chapter VII.

Finding an exact solution of the Schrödinger equation is even more difficult in the case of two particles interacting with each other and with an external field. If the fixed field is considered as a third body, then the problem is equivalent to the three body problem which in general cannot be solved in closed form. To overcome this difficulty it is usual to consider a drastically simplified model. In some connections an entirely satisfactory model of two similar particles subject to the influence of a simple potential field, is a very deep and narrow "hole" at the origin.

If the displacements of the particles from the center of the field are denoted by x and y, then the Schrödinger equation for the problem may be written

$$\left[\frac{\partial^2}{\partial x^2} + \frac{\partial^2}{\partial y^2} + \varepsilon + 2a\delta(x) + 2a\delta(y)\right] \Psi(x, y) = 0, \tag{1.1}$$

where the effects of the field on each particle are characterized by the functions $\delta(x)$ and $\delta(y)$.

The solution of this equation has been obtained in [5, Vol. II, pp. 657—666], where it is shown that in such a system there can be bound, free and partially bound states. In particular, for bound states the solution of (1.1) for quadrants I, II, III, and IV, respectively, has the form

$$\Psi(x, y) = \begin{Bmatrix} N e^{-\alpha(x+y)}; \\ N e^{\alpha(x-y)}; \\ N e^{\alpha(x+y)}; \\ N e^{\alpha(y-x)}, \end{Bmatrix} \tag{1.2}$$

where N is a normalizing coefficient.

If polar coordinates (r, φ) are introduced, then the wave function $\Psi(r, \varphi)$ can be represented by the following Fourier series:

$$\Psi(r, \varphi) = 2 \sum_{k=0}^{\infty} (-1)^k \Omega(k, 0) \cos(4k\varphi) J_{4k}^{(4)}(iar\sqrt{2}), \tag{1.3}$$

where $\Omega(k, 0)$ is a unit step function, equal to $1/2$ for $k = 0$ and 1 for $k > 0$, and $J_\nu^{(n)}(\varphi)$ is the integral

$$\begin{aligned} J_\nu^{(n)}(z) &= \frac{n}{i^\nu \pi} \int_0^{\pi/n} e^{iz\cos u} \cos(\nu u)\, du \\ &= \frac{n}{i^\nu 2\pi} \int_{-\pi/n}^{\pi/n} e^{iz\cos u + i\nu u}\, du. \end{aligned} \tag{1.4}$$

In connection with this and other problems [loc. cit., pp. 681, 699] Ref. [5] treats the properties of (1.4) and of the related function

$$E_\nu^{(n)}(z) = \frac{n}{i^\nu \pi} \int_0^{\pi/n} e^{iz\cos u} \sin(\nu u)\, du. \tag{1.5}$$

In particular, differential equations, recursion relations and asymptotic expansions for large z are obtained. One of the recursion relations for these functions is similar to the recursion relation for cylindrical functions

$$\frac{d}{dz} J_\nu^{(n)}(z) = \frac{1}{2} J_{\nu-1}^{(n)}(z) - \frac{1}{2} J_{\nu+1}^{(n)}(z). \tag{1.6}$$

Therefore they are called "semicylindrical functions" in Ref. [5].

For other properties of these functions the following relation is of interest

$$e^{iz\cos\varphi} = 2 \sum_{k=0}^{\infty} (i)^{kn} \Omega(k, 0) \cos(kn\varphi) J_{kn}^{(k)}(z), \tag{1.7}$$

which is valid for $-\pi/n \leq \varphi \leq \pi/n$. (1.7) represents the Fourier series development of a plane wave. Because the numerical values of the sum in (1.7) repeat themselves n times in the interval $(0, 2\pi)$, we have

$$2 \sum_{k=0}^{\infty} (i)^{kn} \Omega(k, 0) \cos(kn\varphi)\, J_{kn}^{(n)}(z)$$
$$= \begin{cases} e^{iz\cos\varphi}; & -\frac{\pi}{n} \leq \varphi \leq \frac{\pi}{n}; \\ e^{iz\cos(\varphi - 2\pi/n)}; & \frac{\pi}{n} \leq \varphi \leq \frac{3\pi}{n}; \\ \dots\dots\dots & \dots\dots\dots\dots\dots \\ e^{iz\cos(\varphi - 2m\pi/n)}; & \frac{2m-1}{n}\pi \leq \varphi \leq \frac{2m+1}{n}\pi, \end{cases} \tag{1.8}$$

where $m = 0, 1, 2, 3, \ldots, n-1$. Such series developments were encountered earlier in connection with the problem of diffraction by a wedge, (c.f. § 4 of Chapter VII). There also we split up a plane wave in a given angular interval and, as we saw, the problem led to the study of incomplete cylindrical functions of the Bessel form. In fact, taking $u = \pi/2 - \theta$, in (1.4), we have

$$J_\nu^{(n)}(z) = \frac{n}{2\pi} \int_{\pi/2-\pi/n}^{\pi/2+\pi/n} e^{iz\sin\theta - i\nu\theta}\, d\theta. \tag{1.9}$$

On the other hand, the incomplete Anger and Weber functions, by definitions (1.12) and (1.13) of Chapter III, are equal to

$$A_\nu(\beta, z) = \frac{1}{\pi} \int_0^\beta \cos(z \sin\theta - \nu\theta)\, d\theta; \tag{1.10}$$

$$B_\nu(\beta, z) = -\frac{1}{\pi} \int_0^\beta \sin(z \sin\theta - \nu\theta)\, d\theta, \tag{1.11}$$

and from this it follows that

$$\begin{aligned} J_\nu^{(n)}(z) = \frac{n}{2} \Big[A_\nu\Big(\frac{\pi}{2} + \frac{\pi}{n}, z\Big) - A_\nu\Big(\frac{\pi}{2} - \frac{\pi}{n}, z\Big) \Big] \\ - i \Big[B_\nu\Big(\frac{\pi}{2} + \frac{\pi}{n}, z\Big) - B_\nu\Big(\frac{\pi}{2} - \frac{\pi}{n}, z\Big) \Big]. \end{aligned} \tag{1.12}$$

In a similar way it is possible to express the functions $E_\nu^{(n)}(z)$ in terms of incomplete Anger and Weber functions. However, in practical applications they often arise with integer index $\nu = m$, and by (1.5), are simply expressible in terms of elementary functions. Thus, for example

$$\left. \begin{aligned} E_0^{(n)}(z) &= 0; \\ E_1^{(n)}(z) &= \frac{1}{az} (e^{iz\cos\alpha} - e^{iz}), \end{aligned} \right\} \tag{1.13}$$

where $\alpha = \pi/n$.

For the function $J_m^{(n)}(z)$, the recursion relation

$$\frac{m}{z} J_m^{(n)}(z) = \frac{1}{2} J_{m-1}^{(n)}(z) + \frac{1}{2} J_{m+1}^{(n)}(z) - \frac{\sin m\alpha}{(i)^m az} e^{iz\cos\alpha} \tag{1.14}$$

holds (c.f. [5, Vol. II, p. 700]), and therefore we need only know values for $m = 0$ and $m = 1$. But for these values $J_m^{(n)}(z)$ can be expressed in terms of incomplete cylindrical functions of the Poisson form. In fact, from (1.4) for $\nu = 0$ and $\nu = 1$ we have

$$J_0^{(n)}(z) = \frac{n}{\pi} \int_0^\alpha e^{iz\cos u}\, du = \frac{n}{2} E_0^+(\alpha, z); \tag{1.15}$$

$$\begin{aligned} J_1^{(n)}(z) &= \frac{n}{i\pi} \int_0^\alpha e^{iz\cos u} \frac{d \sin u}{du}\, du \\ &= \frac{n}{i\pi} \sin \alpha e^{iz\cos\alpha} + \frac{n}{\pi} z \int_0^\alpha e^{iz\cos u} \sin^2 u\, du \\ &= -i \frac{n}{\pi} \sin \alpha \cdot e^{iz\cos\alpha} + \frac{n}{2} E_1^+(\alpha, z), \end{aligned} \tag{1.16}$$

where $E_0^+(\alpha, z)$ and $E_1^+(\alpha, z)$ are incomplete cylindrical functions of the Poisson form (c.f. § 1 of Chapter II).

The latter formulae allow one to apply all the theory of incomplete cylindrical functions to the analysis of the integrals considered here. In particular, there is no difficulty obtaining asymptotic developments for them for small as well as large values of z. Moreover, since $\alpha = \pi/n$, the expansions with respect to α, (c.f. § 7 of Chapter II, § 5 of Chapter III) may also be used for $n > 3$.

For intermediate values of α and x these functions may be computed from

$$J_0^{(n)}(\varrho) = \frac{n}{2} \left[J_0\left(\frac{\pi}{n}, \varrho\right) + iH_0\left(\frac{\pi}{n}, \varrho\right)\right]; \tag{1.17}$$

and

$$\begin{aligned} J_1^{(n)}(\varrho) &= \frac{n}{2} \left[J_1\left(\frac{\pi}{n}, \varrho\right) + \frac{2}{\pi} \sin\left(\frac{\pi}{n}\right) \sin\left(\varrho \cos \frac{\pi}{n}\right)\right] \\ &+ \frac{in}{2} \left[H_1\left(\frac{\pi}{n}, \varrho\right) - \frac{2}{\pi} \sin\left(\frac{\pi}{n}\right) \cos\left(\varrho \cos \frac{\pi}{n}\right)\right] \end{aligned} \tag{1.18}$$

with the aid of the tables appended below for $J_k(\alpha, \varrho)$ and $H_k(\alpha, \varrho)$, $k = 0, 1$.

For purely imaginary $z = i\varrho$ the evaluation of the integrals under consideration can be effected by the formulae

$$J_0^{(n)}(i\varrho) = nF_0^-\left(\frac{\pi}{n}, \varrho\right); \tag{1.19}$$

$$J_1^{(n)}(i\varrho) = in\left\{F_1^-\left(\frac{\pi}{n}, \varrho\right) - \frac{1}{\pi} \sin \frac{\pi}{n} \exp\left(-\varrho \cos \frac{\pi}{n}\right)\right\}, \tag{1.20}$$

where $F_0^-(\alpha, \varrho)$ and $F_1^-(\alpha, \varrho)$ are tabulated functions.

Returning to (1.3) we see that the general theory of incomplete cylindrical functions applies to the wave function $\psi(r, \varphi)$ and permits quantitative evaluation of the contributions of various harmonics.

2. Some Relations for Evaluating the Transparency Coefficients of Nuclei and the Average Loss of Impulse for Interaction of Particles with their Fields

In studies of the passage of particles through nuclei, it is required to represent the dependence of the nuclear density $n(r)$ and the nuclear potential $V(r)$ on the distance r. Knowledge of these functions is essential for studying nuclear structure, the interactions of particles with nuclei, and also for considering the mechanism of flow-through in nuclear reactions. In early investigations the functions $n(r)$ and $V(r)$ were taken to be constant, with a sharp drop at some constant r_0, i.e.

$$\begin{aligned} n(r) &= n_0 \Omega(1, \varrho); \quad \varrho = \frac{r}{r_0}; \\ V(r) &= V_0 \Omega(1, \varrho). \end{aligned} \tag{2.1}$$

where $\Omega(1, \varrho)$ is a unit step function,

$$\Omega(1, \varrho) = \begin{cases} 1; & 0 < \varrho < 1; \\ 1/2; & \varrho = 1; \\ 0; & \varrho > 1; \end{cases} \tag{2.2}$$

However, in several important cases such representations do not agree well with experimental results — for example in the case of fast electron scattering by nuclei [62]. The use of dependences more representative than (2.1) provides better agreement. Often, instead of function $\Omega(1, \varrho)$ one uses for this purpose a continuous function of two variables, $f(\lambda, \varrho - 1)$ defined either as [63]

$$f(\lambda, \varrho - 1) = [1 + \exp \lambda(\varrho - 1)]^{-1}, \tag{2.3}$$

or as [64, 65]

$$f(\lambda, \varrho - 1) = \frac{1}{2}[1 - \operatorname{th} \lambda(\varrho - 1)], \tag{2.4}$$

which, for $\lambda \to \infty$ reduce to $\Omega(1, \varrho)$. With such potentials, however, it is not possible to solve the Schrödinger equation in closed form. Consequently, we look to approximate methods. The W-K-B method is widely used for this purpose and yields for one of the fundamental characteristics of the scattering process — the phase shift δ_l — the following approximate expression [66]:

$$\delta_l \sim \Psi(x, \lambda) = \int_x^\infty f(\lambda, t - 1)\,(t^2 - x^2)^{1/2}\, t\, dt, \tag{2.5}$$

where $x = kr_0/(l + 1/2)$.

A similar expression arises in the study of the passage of a particle through a nucleus. One of the chief characteristics of this process is the coefficient of transparency T of the nucleus, which by the W-K-B method ([64, 65]) is found to be

$$T = e^{-\beta_0 \Psi(\varrho, \lambda)}. \tag{2.6}$$

In addition to the above characteristics, a very important role is played by the mean value of the impulse loss for each interaction of the particles with the nuclear field. This is equal to

$$\Delta p = -\int_{-\infty}^{+\infty} \frac{dV(r)}{dr} \, dt.$$

Here $r = \sqrt{b^2 + v^2 t^2}$, b is the target parameter of the collision, v is the velocity of the moving particle and t is time. After the substitution $V(r) = V_0 f(\lambda, \varrho - 1)$ and an appropriate change of integration variable the latter expression may be reduced to

$$\Delta p = -\frac{2V_0}{v} \int_{\xi}^{\infty} \frac{x}{\sqrt{x^2 - \xi^2}} \frac{d}{dx} f(\lambda, x - 1) \, dx, \tag{2.7}$$

where $\xi = b/r_0$.

The evaluation of the characteristics just described for either choice of $f(\lambda, \varrho - 1)$ in (2.3) and (2.4) is very awkward, because the integrals which appear are not expressible in terms of known and tabulated functions. But the choice of the distribution f is to some degree arbitrary. We could therefore choose some other distribution, "close" to (2.3) and (2.4); we may take, for example,

$$f(\lambda, \varrho - 1) = \begin{cases} 1 - \frac{1}{2} e^{-\lambda(1-\varrho)}; & \varrho \le 1; \\ \frac{1}{2} e^{\lambda(1-\varrho)}; & \varrho > 1; \end{cases} \tag{2.8}$$

following [39, 67].

We have already dealt with a similar problem in the evaluation of the loss coefficient in the earth's atmosphere, with account taken of the curvature of the atmospheric layers and their inhomogeneity, (c.f. § 2 of Chapter VII). The only difference lies in the fact that in Chap. VII the loss coefficient was calculated in the atmosphere, where the layer density was determined solely by the second part of formula (2.8) for $\varrho > 1$. For the evaluation of the transparency coefficient it is necessary to consider also the possibility of passage of radiation through the region of the nucleus itself, $\varrho < 1$.

We show that for the distribution (2.8), the characteristics of the interaction of a particle with anucleus as described above may be expressed in terms of incomplete cylindrical functions of the Poisson form. As a matter of fact, integrating (2.5) by parts and taking account of (2.8), we find

$$\begin{aligned}\Psi(x,\lambda) &= -\int_x^\infty \sqrt{\sigma^2 - x^2}\,\frac{d}{d\sigma} f(\lambda, \sigma - 1)\,d\sigma \\ &= \frac{\lambda}{2}\int_x^\infty \sqrt{\sigma^2 - x^2}\,e^{-\lambda|1-\sigma|}\,d\sigma.\end{aligned} \tag{2.9}$$

We consider now the cases $x > 1$ and $x < 1$ separately. If $x \geq 1$, then $|1 - \sigma| = \sigma - 1$ and (2.9) becomes

$$\Psi(x,\lambda) = \frac{\lambda}{2} x^2 e^{\lambda} \int_1^\infty \sqrt{t^2 - 1}\,e^{-\lambda x t}\,dt$$

or, by the fundamental definition of the MacDonald function (c.f. expressions (3.27) and (3.28) of Chapter II),

$$\Psi(x,\lambda) = \frac{x}{2} e^{\lambda} K_1(\lambda x). \tag{2.10}$$

If $x < 1$, then we can write (2.9) preliminarily in the form

$$\begin{aligned}\Psi(x,\lambda) &= \frac{\lambda}{2} e^{-\lambda} \int_x^1 \sqrt{\sigma^2 - x^2}\,e^{\lambda\sigma}\,d\sigma \\ &- \frac{\lambda}{2} e^{\lambda} \int_x^1 \sqrt{\sigma^2 - x^2}\,e^{-\lambda\sigma}\,d\sigma + \frac{\lambda}{2} e^{\lambda} \int_x^\infty \sqrt{\sigma^2 - x^2}\,e^{-\lambda\sigma}\,d\sigma.\end{aligned}$$

Setting $\sigma/x = t$,

$$\begin{aligned}\Psi(x,\lambda) &= \frac{x}{2}\Bigg\{e^{-\lambda}\,\lambda x \int_1^{1/x} \sqrt{t^2 - 1}\,e^{\lambda x t}\,dt \\ &- \lambda x e^{\lambda} \int_1^{1/x} \sqrt{t^2 - 1}\,e^{-\lambda x t}\,dt + \lambda x e^{\lambda t} \int_1^\infty \sqrt{t^2 - 1}\,e^{-\lambda x t}\,dt\Bigg\}.\end{aligned}$$

The last term has already been determined, and the first two are expressible by incomplete MacDonald functions, (c.f. § 1 of Chapter II). Thus, we have finally

$$\overline{\Psi}(x,\lambda) = \frac{x}{2} e^{\lambda} K_1(\lambda x) - \frac{x}{2}\Omega(1,x)\,[e^{\lambda} K_1(\beta, \lambda x) + e^{-\lambda} K_1(\beta, -\lambda x)], \tag{2.11}$$

where $\cosh\beta = 1/x$ and $\Omega(1, x)$ is a unit step function. Carrying out similar transformations on (2.7), we find

$$\begin{aligned}\Delta p &= \frac{V_0}{v}\lambda\xi e^{\lambda} K_1(\lambda\xi) \\ &+ \frac{V_0 \lambda\xi}{v}\Omega(1,\xi)\,[e^{-\lambda} K_1(\varrho, -\lambda\xi) - e^{\lambda} K_1(\beta, \lambda\xi)],\end{aligned} \tag{2.12}$$

where $\cosh\beta = 1/z = r_0/b$, and b is the target parameter.

The relations obtained above express the phase shift δ_l in a potential field of the form (1.8), the nuclear transparency T, and the average loss of impulse Δp in a scattering field of the same form, in terms of incomplete cylindrical functions. The general theory of these functions can consequently be used to study the interaction of particles with nuclei. We also note that by appropriate choice of λ, the distribution (2.8) will closely approximate (2.3). Thus the relations obtained have wide application in analysis and interpretation of experimental results.

3. Resonant Absorption of Radiation in Media of Finite Dimensions

For the study of interactions of radiation with various substances a significant characteristic is the fraction of radiation absorbed or transmitted. The weakening of the propagating radiation may be due to inelastic as well as elastic processes. These processes may be either resonant or non-resonant; in the former case the coefficient of absorption or dissipation has sharp peaks for certain values of the incident radiation energy. The resonant case is always dominant near the resonant frequency. Therefore, in practical investigations there is fundamental interest in the resonant absorption in media of finite dimensions [32, 33, 68, 69]. Knowledge of this quantity permits determination of the absorption intensity, i.e. the section in resonance, and also the width of this resonance, which in turn allows the determination of the temperature and density of the medium. Studies of this type arise in connection with investigations of plasma properties, of the Earth's atmosphere [32, 34], and in the solution of many other problems of atomic and nuclear physics.

If we denote the coefficient of absorption by $k(\nu)$, then the integrated absorption is determined by

$$A = \int_{-\infty}^{\infty} (1 - e^{-wk(\nu)})\, d\nu, \tag{3.1}$$

where w is the per-unit absorption of the gas. When the loss curve has the usual resonant shape, (c.f. § 1 of Chapter VIII), the dependence of the absorption coefficient $k(\nu)$ on the frequency ν has the form

$$k(\nu) = \frac{s}{\pi} \cdot \frac{\Gamma}{2} \frac{1}{(\nu - \nu_0)^2 + \left(\frac{\Gamma}{2}\right)^2}, \tag{3.2}$$

where $\Gamma/2$ is the half-width of the line and s is the absorption intensity, proportional to the section. For convenience in the following analysis we write (3.1) in the somewhat different form

$$A = \int_0^w dx \int_{-\infty}^{\infty} e^{-xk(\nu)}\, k(\nu)\, d\nu. \tag{3.3}$$

Substituting $k(\nu)$ form (3.2) into this the expression, we have

$$A = \frac{s\Gamma}{2\pi} \int_0^w dx \int_{-\infty}^{\infty} \frac{dt}{t^2 + \left(\frac{\Gamma}{2}\right)^2} \exp\left(-\frac{x\Gamma s}{2\pi} \frac{1}{t^2 + \left(\frac{\Gamma}{2}\right)^2}\right). \tag{3.4}$$

Introducing a new variable of integration by $2t/\Gamma = \tan\theta$, we find

$$A = \frac{s}{\pi} \int_0^w dx \int_{-\pi/2}^{\pi/2} \exp\left(-\frac{2xs}{\pi\Gamma} \sin^2\theta\right) d\theta$$

$$= \frac{s}{2\pi} \int_0^w \mathrm{e}^{-xs/\pi\Gamma} dx \int_{-\pi}^{\pi} \mathrm{e}^{xs/\pi\Gamma \cos\psi} d\psi = s \int_0^w \mathrm{e}^{-xs/\pi\Gamma} I_0\left(\frac{xs}{\pi\Gamma}\right) dx$$

or

$$A = \pi\Gamma \int_0^\varrho \mathrm{e}^{-t} I_0(t)\, dt, \tag{3.4'}$$

where $\varrho = ws/\pi\Gamma$. The latter integral is the incomplete Lipschitz-Hankel integral $Ie_0(1, \varrho)$ (c.f. (5.41) of Chapter II), so that A takes the form

$$A = \pi\Gamma\varrho \mathrm{e}^{-\varrho} [I_0(\varrho) + I_1(\varrho)]. \tag{3.5}$$

Thus if there is just one isolated resonant line in (3.2) the absorption integral is expressible in closed form by modified Bessel functions. Actually there is not one isolated absorption line, but an entire group of lines, overlapping one another. In this situation, as we have already seen in Sections 1 and 2 of Chapter VIII, the form of the absorption lines is more complicated. An approximate expression for them may be obtained from (1.21) of Chapter VIII,

$$J \approx \frac{c(\nu_0) \sinh\left(\frac{\pi\Gamma}{\nu_0}\right)}{\cosh^2 \frac{\pi\Gamma}{2\nu_0} - \cos^2\left(\frac{\nu - \nu_0}{\nu_0}\pi\right)} = \frac{c(\nu_0) \sinh\left(\frac{\pi\Gamma}{\nu_0}\right)}{\cosh \frac{\pi\Gamma}{\nu} - \cos \frac{2\pi(\nu - \nu_0)}{\nu_0}}. \tag{3.6}$$

Consequently, the absorption coefficient $k(\nu)$ is determined by

$$k(\nu) = \frac{s}{\nu_0} \frac{\sinh\beta}{\cosh\beta - \cos t}, \tag{3.7}$$

where $\beta = \pi\Gamma/\nu_0$ and $t = (1/\nu_0)\, 2\pi(\nu - \nu_0)$.

Such a description of the loss curve was first suggested by Elsasser, and it is customary in the literature to call this the Elsasser model (c.f. [32, p. 65] and [34]). Despite its idealized nature several investigations have shown that such a model does lead to the correct functional relation between absorption, pressure, and optical density.

Substituting (3.7) into (3.3) and taking account of the periodicity of $k(\nu)$, we get

$$A = \sum_k A_k, \tag{3.8}$$

where the summation is carried out over all lines of the given zone, and the quantities A_k are determined by

$$A_k = \frac{s_k}{\pi} \int_0^w dx \int_0^\pi \frac{\sinh\beta}{\cosh\beta - \cos t} \exp\left(-\frac{x s_k}{v_0} \frac{\sinh\beta}{\cosh\beta - \cos t}\right) dt. \quad (3.9)$$

Introduce now a new variable of integration by

$$\cos\varphi = \frac{1 - \cosh\beta\cos t}{\cosh\beta - \cos t},$$

so that

$$d\varphi = -\frac{\sinh\beta}{\cosh\beta - \cos t}\, dt.$$

After some simplifications:

$$A_k = s_k \int_0^w e^{-x s_k/v_0 \cosh\beta/\sinh\beta}\, I_0\left(\frac{s_k x}{v_0 \sinh\beta}\right) dx \quad (3.10)$$

or

$$A_k = v_0 \sinh\beta \int_0^\varrho e^{-t\cosh\beta} I_0(t)\, dt = v_0 \sinh\beta\, Ie_0(\cosh\beta, \varrho), \quad (3.11)$$

where $\varrho = \omega s_k/v_0 \sinh\beta$, and $Ie_0(a, \varrho)$ is the incomplete Lipschitz-Hankel integral of the modified Bessel function, (c.f. § 5 of Chapter II).

We want to consider simultaneously a similar problem, which is important in the theory of nuclear reactions with slow neutrons. The interaction cross-sections of neutrons with nuclei at medium and low energies has a resonant structure as is well known. Therefore, one of the principal parameters in the theory and design of nuclear reactors is the probability of resonant capture of neutrons in the filled reactor. This probability is also of great significance in the study of the interaction of neutrons with nuclei and of their passage through substances. This quantity is usually called the resonance integral, which in the case of a region of finite dimension has the form (c.f., e.g. [33])

$$\begin{aligned} I_{\text{ef}} = \frac{\Gamma}{2E} \Bigg[& \frac{1}{L} \int_0^L dt \int_{-\infty}^\infty e^{-\sigma_t l} \sigma_a(x)\, dx \\ & + \frac{\sigma_p}{L} \int_0^L dl \int_0^l dl' \int_{-\infty}^\infty e^{-\sigma_t l'} \sigma_a(x)\, dx \Bigg], \end{aligned} \quad (3.12)$$

where E is the energy, Γ the total width, $L = 4VN/F$, V and F are the volume and surface area of the region, N is the number of absorbing nuclei per cm^3, σ_α, σ_p and σ_t are, respectively the absorption cross section, the scattering potential and the complete scattering cross section. The dependence of these cross-sections on the energy is given by

$$\left.\begin{aligned} \sigma_a &= \frac{\sigma_0 \Gamma_a}{\Gamma} \frac{1}{1+x^2}; \quad x = \frac{2(E-E_0)}{\Gamma}; \\ \sigma_t &= \sigma_p + \frac{\sigma_0}{1+x^2}(\cos 2\varphi - x \sin 2\varphi). \end{aligned}\right\} \tag{3.13}$$

We show now that I_{ef} may be expressed in terms of incomplete Lipschitz-Hankel integrals. To this end we write (3.12) as

$$\begin{aligned} I_{\text{ef}} &= \frac{\Gamma}{2E}\left[\frac{1}{L} D(L) + \frac{\sigma_p}{L}\int_0^L D(l)\, dl\right] \\ &= \frac{\Gamma}{2E}\left[\left(\frac{1}{L} + \sigma_p\right) D(L) - \frac{\sigma_p}{L}\int_0^L l \frac{dD(l)}{dl}\, dl\right], \end{aligned} \tag{3.14}$$

where we have introduced

$$D(y) = \int_0^y dl \int_{-\infty}^{\infty} e^{-\sigma_t(x) l} \sigma_a(x)\, dx. \tag{3.15}$$

Substituting in (3.15) the expressions for $\sigma_t(x)$ and $\sigma_a(x)$ from (3.13), we find

$$D(y) = \frac{\sigma_0 \Gamma_a}{\Gamma} \int_0^y e^{-\sigma_p l}\, dl \int_{-\infty}^{\infty} \frac{\exp\left[-\dfrac{\sigma_0 l}{1+x^2}(\cos 2\varphi - x \sin 2\varphi)\right]}{1+x^2}\, dx. \tag{3.16}$$

Introducing a new variable by $\tan(\theta/2) = x$, we obtain after some simplifications

$$D(y) = \frac{\sigma_0 \Gamma_a}{2\Gamma} \int_0^y dl e^{-(\sigma_p + \sigma_0 \cos 2\varphi/2) l} \int_{-\pi}^{\pi} e^{-\sigma_0 l/2 \cos(\theta + 2\varphi)}\, d\theta.$$

Here the inner integral can be expressed in terms of the modified Bessel function $I_0(z)$. We therefore have

$$D(y) = \frac{2\pi \Gamma_a}{\Gamma} \int_0^{\sigma_0 y/2} e^{-at} I_0(t)\, dt, \tag{3.17}$$

where we have put $a = \cos 2\varphi + \sigma_p/\sigma_0$, and assumed this quantity is always larger than unity [33]. Now it is not difficult to find an expression for the second term in (3.14). We have

$$\begin{aligned} \frac{\sigma_p}{L}\int_0^L l \frac{d}{dl} D(l)\, dl &= \frac{\sigma_p}{L} \frac{2\pi \Gamma_a}{\Gamma} \frac{2}{\sigma_0} \int_0^{L\sigma_0/2} e^{-at} t I_0(t)\, dt \\ &= \frac{4\pi}{L} \frac{\Gamma_a}{\Gamma} \frac{\sigma_p}{\sigma_0}\left[\frac{\sigma_0 L}{2} I_1\left(\frac{\sigma_0 L}{2}\right) + a \int_0^{\sigma_0 L/2} t I_1(t)\, e^{-at}\, dt\right]. \end{aligned} \tag{3.18}$$

Substituting (3.17) and (3.18) intro (3.14) and recalling the definition of the incomplete Lipschitz-Hankel integral $Ie_\nu(a, \varrho)$ (c.f. Sections 5 and 6 of Chapter II), we obtain finally

$$I_{\text{ef}} = \frac{\pi \sigma_0 \Gamma_a}{2E} \left[\left(\frac{1}{\varrho} + \frac{2\sigma_p}{\sigma_0} \right) Ie_0(a, \varrho) - \left(\frac{2\sigma_p}{\sigma_0} \right)^2 \varrho I_1(\varrho) - \left(\frac{2\sigma_\varrho}{\sigma_0} \right)^2 a Ie_1(a, \varrho) \right], \tag{3.19}$$

where $\varrho = \sigma_0 L/2 = 2VN\sigma_0/F$.

Thus the integrated absorption coefficient in the Elsasser model and the resonance absorption integral for neutrons in media of finite dimensions can be expressed very simply in terms of incomplete Lipschitz-Hankel integrals. The detailed theory of these integrals and their relation to incomplete cylindrical functions of Poisson form given in Sections 5 and 6 of Chapter II may prove to be of considerable help in the evaluation of these physical characteristics, both in limiting cases with the aid of asymptotic expansions as well as in numerical evaluation for fixed values of the appropriate parameters. In this connection we note that in practical cases the quantity $(2\sigma_\varrho/\sigma_0)^2$ is small, and the first two terms of (3.19) are dominant, so that the computation of A and I_{ef} leads to the evaluation of the incomplete Lipschitz-Hankel integral of zero order, $Ie_0(\alpha, \varrho)$. Assuming that $a \geq 1$, numerical values of this integral may be determined from tabulated cylindrical functions of two purely imaginary arguments [22] by means of (6.15) of Chapter IV:

$$Ie_0(a, \varrho) = \frac{1}{\sqrt{a^2 - 1}} \{ e^{-a\varrho} [I_0(\varrho) + 2Y_2(\lambda\varrho, \varrho) + Y_1(\lambda\varrho, \varrho)] - 1 \}, \tag{3.20}$$

where $\lambda = a + \sqrt{a^2 - 1}$, and $Y_1(y, \varrho)$, $Y_2(y, \varrho)$ are tabulated functions.

We remark that for the analysis of limiting cases it is convenient to use the asymptotic expressions for $Ie_0(a, \varrho)$ obtained in [6] and [22].

To conclude this section we show that the widening of resonance curves of the type (3.2) due to thermal motion of the radiators or absorbers may also be expressed in terms of incomplete cylindrical functions in Bessel form. As is well known, the spreading of the curve (3.2) by a Maxwellian distribution leads one to the evaluation of

$$\Psi(x, \xi) = \frac{\xi}{2\sqrt{\pi}} \int_{-\infty}^{\infty} \frac{e^{-\xi^2/4\,(x-y)^2}}{1 + y^2}\, dy, \tag{3.21}$$

where $x = 2(E - E_0)/\Gamma$ and $\xi = \Gamma/2\sqrt{M/(mE_0 kT)}$ (c.f., e.g. [68, p. 86]). The integral on the right side of (3.21) is equal to the real part of the probability integral with complex argument [17]. It is appropriate to express it in terms of incomplete cylindrical functions. To this end we substitute the obvious identity

$$\frac{1}{1 + y^2} = \int_0^\infty e^{-\alpha(1+y^2)}\, d\alpha \tag{3.22}$$

into (3.21). After interchanging the order of integration we obtain

$$
\begin{aligned}
\Psi(x, \xi) &= \frac{\xi}{2\sqrt{\pi}} \int_0^\infty e^{-\alpha}\, d\alpha \int_{-\infty}^{\infty} e^{-\xi^2/4\,(x-y)^2 - \alpha y^2}\, dy \\
&= \frac{\xi e^{-\xi^2 x^2/4}}{2\sqrt{\pi}} \int_0^\infty e^{-\alpha}\, d\alpha \int_{-\infty}^{\infty} e^{-(\alpha+\xi^2/4)y^2 + \xi^2 x/2y}\, dy \\
&= \frac{1}{2} e^{-\xi^2 x^2/4} \int_0^\infty e^{-\alpha + \xi^4 x^2/16(\xi^2/4+\alpha)} \frac{d\alpha}{\sqrt{\frac{\xi^2}{4} + \alpha}}.
\end{aligned}
\tag{3.23}
$$

Introducing a new variable of integration by $\alpha = \xi^2/4\,(|x|/t - 1)$, we find

$$
\Psi(x, \xi) = \frac{\sqrt{|x|}\,\xi^2}{4} e^{\xi^2/4(1-x^2)} \int_0^{|x|} e^{\xi^2|x|/4(t-1/t)} \frac{dt}{t^{3/2}}. \tag{3.24}
$$

The latter integral is not difficult to express in terms of the incomplete cylindrical function of Bessel form with index $\nu = 1/2$ by putting

$$
\Psi(x, \xi) = \pi i \frac{\sqrt{|x|}}{4} \xi^2 e^{\xi^2/4(1-x^2)}\, \Phi_{1/2}\left(\ln |x|, \frac{\xi^2\,|x|}{2}\right). \tag{3.25}
$$

Using now (6.29) of Chapter III for Im $(b) = 0$, we find the following expansion for $\psi(x, \xi)$, when $z \cosh(\omega) = \xi^2 (1 + x^2)/4 > 1$:

$$
\begin{aligned}
\Psi(x, \xi) &= \frac{1}{1 + x^2} \sum_{n=0}^{\infty} (-1)^n \left\{\frac{4}{\xi^2 (1 + x^2)}\right\}^n \\
&\times \frac{\Gamma\left(n + \frac{1}{2}\right)}{\sqrt{\pi}} F\left(-n, n + 1; \frac{1}{2}; \frac{x^2}{1 + x^2}\right).
\end{aligned}
\tag{3.26}
$$

This equation may be written in somewhat different form by using the relation (c.f. [11, p. 1056])

$$
F\left(-n, n + 1; 1/2; \frac{x^2}{1 + x^2}\right) = \sqrt{1 + x^2} \cos\left[(2n + 1) \cot^{-1} x\right]. \tag{3.27}
$$

One finds

$$
\begin{aligned}
\Psi(x, \xi) &= \frac{1}{\sqrt{1 + x^2}} \sum_{n=0}^{\infty} (-1)^n \left[\frac{4}{\xi^2 (1 + x^2)}\right]^n \frac{\Gamma\left(n + \frac{1}{2}\right)}{\sqrt{\pi}} \\
&\times \cos\left[(2n + 1) \cot^{-1} x\right].
\end{aligned}
\tag{3.28}
$$

It is easy to obtain the first term of (3.28):

$$
\Psi(x, \xi) \approx \frac{1}{1 + x^2} \left[1 - \frac{2(1 - 3x^2)}{\xi^2 (1 + x^2)^2}\right]. \tag{3.29}
$$

We recall that for validity of (3.29) it was required only that $\xi^2 (1 + x^2)/4 > 1$. The result obtained is therefore correct for small as well as for large values of x when $\xi > 2$. In this case the Doppler widening is practically insignificant, i.e. (3.29) is close to the usual resonance curve.

4. Application of Incomplete Cylindrical Functions in the Study of a Heterogeneous Reactor with Small Number of Blocks

Computations in the study of nuclear reactors, such as the determination of critical dimensions and the distribution of neutron density, in addition to estimates of the effects of the controlling system, are usually carried out for homogeneous models. To this end the actual medium, consisting of a lattice of separate fuel blocks, and situated in a moderator, is replaced by an equivalent homogeneous medium. Such an approximation gives entirely satisfactory results when large numbers of blocks are present. When the number of blocks is small more precise models have been considered, in which account has been taken of the discrete distribution of neutron sources. Here we shall not discuss all the details in the hetrogeneous computation, which has been the subject of a number of theoretical and experimental studies [68, 70—73], but shall present only the fundamental steps.

The differential equation for the density of thermal neutrons $N(\boldsymbol{r})$ in the moderator for the stationary case has the form

$$
\begin{aligned}
D\nabla^2 N(\boldsymbol{r}) - \frac{N(\boldsymbol{r})}{T_0} + \eta\gamma \sum_k W(\boldsymbol{r}, \boldsymbol{r}_k)\, N(\boldsymbol{r}_k) & \\
- \gamma \sum_k \delta(\boldsymbol{r} - \boldsymbol{r}_k)\, N(\boldsymbol{r}_k) = 0, &
\end{aligned}
\tag{4.1}
$$

where D and T_0 are the coefficients of diffusion and the lifetime of thermal neutrons in the moderator, η is the number of secondary neutrons, γ is the thermal constant of the block, and $W(\boldsymbol{r}, \boldsymbol{r}_k)$ is the probability that a fast neutron, originating in block k, becomes thermalized at point $\boldsymbol{r}$, which for the case of linear blocks is equal to

$$
W(\boldsymbol{r}, \boldsymbol{r}_k) = \frac{1}{4\pi\tau} \exp\left[-\frac{(\boldsymbol{r} - \boldsymbol{r}_k)^2}{4\tau}\right]. \tag{4.2}
$$

It is convenient to carry out the solution of (4.1) by means of Fourier transforms. To this end we multiply the (4.1) by exp. $(i\boldsymbol{s} \cdot \boldsymbol{r})$ and integrate over all space. Taking into account the fact that $N(\boldsymbol{r}) \to 0$ and grad $N(\boldsymbol{r}) \to 0$ as $\boldsymbol{r} \to \infty$, so that

$$
\iint_{-\infty}^{\infty} e^{i\boldsymbol{s}\boldsymbol{r}} \nabla^2 N(\boldsymbol{r})\, d\boldsymbol{r} = -s^2 \iint_{-\infty}^{\infty} e^{i\boldsymbol{s}\boldsymbol{r}} N(\boldsymbol{r})\, d\boldsymbol{r},
$$

we have

$$
\begin{aligned}
-\left(s^2 + \frac{1}{L^2}\right)\Phi(\boldsymbol{s}) + \frac{\gamma}{D}\Big[\eta \sum_k N(\boldsymbol{r}_k)\, \Psi(\boldsymbol{s}, \boldsymbol{r}_k) & \\
- \sum_k e^{i\boldsymbol{s}\boldsymbol{r}_k} N(\boldsymbol{r}_k)\Big] = 0, &
\end{aligned}
\tag{4.3}
$$

where $L^2 = DT_0$ and we have introduced the notations

$$\Phi(\mathbf{s}) = \iint\limits_{-\infty}^{\infty} e^{i\mathbf{s}\mathbf{r}} N(\mathbf{r})\, d\mathbf{r}; \tag{4.4}$$

$$\Psi(\mathbf{s}, \mathbf{r}_k) = \iint\limits_{-\infty}^{\infty} e^{i\mathbf{s}\mathbf{r}} W(\mathbf{r}, \mathbf{r}_k)\, d\mathbf{r}. \tag{4.5}$$

Solving (4.3) for $\Phi(s)$ and inverting the Fourier transform, we have

$$N(\mathbf{r}) = \frac{\gamma}{(2\pi)^2 D} \sum_k N(\vec{\mathbf{r}}_k) \left(\eta \iint\limits_{-\infty}^{\infty} \frac{e^{-i\mathbf{s}\mathbf{r}} \Psi(\mathbf{s}, \mathbf{r}_k)}{s^2 + \frac{1}{L^2}}\, d\mathbf{s} - \iint\limits_{-\infty}^{\infty} \frac{e^{i\mathbf{s}\mathbf{r}_k - i\mathbf{s}\mathbf{r}}}{s^2 + \frac{1}{L^2}}\, d\mathbf{s} \right). \tag{4.6}$$

To simplify this expression we first evaluate $\Psi(s, r_k)$. From (4.5) and (4.2) we find

$$\begin{aligned} \Psi(\mathbf{s}, \mathbf{r}_k) &= \frac{1}{4\pi\tau} \iint\limits_{-\infty}^{\infty} \exp\left[i\mathbf{s}\mathbf{r} - \frac{(\mathbf{r} - \mathbf{r}_k)^2}{4\tau} \right] d\mathbf{r} \\ &= \frac{e^{i\mathbf{s}\mathbf{r}_k}}{4\pi\tau} \iint\limits_{-\infty}^{\infty} \exp\left(i s \varrho - \frac{\varrho^2}{4\tau} \right) d\varrho \\ &= \frac{e^{i\mathbf{s}\mathbf{r}_k}}{4\pi\tau} \int\limits_{-\infty}^{\infty} \exp\left(i s_x x - \frac{x^2}{4\tau} \right) dx \int\limits_{-\infty}^{\infty} \exp\left(i s_y y - \frac{y^2}{4\tau} \right) dy \\ &= e^{i\mathbf{s}\mathbf{r}_k - \tau s^2}. \end{aligned} \tag{4.7}$$

In this case the first integral in (4.6) can be written

$$\begin{aligned} \iint\limits_{-\infty}^{\infty} \frac{e^{-i\mathbf{s}\mathbf{r}} \Psi(\mathbf{s}, \mathbf{r}_k)}{s^2 + \frac{1}{L^2}}\, d\mathbf{s} &= \iint\limits_{-\infty}^{\infty} \frac{e^{i\mathbf{s}(\mathbf{r}_k - \mathbf{r}) - \tau s^2}}{s^2 + \frac{1}{L^2}}\, d\mathbf{s} \\ &= \int\limits_0^{\infty} \frac{e^{-\tau s^2}}{s^2 + \frac{1}{L^2}}\, s\, ds \int\limits_0^{2\pi} e^{is|\mathbf{r}_k - \mathbf{r}| \cos\varphi}\, d\varphi. \end{aligned}$$

Or, recalling the definition of the Bessel functions, we have

$$\iint\limits_{-\infty}^{\infty} \frac{e^{i\mathbf{s}(\mathbf{r}_k - \mathbf{r}) - \tau s^2}}{s^2 + \frac{1}{L^2}}\, d\mathbf{s} = 2\pi \int\limits_0^{\infty} \frac{e^{-\tau s^2} J_0(s\,|\mathbf{r}_k - \mathbf{r}|)}{s^2 + \frac{1}{L^2}}\, s\, ds. \tag{4.8}$$

But the integral on the right side of (4.8) is a particular case of a more general Bessel function integral which we have considered earlier, and by (4.21) of Chapter V, it can be expressed in terms of the incomplete

MacDonald function of zero order. The last formula is therefore

$$\iint_{-\infty}^{\infty} \frac{e^{is(r_k - r) - \tau s^2}}{s^2 + \frac{1}{L^2}} ds$$
$$= \pi e^{\tau/L^2} \left\{ K_0\left(\frac{|r - r_k|}{L}\right) - K_0\left(\beta, \frac{|r - r_k|}{L}\right) \right\}, \tag{4.9}$$

where $\beta = \ln 2\pi/L\,|r - r_k|$, and $K_0(\beta, \varrho)$ is the incomplete MacDonald function

$$K_0(\beta, \varrho) = \int_0^\beta e^{-\varrho \cosh t}\, dt. \tag{4.10}$$

To evaluate the second integral in (4.6) it is sufficient to consider (4.9) in the limit as $\tau \to 0$. Assuming that as $\tau \to 0$

$$\beta = \ln \frac{2\tau}{L(r - r_k)} \to -\infty,$$

i.e., that

$$\lim_{\tau \to 0} K_0(\beta, \varrho) = \int_0^{-\infty} e^{-\varrho \cosh t}\, dt = -K_0(\varrho), \tag{4.11}$$

we find

$$\iint_{-\infty}^{\infty} \frac{e^{is(r_k - r)}}{s^2 + \frac{1}{L^2}} dr = 2\pi K_0\left(\frac{|r_k - r|}{L}\right). \tag{4.12}$$

Substituting (4.9) and (4.12) into (4.6), we obtain for $N(r)$:

$$N(r) = \frac{\gamma}{2\pi D} \sum_k N(r_k) \left[\left(\frac{\eta}{2} e^{\tau/L^2} - 1\right) K_0\left(\frac{|r - r_k|}{L}\right) - \frac{\eta}{2} e^{\tau/L^2} K_0\left(\beta, \frac{|r - r_k|}{L}\right) \right]. \tag{4.13}$$

Here, as above, the summation in the right side is carried out over all blocks belonging to the system. Therefore, taking successively $r = r_k$, $k = 1, 2, \ldots$, in (4.13) we find a system of homogeneous linear algebraic equations for determining the values of $N(r_k)$:

$$N(r_k) = \frac{\gamma}{2\pi D} \sum_k N(r_k)\, A_{k,n}; \tag{4.14}$$

$$A_{k,n} \equiv A(d_{nk}) = \left(\frac{\eta}{2} e^{\tau/L^2} - 1\right) K_0\left(\frac{d_{kn}}{L}\right) - \frac{\eta}{2} e^{\tau/L^2} K_0\left(\beta_{kn} \frac{d_{kn}}{L}\right), \tag{4.15}$$

where $d_{kn} = r_k - r_n$, and $\beta_{kn} = \ln 2\tau/L d_{kn}$. From the solvability condition of system (4.14) — namely, the vanishing of its determinant — one can evaluate any one of the quantities η, γ, δ_{kn}, L and τ, given the others. Thus description of a heterogeneous reactor containing m linear fuel blocks leads to the solution of a system of linear algebraic equations, the coefficients of which are expressed in terms of the incomplete MacDonald function $K_0(\beta, \varrho)$. This tunction, as was shown in § 1 of Chapter VII, can be simply expressed in terms of the tabulated generalized exponential integrals, (c.f. (1.24) of Chapter VII). The solution of the characteristic equation and subsequent evaluation of $N(\boldsymbol{r})$ are thereby substantially simplified.

In limiting cases the appropriate asymptotic expansion of $K_0(\beta, \varrho)$ facilitates the analysis. As we showed in § 10 of Chapter II, the parameter in this expansion is $|\beta \sinh \beta|$, which in the present case is equal to

$$|\varrho \sinh \beta| = \left|\frac{\tau}{L^2} - \frac{d^2}{4\tau}\right|. \tag{4.16}$$

If this quantity is much larger than one, then, by (10.28) of Chapter II,

$$K_0(\beta, \varrho) = K_0(\varrho) - (\tau/L^2 - d^2/4\tau)^{-1} \exp(-\tau/L^2 - d^2/4\tau). \tag{4.17}$$

Thus for $|\tau/L^2 - d^2/4\tau| \gg 1$, the coefficients A_{kn} may be written approximately as

$$A(d) \approx \frac{\eta}{2}\left(\frac{\tau}{L^2} - d^2/4\tau\right)^{-1} \exp(-d^2/4\tau) - K_0\left(\frac{d}{L}\right). \tag{4.18}$$

Conversely, if (4.16) is much smaller than one, then, by the (1.28) and (7.5)—(7.9) of Chapter II, the first term in expansion of $K_0(\beta, \varrho)$ is

$$K_0(\beta, \varrho) \approx e^{-\varrho}[\beta(1+\varrho) - \varrho \sinh \beta] \tag{4.19}$$

and an approximate expression for the A_{kn} can be written as

$$\begin{gathered} A(d) \approx \left(\frac{\eta}{2} e^{\tau/L^2} - 1\right) K_0\left(\frac{L}{d}\right) - \frac{\eta}{2} e^{\tau/L^2} \\ \times \left[\left(1 + \frac{d}{L}\right) \ln \frac{2\tau}{Ld} - \frac{\tau}{L^2} + \frac{d^2}{4\tau}\right]. \end{gathered} \tag{4.20}$$

Other asymptotic expansions can be found if we employ the relation between $K_0(\beta, \varrho)$ and the incomplete Weber integrals. Thus, from (4.21) of Chapter V

$$\begin{gathered} K_0\left(\beta, \frac{d}{L}\right) = K_0\left(\frac{d}{L}\right) \\ - \alpha e^{-\tau/L^2}\left[K_0\left(\frac{d}{L}\right) \tilde{Q}_0\left(\frac{\tau}{L^2}, \frac{d}{L}\right) + I_0\left(\frac{d}{L}\right) \tilde{P}_0\left(\frac{\tau}{L^2}, \frac{d}{L}\right)\right] \end{gathered} \tag{4.21}$$

If now $dL/2\tau \ll 1$, then, by (3.24) and (3.46) of Chapter IV and the relation $K_0(x) I_1(x) + K_1(x) I_0(x) = 1/x$, we find

$$K_0\left(\beta, \frac{d}{L}\right) \approx K_0\left(\frac{d}{L}\right) - \frac{L^2}{\tau} e^{-d^2/4\tau}. \tag{4.22}$$

Thus the A_{kn} can be written approximately as

$$A(d) = \frac{L^2}{\tau} \exp(\tau/L^2 - d^2/4\tau) - K_0(d/L). \tag{4.23}$$

If $dL/2\tau \gg 1$, then again from (3.24) and (3.46) of the same Chapter we find

$$K_0\left(\beta, \frac{d}{L}\right) \approx (1 - 2e^{\tau/L^2}) K_0\left(\frac{d}{L}\right), \tag{4.24}$$

and the corresponding expression for the $A_{kn} = A(d_{kn})$ is

$$A(d) = \left[\eta \exp\left(\frac{2\tau}{L^2}\right) - 1\right] K_0\left(\frac{d}{L}\right). \tag{4.25}$$

To conclude this section we point out still another problem connected with the study of moderated neutrons. In [74], which is concerned with the theory of neutron moderation by elastic collisions with heavy water nuclei, the method of moments is applied to find an approximate solution (c.f. [75]). We shall show that the spectrum of the moderated neutrons may be described by a function of the form

$$F_1(x) = \frac{1}{a} x^{1/a} \int_x^\infty \frac{F_0(t)}{x^{1+1/a}} dt, \tag{4.26}$$

where $x = tv/l_0$; v is the velocity of the neutrons, t is time, l_0 is the mean free path, and $F_0(x)$ is

$$F_0(x) = Ax^{2/1-r^2} \exp\left(-\frac{b}{x} - x\right). \tag{4.27}$$

The parameters a, b and $r = (M-1)/(M+1)$ depend only on the mass number of the moderating medium (c.f. [74]).

$F_1(x)$ plays a major role in neutron spectrometry as well as in reactor theory, (c.f. [76]). It is not difficult to verify that this function is expressible in terms of incomplete cylindrical functions. From (4.27) we have

$$F_1(x) = \frac{A}{a} x^{1/a} \int_x^\infty \frac{\exp\left(-\frac{b}{t} - t\right)}{t^{\nu+1}} dt, \tag{4.28}$$

where $\nu = 1/a - 2/(1-r^2)$. Introducing a new variable of integration by $t = \sqrt{b}\, e^u$, we find

$$F_1(x) = \frac{A}{a} x^{1/a} b^{-\nu/2} \int_{\ln x/\sqrt{b}}^\infty e^{-2\sqrt{b}\cosh u - \nu u} du. \tag{4.29}$$

For half-odd ν, such integrals can be expressed in terms of the probability integral as was shown in § 4 of Chapter III. For arbitrary ν the integral on the right in (4.29) is similar to the integral in (4.8). Comparing (4.29) with (4.12) of Chapter V, one sees that $F_1(x)$ can be expressed either by incomplete cylindrical functions of Bessel form, or by incomplete Weber integrals, or by incomplete cylindrical functions of Poisson form for integer $\nu = n$. Thus, from (4.11) and (4.12) of Chapter V, with $\alpha = 1$, $p^2 = x$, and b replaced by $2\sqrt{b}$, we have

$$F_1(x) = 2\,\frac{A}{a}\,b^{-\nu/2}x^{1/a}e^{-2x}\,[K_\nu(2\sqrt{b})\,\tilde{Q}_\nu(x, 2\sqrt{b}) + I_\nu(2\sqrt{b})\,\tilde{P}_\nu(x, 2\sqrt{b})]. \tag{4.30}$$

This formula may be convenient for analytic study of limiting cases, when $\sqrt{b}/x$ is either large or small compared with one. If $\sqrt{b}/x < 1$, then by (3.24) and (3.46) of Chapter IV and the relation $K_\nu(z)\,I_{\nu+1}(z) + K_{\nu+1}(z)\,I_\nu(z) = 1/z$, we have

$$F_1(x) \approx \frac{A}{a}\,x^{1/a-\nu-1}\exp\left(-x - \frac{b}{x}\right). \tag{4.31}$$

If, on the other hand $\sqrt{b}/x > 1$, then, using (3.34) and (3.48) of the same Chapter, we find

$$F_1(x) \approx 2b^{-\nu/2}\,\frac{A}{a}\,x^{1/a}K_\nu(2\sqrt{b}). \tag{4.32}$$

Formula (4.30) may also facilitate numerical computations, particulary for integer $\nu = n$, since in this case the incomplete weber integrals $\tilde{Q}_n(x, z)$ and $\tilde{P}_n(x, z)$, are simple to express in terms of $\tilde{Q}_0(x, z)$ and $\tilde{P}_0(x, z)$ (c.f. (3.17) and (3.19) of Chapter II). The latter can then be evaluated with available tables, (c.f. Chapter XI).

5. Application of Incomplete Cylindrical Functions to the Solution of the Diffusion Equation

One of the most extensively applied methods for solution of differential equations in mathematical physics is that of separation of variables. If for the given boundary conditions it is possible to find a system of coordinates in which the partial differential equation under consideration is separable, then this method readily permits one to construct the desired solution. This solution, however, is usually obtained as an infinite series, which often converges very slowly. This makes it difficult to analyze the behavior of such a solution, i.e. singularities, etc. It is therefore appropriate to seek closed form solutions if only in the form of integrals containing known functions. As is well known, such solutions can be obtained by Green's function methods. Thus, the solution of the diffusion equation

$$\nabla^2\Psi(\mathbf{r}, t) - a^2\frac{\partial\Psi}{\partial t} = -4\pi\varrho(\mathbf{r}, t), \tag{5.1}$$

satisfying inhomogeneous boundary conditions can be written in the integral form [5, Vol. I, p. 795]:

$$\Psi(r, t) = \int_0^t dt_0 \int \varrho(\boldsymbol{r}_0, t_0)\, G(\boldsymbol{r}, t \,|\, \boldsymbol{r}_0, t_0)\, d\boldsymbol{r}_0$$
$$+ \frac{1}{4\pi} \int_0^t dt_0 \int [G \operatorname{grad}_0 \Psi - \Psi \operatorname{grad}_0 G]\, ds_0 + \frac{a^2}{4\pi} \int [\Psi G]_{t_0=0}\, d\boldsymbol{r}_0. \tag{5.2}$$

The Green's function $G(\boldsymbol{r}, t; \boldsymbol{r}_0, t_0)$ satisfies the equation

$$\nabla^2 G - a^2 \frac{\partial G}{\partial t_0} = -4\pi\delta(\boldsymbol{r} - \boldsymbol{r}_0)\, \delta(t - t_0) \tag{5.3}$$

and is chosen so as to satisfy the appropriate homogeneous boundary conditions for the problem. The first term in (5.2) characterizes the effect of sources, the second the effects of the boundaries, and the third, the influence of the initial conditions.

In a number of cases the solution in integral form, Eq. (5.2), may be expressed in terms of incomplete cylindrical functions. We limit ourselves here to fairly simple cases of unbounded spatial regions and zero n itial and boundary conditions. Then Eq. (5.2) takes the form

$$\Psi(\boldsymbol{r}, t) = \int_0^t dt_0 \int \varrho(\boldsymbol{r}_0, t_0)\, G(\boldsymbol{r} - \boldsymbol{r}_0, t - t_0)\, d\boldsymbol{r}_0, \tag{5.4}$$

and the Green's function is [5, Vol. I., p. 796]

$$G(R, \tau) = \frac{4\pi}{a^2} \left(\frac{a}{2\sqrt{\pi\tau}}\right)^n \exp\left(-\frac{a^2 R^2}{4\tau}\right), \tag{5.5}$$

where n is equal to one, two, or three, depending on the number of spatial dimensions. If $\varrho(\nu, t)$, which characterizes the distribution of sources, is taken to be of the form

$$\varrho(\boldsymbol{r}, t) = T\delta(\boldsymbol{r})\, f(t), \tag{5.6}$$

then (5.4), can be written in view of (5.5) as

$$\Psi(\boldsymbol{r}, t) = \frac{4\pi T}{a^2} \left(\frac{a}{2\sqrt{\pi}}\right)^n \int_0^t \frac{f(t_0)}{(\sqrt{t - t_0})^n} \exp\left(-\frac{a^2 r^2}{4(t - t_0)}\right) dt_0. \tag{5.7}$$

From this it is evident that for an instantaneously acting source, i.e. for $f(t) = \delta(t)$), $\Psi(\boldsymbol{r}, t)$ is the influence function of an impulsive source with total intensity T:

$$\Psi(\boldsymbol{r}, t) = \frac{4\pi}{a^2}\, T \left(\frac{a}{2\sqrt{\pi}}\right)^n \frac{1}{t^{n/2}} \exp\left(-\frac{a^2 r^2}{4t}\right) \tag{5.8}$$

In a number of cases $f(t)$ varies exponentially

$$f(t) = \exp(-\sigma t), \tag{5.9}$$

where σ is some complex number. Such an f arises for example in the study of particle diffusion from a linear radioactive source. Substituting (5.9) into (5.7), we obtain

$$\Psi(\mathbf{r}, t) = \frac{4\pi}{a^2} T \left(\frac{a}{2\sqrt{\pi}}\right)^n \int_0^t e^{-\sigma t_0 - a^2 r^2/4(t-t_0)} \frac{dt_0}{(t - t_0)^{n/2}}. \tag{5.10}$$

Introducing a new variable of integration by $t - t_0 = ar/(2\sqrt{\sigma})\, e^{\xi}$, we find

$$\Psi(\mathbf{r}, t) = \left(\frac{a\sqrt{\sigma}}{2\pi r}\right)^{n/2-1} T e^{-\sigma t} \int_{-\infty}^{\beta} e^{\varrho \sinh \xi - (n/2-1)\xi}\, d\xi, \tag{5.11}$$

where $\varrho = ra\sqrt{\sigma}$ and $\beta = \ln\,[2\sqrt{\sigma}t/(ar)]$. Recalling the definition of the incomplete cylindrical function of Bessel form, we have

$$\Psi(\mathbf{r}, t) = \pi i \left(\frac{a\sqrt{\sigma}}{2\pi r}\right)^{n/2-1} T e^{-\sigma t} [\varepsilon_{n/2-1}(\beta, \varrho) - \varepsilon_{n/2-1}(-\infty, \varrho)]. \tag{5.12}$$

In the one and three dimensional cases, $n = 1$ and $n = 3$, $\Psi(\mathbf{r}, t)$ can be expressed by $\varepsilon_{1/2}(\omega, z)$ and $\varepsilon_{1/2}(\omega, z)$, respectively, which in turn, as was shown in § 4 of Chapter III, can be expressed by the probability integral. Thus, from (4.5) and (4.12) of Chapter III, we have for $n = 3$

$$\Psi(\mathbf{r}, t) = \frac{T}{r} \left\{ e^{-\sigma t} \cos \varrho - \frac{2}{\sqrt{\pi}} \exp\left(-\frac{a^2 r^2}{4t}\right) \operatorname{Im}\,[W(z)] \right\}, \tag{5.13}$$

where Im $(W(z))$ denotes the imaginary part of the probability integral

$$W(z) = e^{-z^2} \int_0^z e^{t^2}\, dt \tag{5.14}$$

and

$$z = \sqrt{\sigma t} + i \sqrt{\frac{a^2 r^2}{4t}}. \tag{5.15}$$

Similarly, for the one dimensional case, we get from (4.6) and (4.14) of Chapter III,

$$\Psi(x, t) = \frac{2\pi}{a\sqrt{\sigma}} T \left\{ \frac{2}{\sqrt{\pi}} e^{-a^2 x^2/4t} \operatorname{Re}[W(z_0)] - e^{-\sigma t} \sin ax\sqrt{\sigma} \right\}, \tag{5.16}$$

where

$$z_0 = \sqrt{\sigma t} + i \sqrt{\frac{a^2 x^2}{4t}}. \tag{5.17}$$

If either $|z|$ or $|z_0|$ is large compared with unity, which corresponds physically either to large times or large distances from the source, then, using the asymptotic properties of $W(z)$,

$$W(z) \approx i \frac{\sqrt{\pi}}{2} e^{-z^2} + \frac{1}{2z}, \tag{5.18}$$

we find the following approximations for $\Psi(\boldsymbol{r}, t)$:
in the three dimensional case:

$$\Psi(\boldsymbol{r}, t) \approx \frac{aT}{2\sqrt{\pi t}} \cdot \frac{1}{\sigma t + \frac{a^2 r^2}{4t}} \exp\left(-\frac{a^2 r^2}{4t}\right) \tag{5.19}$$

and in the one dimensional case:

$$\Psi(x, t) \approx T\frac{2\sqrt{\pi t}}{a} \frac{1}{\sigma t + \frac{a^2 x^2}{4t}} \exp\left(-\frac{a^2 x^2}{4t}\right). \tag{5.20}$$

In the two dimensional case, which corresponds to a linear source of unbounded dimension, the solution $\Psi(\boldsymbol{r}, t)$, by (5.12), has the form

$$\Psi(\boldsymbol{r}, t) = \pi i T \mathrm{e}^{-\sigma t}\left[\varepsilon_0(\beta, \varrho) - \varepsilon_0(-\infty, \varrho)\right], \tag{5.21}$$

i.e. it can be expressed in terms of incomplete cylindrical functions of zero order.

Using the relations between $\varepsilon_{n/2-1}(w, z)$ and the other incomplete cylindrical functions, one can obtain various integral representations of the desired solution, which may prove to be more advantageous in particular cases. When finding asymptotic expansions, for example, it is convenient to use (6.7) of Chapter III and to write Ψ as

$$\Psi(\boldsymbol{r}, t) = i\pi T\left(\frac{a\sqrt{\sigma}}{2\pi r}\right)^{n/2-1} \mathrm{e}^{-\sigma t}\Phi_{n/2-1}(\beta, \varrho), \tag{5.22}$$

where $n = 1, 2, 3$; $\beta = \ln 2\sqrt{\sigma t}/(ra)$; and $\Phi_\nu(\beta, \varrho)$ is an incomplete cylindrical function:

$$\Phi_\nu(\beta, \varrho) = \frac{1}{\pi i}\int_{-\infty}^{\beta} \mathrm{e}^{\varrho \sinh u - \nu u}\, du. \tag{5.23}$$

Using now the relation (6.22) of Chapter III for $\operatorname{Im}(b) = 0$, we find the following asymptotic development of $\Psi(r, t)$ for $\varrho \cosh\beta = \sigma t + a^2 r^2/yt \gg 1$

$$\Psi(\boldsymbol{r}, t) \approx T\left(\frac{a^2}{4\pi t}\right)^{n/2-1} \frac{\exp\left(-\frac{a^2 r^2}{4t}\right)}{\sigma t + \frac{a^2 r^2}{4t}}$$

$$\times \sum_{k=0}^{\infty}(-1)^k\left(\sigma t + \frac{a^2 r^2}{4t}\right)^{-k} \frac{\Gamma\left(k + 2 - \frac{n}{2}\right)}{\Gamma\left(2 - \frac{n}{2}\right)} \tag{5.24}$$

$$\times F\left(-k, k+1; 2 - \frac{n}{2}; \frac{\sigma t}{\sigma t + \frac{a^2 r^2}{4t}}\right).$$

Limiting ourselves to the first term of the expansion, we find from (5.24)

$$\Psi(\mathbf{r}, t) \approx T\left(\frac{a^2}{4\pi t}\right)^{n/2-1} \frac{\exp\left(-\frac{a^2 r^2}{4t}\right)}{\sigma t + \frac{a^2 r^2}{4t}}. \tag{5.25}$$

Taking here $n = 3$ and $n = 1$, we obtain (5.19) and (5.20). For the two dimensional case $n = 2$, the asymptotic representation for $\Psi(r, t)$ has the form

$$\Psi(\mathbf{r}, t) \approx T \frac{\exp\left(-\frac{a^2 r^2}{4t}\right)}{\sigma t + \frac{a^2 r^2}{4t}}. \tag{5.26}$$

In addition to the expressions just obtained it is also of interest to express the solution by the incomplete Weber integrals $Q_\nu(x, z)$ and $P_\nu(x, z)$. For this purpose we use (4.2) of Chapter IV; we have

$$\Psi(\mathbf{r}, t) = \pi T\left(\frac{a\sqrt{\sigma}}{2\pi r}\right)^{n/2-1} \mathrm{e}^{-\sigma t}\{N_{n/2-1}(\varrho)\,[Q_{n/2-1}(\sigma t, \varrho)$$
$$- Q_{n/2-1}(0, \varrho)] + N_{n/2-1}(\varrho)\,[P_{n/2-1}(\sigma t, \varrho) - P_{n/2-1}(0, \varrho)]\}.$$

But, by (3.32) and (3.43) of Chapter IV, $Q_\nu(0, \varrho) = 1$ and $P_\nu(0, \varrho) = 0$, so that

$$\begin{aligned}\Psi(\mathbf{r}, t) = \pi T\left(\frac{a\sqrt{\sigma}}{2\pi r}\right)^{n/2-1} \mathrm{e}^{-\sigma t}\{N_{n/2-1}(\varrho)\cdot P_{n/2-1}(\sigma t, \varrho) \\ - N_{n/2-1}(\varrho)\,[1 - Q_{n/2-1}(\sigma t, \varrho)]\},\end{aligned} \tag{5.27}$$

where as before $\varrho = ra\sqrt{\sigma}$. This formula also permits one to obtain developments of $\Psi(r, t)$ with respect to the parameter $\varrho/(2\sigma t) = ar/(2\sqrt{\sigma t})$. Thus, if this parameter is small compared to unity, then from (3.23), (3.39) of Chapter IV we find

$$\Psi(\mathbf{r}, t) \approx \pi T\left(\frac{a\sqrt{\sigma}}{2\pi r}\right)^{n/2-1} \left[\left(\frac{ar}{2\sqrt{\sigma t}}\right)^{n/2} \frac{2}{\pi\sqrt{\sigma}\, ar} \mathrm{e}^{-a^2r^2/4t} - \mathrm{e}^{-\sigma t} N_{n/2-1}(\varrho)\right]. \tag{5.28}$$

In particular, we have in the two dimensional case

$$\Psi(\mathbf{r}, t) \approx T\left[\frac{1}{\sigma t}\mathrm{e}^{-a^2r^2/4t} - \pi \mathrm{e}^{-\sigma t} N_0(\varrho)\right]. \tag{5.29}$$

We give still another example which illustrates the application of incomplete cylindrical functions in the solution of the diffusion equation. We consider the problem of time-dependent diffusion of a gas with dissociation [77]. Let there be a linear source of strength T located at the origin, which is activated at $t = 0$. The diffusion equation, with account of loss, then takes the form

$$\frac{1}{a^2}\nabla^2\Psi - \gamma\Psi - \frac{\partial\Psi}{\partial t} = -4\pi T_0\delta(\boldsymbol{r}), \tag{5.30}$$

where γ is the loss constant. We suppose further that at $t = 0$ the concentration of gas was everywhere zero. Then the solution of (5.30), which vanishes at infinity according to (5.4), takes the form

$$\Psi(\boldsymbol{r}, t) = T_0 \int_0^t \tilde{G}(r, t - t_0)\, dt_0, \tag{5.31}$$

where now the Green's function $\tilde{G}(\boldsymbol{r}, t)$ satisfies the homogeneous differential equation

$$\frac{1}{a^2}\nabla^2\tilde{G} - \gamma\tilde{G} - \frac{\partial\tilde{G}}{\partial t} = 0. \tag{5.32}$$

To solve (5.32), we set $\tilde{G} = G\mathrm{e}^{-\gamma t}$, and obtain the usual two-dimensional homogeneous diffusion equation for G

$$\nabla^2 G - a^2\frac{\partial G}{\partial t} = 0, \tag{5.33}$$

whose Green's function, according to (5.5), is

$$G = \frac{1}{\tau}\exp\left(-\frac{a^2r^2}{4\tau}\right). \tag{5.34}$$

Consequently, the Green's function for (5.5) takes the form

$$\tilde{G} = \frac{1}{\tau}\exp\left(-\frac{a^2r^2}{4\tau} - \gamma\tau\right) \tag{5.35}$$

and substitution of (5.35) in (5.31) gives

$$\Psi(\boldsymbol{r}, t) = T_0 \int_0^t \mathrm{e}^{-r^2a^2/4\tau - \gamma\tau}\frac{d\tau}{\tau}. \tag{5.36}$$

Introducing a new variable of integration according to $\tau = [ar/(2\sqrt{\gamma})]\,\mathrm{e}^u$, we find

$$\Psi(\boldsymbol{r}, t) = T_0 \int_{-\infty}^{\beta} \mathrm{e}^{-ar\sqrt{\gamma}\cosh u}\, du = T_0\left(\int_0^{\beta} \mathrm{e}^{-ar\sqrt{\gamma}\cosh u}\, du - \int_0^{-\infty} \mathrm{e}^{-ar\sqrt{\gamma}\cosh u}\, du\right),$$

where $\beta = \ln[2\sqrt{\gamma t}/(ar)]$. The second term represents the MacDonald function, and the first term the incomplete MacDonald function $K_0(\beta, ar\sqrt{\gamma})$. Therefore we get finally

$$\Psi(\boldsymbol{r}, t) = T_0[K_0(a\,r\sqrt{\gamma}) + K_0(\beta, ar\sqrt{\gamma})]. \tag{5.37}$$

Formula (5.37) lends itself to immediate numerical evaluation because the incomplete MacDonald functions are connected with tabulated integrals, (c.f. § 1 of Chapter VII, and § 2 of Chapter XI).

From (5.37) for $t \to \infty$ ($\beta \to \infty$) we obtain the well known steady state result for the diffusion of a gas with losses [68, 77].

$$\Psi(\mathbf{r}, t)_{t\to\infty} = 2T_0 K_0(ra\sqrt{\gamma}). \tag{5.38}$$

In this section we have shown by means of simple examples how incomplete cylindrical functions appear naturally in the solution of the diffusion equation for sources which vary exponentially with time, or for situations where there are either losses or capture associated with the diffusion process. Incomplete cylindrical functions may also appear in the search for closed form solutions of the diffusion or heat conduction equation, in the more general case where volume or surface sources vary exponentially with time; In fact, since the Green's function for the diffusion equation varies with time as

$$G(R, \tau) = \frac{1}{\tau^n} \exp\left(-\frac{R}{\tau}\right),$$

it follows that the integral with respect to time in (5.2) leads immediately to incomplete cylindrical functions.

We remark that incomplete cylindrical functions often appear in the solutions of problems of mathematical physics dealing with unsteady (transient) process. This statement is supported by the examples considered above, and also by the results given in § 4 of Chapter VII and §§ 1, 2, and 4 of Chapter VIII (c.f. also [6, 7, 21, 30, 31, 37, 41, and 44].)

Chapter X

Other Applied Problems Leading to Incomplete Cylindrical Functions

In this chapter we present in outline form some further problems which lend themselves to analysis with the theory of incomplete cylindrical functions. Because they come from diverse fields, we will not attempt to present them systematically.

1. Excitation of Betatron Oscillations

In the study of excitation of betatron oscillations synchronized with an oscillation impulse in accelerators with hard focussing, it has been shown [78] that the oscillation amplitude is characterized by th eintegral

$$\int_0^t e^{i\Delta_0 t + i\,\Delta_1/\Omega \cos\,\Omega t}\,dt. \tag{1.1}$$

This integral is immediately expressible by means of the function $\varepsilon_\mu(w, z)$ Taking $\Omega t = -iu - \pi/2$, we find

$$\int_0^t e^{i\Delta_0 t + i\,\Delta_1/\Omega\,\cos\Omega t}\,dt = \frac{\pi}{\Omega}e^{-\,i\mu\pi/2}\left[\varepsilon_{-\mu}\left(i\Omega t + \frac{i\pi}{2}, \varrho\right) - \varepsilon_{-\mu}\frac{i\pi}{2}, \varrho)\right],$$

where $\mu = \Delta_0/\Omega$ and $\varrho = \Delta_1/\Omega$. Further analysis follows the theory of these functions.

2. Transient Processes in Electrical and Microphonic Circuits

As is well known, most of the processes in electrical, microphonic, and other circuits can be described by linear differential equations. There is considerable interest in the solution of these equations in the transient regime. In [79] it is shown that the solution of the appropriate differential equation for this regime leads to analysis of the function

$$Y = \exp\,(-\Delta t + \varrho\,\sin\,\omega t)\int_0^t e^{\Delta\tau - \varrho\sin\omega\tau}\,d\tau, \tag{2.1}$$

where Δ and ϱ are parameters characterizing the given circuit. The integral on the right side of this formula can, with the aid of the obvious change of variable $u = i\omega\tau$, be expressed in terms of the functions $\varepsilon_\nu(w, z)$:

$$Y = \frac{\pi}{\omega} e^{-\Delta t + \varrho \sin \omega t} \varepsilon_\mu(i\omega t, i\varrho), \tag{2.2}$$

where $\mu = i\Delta/\omega$.

We have already dealt with similar expressions in the solution of problems of absorption and scattering of light by atomic systems and the motion of charged particles in electromagnetic fields, c.f. § 1, 2, and 4 of Chapter VIII. These problems are also of non-steady character, and the appearance of incomplete cylindrical functions characterizes this transience .

3. Influence of Optical Systems on the Amplitude of Transmitted Waves

In optical systems attenuation and phase changes of the incident waves take place. Problems in this area were considered in § 6 of Chapter VII, where the diffractive properties of optical systems were treated. References [54] and [80] study in detail methods of computing the complex amplitude of the transmitted waves in the focal and image planes by means of a complex transparency coefficient, which embodies the optical properties of the system (abberation, focal defect, diffraction, etc.).

In some cases of weak absorption, the amplitude in the focal plane is given (up to a multiplicative constant) by the formulae [80]:

$$u_1(\xi) = \int_{-x_0/2}^{x_0/2} e^{-i(k\xi/f - b)t + i\alpha_n \cos n\varepsilon t}\, dt; \tag{3.1}$$

$$u_2(\xi) = \int_{-x_0/2}^{x_0/2} e^{-i(k\xi/f - b)t + i\beta_n \sin n\varepsilon t}\, dt: \tag{3.2}$$

$$u_3(\xi) = \int_{-x_0/2}^{x_0/2} e^{-i(k\xi/f - b)t + i\alpha_n \cos n\varepsilon t + i\beta_m \sin m\varepsilon t}\, dt. \tag{3.3}$$

In [80] all these integrals are analyzed by expanding $\exp(i\alpha_n \cos n\varepsilon t)$ and $\exp(i\beta_m \sin m\varepsilon t)$ in series of Bessel functions. However, using the methods considered in detail above, they can also be expressed without difficulty in terms of incomplete cylindrical functions.

4. Density Perturbation in a Gas due to a Rapidly Moving Body

In the study of rapidly moving bodies in very diffuse gases, and in particular, for the study of the motion of artificial satellites in the upper layers of the atmosphere, there is considerable interest in determining the density distribution of gas in the neighborhood of a moving body.

A large number of works has been devoted to this question; see especially [38]. Here we shall touch only on the formulation of the problem and on those of its aspects which can be conveniently analyzed with the the theory of incomplete cylindrical functions.

Since in a rarified gas the mean free path of a particle is long in comparison with its characteristic dimensions, the solution of rare gas problems requires a kinetic theory approach. The distribution function $f(\boldsymbol{r}, \boldsymbol{v}, t)$ for neutral particles satisfies the kinetic equation

$$\frac{\partial f}{\partial t} + \boldsymbol{v}\frac{\partial f}{\partial r} - \frac{1}{M}\frac{\partial U}{\partial \boldsymbol{r}}\frac{\partial f}{\partial \boldsymbol{v}} = 0, \tag{4.1}$$

where M is the mass of the particle, and $U = U(\boldsymbol{r}, t)$ is the potential energy of the interaction of particles with the surface of the body. If the latter is moving uniformly with velocity $\boldsymbol{v}_0$, then $U = U(\boldsymbol{r} - \boldsymbol{v}_0 t)$. In this case it is convenient to use a system of coordinates fixed on the moving body. The distribution of particles will be stationary in this system, and (4.1) takes the form

$$\boldsymbol{v}\frac{\partial f}{\partial \boldsymbol{r}} - \frac{1}{M}\frac{\partial U}{\partial \boldsymbol{r}}\frac{\partial f}{\partial \boldsymbol{u}} = 0, \tag{4.2}$$

where $\boldsymbol{u} = \boldsymbol{v} + \boldsymbol{v}_0$.

In the system of coordinates described, the stream of particles has an incident velocity $-\boldsymbol{v}_0$. This indicates that at large distances from the body one should expect find the usual Maxwellian distribution, i.e.:

$$\lim_{r\to\infty} f(\boldsymbol{r}, \boldsymbol{u}, t) = f_0(u) = n_0\left(\frac{M}{2\pi k T}\right)^{3/2}\exp\left(-\frac{Mu^2}{2kT}\right). \tag{4.3}$$

Here n_0 is the undistributed density of particles and T is their temperature.

The interaction of particles with the surface of the body may be described either by giving the particular form of the potential $U(\boldsymbol{r})$, as was done in our study of diffraction by a screen of given form, or by introducing definite boundary conditions for the distribution function on the surface of the body. For example, if all particles are absorbed by the body, then $f(\boldsymbol{r}, \boldsymbol{u}) = 0$ for $\boldsymbol{n}\cdot\boldsymbol{v} > 0$, where $\boldsymbol{n}$ is the outer unit normal to the body.

Equation (4.2) for charged particles — writing f_e for electrons, and f_i for ions, takes the form

$$\left.\begin{aligned} \vec{\boldsymbol{v}}\frac{\partial f_e}{\partial \boldsymbol{r}} - \left[\left(\frac{e}{m}\frac{\partial\varphi}{\partial \boldsymbol{r}} + \frac{1}{m}\frac{\partial U}{\partial \boldsymbol{r}}\right) - \frac{e}{mc}\,[\boldsymbol{H}\boldsymbol{u}]\right]\frac{\partial f_e}{\partial \boldsymbol{u}} = 0;\\ \boldsymbol{v}\frac{\partial f_i}{\partial \boldsymbol{r}} - \left[\left(\frac{e}{M_i}\frac{\partial\varphi}{\partial \boldsymbol{r}} + \frac{1}{M_i}\frac{\partial U}{\partial \boldsymbol{r}}\right) + \frac{e}{M_i c}\,[\boldsymbol{H}\boldsymbol{u}]\right]\frac{\partial f_i}{\partial \boldsymbol{u}} = 0, \end{aligned}\right\} \tag{4.4}$$

where e is the ionic charge, m is the electron mass, M_i the ionic mass, $\boldsymbol{H}$ is the magnetic field (in the ionosphere, $\boldsymbol{H}$ is the Earth's magnetic field) and φ is the potential of the electric field, determined from the Poisson equation

$$\nabla^2\varphi = 4\pi e\,[\int f_e(\boldsymbol{r}, \boldsymbol{u})\, d\boldsymbol{u} - \int f_i(\boldsymbol{r}, \boldsymbol{u})\, d\boldsymbol{u}], \tag{4.5}$$

We assume that f_i and f_e satisfy (4.3) at infinity, and thus that $\varphi(r) \to 0$ as $r \to \infty$.

It is convenient to write f as the sum

$$f(\boldsymbol{r}, \boldsymbol{u}) = f_1(\boldsymbol{r}, \boldsymbol{u}) + f_2, (\boldsymbol{r}, \boldsymbol{u}), \tag{4.6}$$

where $f_1(\boldsymbol{r}, \boldsymbol{u})$ is the distribution function of particles which do not collide with the body, and f_2 is that of the reflected particles.

At sufficiently large distances from the body we can take $f \approx f_1$. In this approximation the actual form of the body is immaterial. It is necessary only to know its maximum cross section S perpendicular to the direction of motion. In other words, in this approximation the body may be regarded as a plane screen.

Moreover, if v_0 is much larger than the thermal speed, then in the direction of the z-axis (parallel to $\boldsymbol{v}_0$) all the particles will have roughly the same velocity v_0. The problem of determining the density is thus a two dimensional one, and the distribution function in the (x, y)-plane at the initial moment of time will equal:

$$f(x, y; u_x, u_y; 0) = \Omega(D, D_s)\, n_0 \frac{M}{2\pi kT} \exp\left[-\frac{M(u_x^2 + u_y^2)}{2kT}\right], \tag{4.7}$$

where $\Omega(D, D_s)$ is a unit step function, equal to zero for all points D interior to the screen and to one outside of it. At all other times the distribution function of the freely moving particles, will have according to (4.2), the form

$$f(x, y; u_x, u_y; t) = f(x_0, y_0; u_x, u_y;\ 0), \tag{4.8}$$

where $x_0 = x - u_x t$, $y_0 = y - u_y t$, $t = z/v_0$, and the expression on the right is determined from Eq. (4.7).

It is evident that the particle density and density perturbation will be equal to

$$\begin{aligned} n(x, y, z) &= \int f\left(x, y; u_x, u_y; \frac{z}{v_0}\right) du_x\, du_y \\ &= \left(\frac{v_0}{z}\right)^2 \int f\left(x_0, y_0; \frac{x - x_0}{z} v_0, \frac{y - y_0}{z} v_0; 0\right) dx_0 dy_0 \\ &= \frac{n_0 a}{z^2 \pi} \int \Omega(D, D_s) \exp\left[-\frac{a}{z^2}(\varrho - \varrho_0)^2\right] d\boldsymbol{\varrho}_0, \end{aligned} \tag{4.9}$$

and

$$\Delta n(x, y, z) = n_0 - n = \frac{n_0 a}{\pi} \int \Omega(D_s, D) \exp\left[-\frac{a}{z^2}(\varrho - \varrho_0)^2\right] d\varrho_0, \quad (4.10)$$

where the non-dimensional quantity $\varphi = E/kT$ is the ratio of the kinetic energy of the incident particles to their thermal energy, n_0 is the undisturbed density of the gas, and $\varrho = \sqrt{x^2 + y^2}$. Here the z-axis is chosen in the direction of motion of the body, and the unit step function $\Omega(D_s, D) = 1 - \Omega(D, D_s)$ is introduced so that the integration in (4.10) is carried out over the entire maximal perpendicular section.

If this section is a circle of radius R_0, then (4.10) takes the form

$$\begin{aligned} \Delta n(x, y, z) = {} & \frac{n_0 a}{\pi z^2} \int_0^{R_0} \varrho_0 \exp\left[-\frac{a\varrho^2}{z^2} - \frac{a\varrho_0^2}{z^2}\right] d\varrho_0 \\ & \times \int_0^{2\pi} \exp\left(\frac{2a\varrho\varrho_0 \cos\varphi}{z^2}\right) d\varphi. \end{aligned} \quad (4.11)$$

The inner integral with respect to φ is immediately expressible in terms of the Bessel function of zero order and purely imaginary argument, so that after changing the integration variable according to $\varrho_0 \sqrt{a} = zt$ we find

$$\Delta n(x, y, z) = 2n_0 \exp\left[-\left(\frac{\varrho}{z}\right)^2 a\right] \int_0^{R_0\sqrt{a}/z} e^{-t^2} I_0\left(2\frac{\varrho\sqrt{a}}{z} t\right) t\, dt. \quad (4.12)$$

If the maximal perpendicular section of the body is a circular sector of radius R_0 and central angle α, then (4.10) can be written as

$$\begin{aligned} \Delta n(x, y, z) = {} & \frac{n_0 a}{\pi z^2} \int_0^{R_0} \varrho_0 \exp\left(-a\frac{\varrho^2 + \varrho_0^2}{z^2}\right) d\varrho_0 \\ & \times \int_0^{\alpha} \exp\left[\frac{2a\varrho_0}{z^2}(x\cos\varphi + y\sin\varphi)\right] d\varphi. \end{aligned} \quad (4.13)$$

In this case the inner integral is expressible in terms of the incomplete cylindrical function $F_0^+(w, z)$ (c.f. § 1 of Chapter II). Therefore, after the change of integration variable $\varrho_0 \sqrt{a} = zt$:

$$\begin{aligned} \Delta n(x, y, z) = n_0 \exp\left(-\frac{a\varrho^2}{z^2}\right) \int_0^{R_0\sqrt{a}/z} t e^{-t^2} \Bigg[& F_0^+\left(\psi - \alpha, \frac{2\varrho\sqrt{a}}{z} t\right) \\ & - F_0^+\left(\psi, \frac{2\varrho\sqrt{a}}{z} t\right)\Bigg] dt, \end{aligned} \quad (4.14)$$

where $\varrho = \sqrt{x^2 + y^2}$ and $\tan\psi = x/y$.

Thus in the case considered the perturbation of density behind a moving body has been expressed by means of incomplete cylindrical functions. It is useful to note that the analysis of the present problem is reminescent of that for the problem of difraction of plane waves by a circular sector, (c.f. § 5 of Chapter VII). The only difference is that in the diffraction problem a given screen cut out a bundle of plane waves, while in the present problem a moving body with the same perpendicular cross section cuts out a "valley" of gas particles when the latter is subject to a maxwellian distribution.

Here we note that even in the more important case when the moving body has a circular perpendicular section, the density perturbation can be expressed with incomplete cylindrical functions. This holds for neutral as well as for charged particles interacting in an external magnetic field [38]. For neutral particles the density perturbation is determined by (4.12). But the integral appearing in (4.12) represents an incomplete Weber integral. Therefore, the expression for the density disturbance takes the form

$$\Delta n(x, y, z) = n_0 e^{-u/c} \tilde{Q}_{0'}\left(\frac{u}{2c}, u\right), \tag{4.15}$$

where $u = 2\varrho R_0 a/z^2$; $c = R_0/\sqrt{x^2 + y^2}$. This formula is convenient for analysis of the limiting cases $c \gg 1$, $c \ll 1$ or $u \ll 1$, $u \gg 1$. In intermediate cases, it may be used for numerical computations with the aid of available tables, (see Chapter XI).

5. Computation of the Sonic Field in Closed Domains

For computing the sonic field set up by some radiator in a closed space, it is necessary to evaluate a function of the form

$$T(\varrho, \alpha) = \int_{\alpha}^{\alpha+\pi/2} e^{-\varrho \sin u}\, du. \tag{5.1}$$

By means of the obvious transformation $u = \pi/2 + \theta$ we find

$$T(\varrho, \alpha) = \int_{\alpha-\pi/2}^{\alpha} e^{-\varrho \cos\theta}\, d\theta = \pi\left[F_0^-(\alpha, \varrho) + F_0^-\left(\frac{\pi}{2} - \alpha, \varrho\right)\right]. \tag{5.2}$$

From this it is clear that $T(\varrho, \alpha)$ can be expressed by means of the incomplete cylindrical function $F_0^-(\omega, z)$.

From (5.2) one sees that

$$T(\varrho, \alpha) = T\left(\varrho, \frac{\pi}{2} - \alpha\right), \tag{5.3}$$

so that T is determined by knowledge of its values in the interval $0 \leq \alpha \leq \pi/4$. In [81] some formulae for estimating this function are given, which were obtained by means of the following approximations for $\exp(-\varrho \sin u)$:

$$\exp(-\varrho \sin u) \approx \begin{cases} e^{-\varrho u}, & 0 \leq u \leq \frac{\pi}{6}; \\ e^{-\varrho} + \frac{9}{\pi^2}(e^{-\varrho/2} - e^{-\varrho})\left(u - \frac{\pi}{2}\right)^2, & \frac{\pi}{6} < u < \frac{\pi}{4}. \end{cases} \tag{5.4}$$

In addition, short tables of T are given, there covering the ranges $\alpha = 0\ (0, 1)\ \pi/4$ and $\varrho = 0\ (0.1)\ 1;\ 1\ (1)\ 10;\ 10\ (10)\ 100$.

6. Exchange Processes between Liquid and Solid Phases

In this section we touch briefly on the mathematical analysis of exchange processes which take place during the flow of a liquid through pores or voids in a column filled with crushed solid particles.

Let two sorts of ions be contained in the column having respective concentrations c and $c_0 - c$ in the fluid and q, $Q - q$ in the solid phase. If we denote by R the rate of flow of liquid along the column and by α the fraction of the liquid phase in the volume, then the conservation equation can be written [7]

$$\frac{\partial c}{\partial t} + R \cdot \frac{\partial c}{\partial X} + \frac{1-\alpha}{\alpha}\frac{\partial q}{\partial t} = 0. \tag{6.1}$$

For this process, changes in concentration in the solid phase can be described by

$$\frac{\partial q}{\partial t} = K\left[c(Q-q) - \frac{1}{K}q(c_0 - c)\right]. \tag{6.2}$$

For the case of absorption on the surface of the solid, (6.2) takes the form:

$$\frac{\partial q}{\partial t} = K'\left[c(Q-q) - \frac{1}{K'}q\right]. \tag{6.3}$$

Here K, K' are physical characteristics which determine the speed of flow of the exchange processes, and are usually taken to be independent of time.

If in (6.1) and (6.2) the non-dimensional quantities

$$\left.\begin{aligned} x &= \frac{KQ(1-\alpha)}{\alpha R}X; \quad y = \frac{Kc_0}{R}(Rt - X); \\ u &= c/c_0; \quad v = q/Q; \quad r = 1/K \end{aligned}\right\} \tag{6.4}$$

are introduced, and if in (6.1) and (6.3) we put

$$\left.\begin{aligned} x &= \frac{K'Q(1-\alpha)}{\alpha R} X; \quad y = \frac{K'(1+K'c_0)}{K'R}(Rt - X); \\ u &= \frac{c}{c_0}; \quad v = \frac{q(1+K'c_0)}{(K'c_0Q)}; \quad r = \frac{1}{(1+K'c_0)}, \end{aligned}\right\} \tag{6.5}$$

respectively, then the equations may be written in the form

$$\left.\begin{aligned} &\frac{\partial u}{\partial x} + \frac{\partial v}{\partial y} = 0; \\ &\frac{\partial v}{\partial y} = u - rv + (r-1)\,uv. \end{aligned}\right\} \tag{6.6}$$

We shall seek a solution of this system which satisfies the boundary conditions

$$\left.\begin{aligned} v &= 0, && \text{for} \quad y = 0, \quad x \geq 0; \\ u &= u_0(y) && \text{for} \quad x = 0, \quad y > 0. \end{aligned}\right\} \tag{6.7}$$

If a new function $\Psi(x, y)$ is introduced so that

$$\left.\begin{aligned} u &= \frac{\partial \Psi}{\partial y} = \Psi_y; \\ v &= -\frac{\partial \Psi}{\partial x} = -\Psi_x, \end{aligned}\right\} \tag{6.8}$$

then (6.6) yields for $\Psi(x, y)$:

$$\Psi_{xy} + \Psi_y + r\Psi_x + (1-r)\,\Psi_x\Psi_y = 0. \tag{6.9}$$

The boundary conditions (6.7) take the forms

$$\left.\begin{aligned} &\Psi_x(x, 0) = 0; \quad \Psi(x, 0) = 1; \quad x \geq 0; \\ &\Psi_y(0, y) = u_0(y); \quad \Psi(0, y) = 1 + \int_0^y u_0(\tau)\,d\tau; \quad y \geq 0. \end{aligned}\right\} \tag{6.10}$$

Finally, introducing the auxiliary function $F(x, y)$ by

$$\Psi(x, y) = \frac{\ln F(x, y)}{1-r} + (1 + y - x), \tag{6.11}$$

we obtain

$$F_{xy} + F_x + rF_y = 0; \tag{6.12}$$

$$F(x, 0) = \exp(1-r)\,x; \quad F(0, y) = \exp\left\{(1-r)\int_0^y [u_0(\tau) - 1]\,d\tau\right\}. \tag{6.13}$$

The desired solutions u and v are related to this function by

$$u = 1 + \frac{F_y}{1-r}\frac{1}{F}; \quad v = 1 - \frac{F_x}{1-r}\frac{1}{F}. \tag{6.14}$$

for $r \neq 1$.

Now the solution of (6.12) and (6.13) presents no difficulty. It can be obtained by various methods. For example, application of the Laplace transformation yields the following expression for $F(x, y)$ in the case $u(\tau) = \text{const} = u_0$,

$$\begin{aligned} F(x, y) &= [1 - J(x, ry)]\, e^{(1-r)(x-y)} \\ &+ J\left(\frac{rx}{\varrho}, \beta y\right) \exp\left[\frac{1-\beta}{\beta}(rx - \beta y)\right], \end{aligned} \tag{6.15}$$

where $\beta = 1 - (1 - r)(1 - u_0)$, and $J(x, y)$ is

$$J(x, y) = 1 - e^{-y} \int_0^x e^{-\tau} I_0\left(2\sqrt{y\tau}\right) d\tau. \tag{6.16}$$

The desired solution is completely determined by $J(x, y)$ whose properties were studied in detail in [7]. It is not difficult to find a relation between this function and the incomplete Weber integral $\tilde{Q}_0(y, z)$ of the modified Bessel function. This can be done by introducing in (6.16) a new integration variable by $2\sqrt{y\tau} = t$, from which we get

$$\begin{aligned} J(x, y) &= 1 - \frac{e^{-y}}{2y} \int_0^{2\sqrt{xy}} t\, I_0(t) \exp\left(-\frac{t^2}{4y}\right) dt \\ &= 1 - e^{-2y} \tilde{Q}_0\left(y, 2\sqrt{xy}\right). \end{aligned} \tag{6.17}$$

7. Drift of Non-Fundamental Carriers in Semiconductors

For the analysis of drift of carriers in semiconductors in the presence of capture (in the linear approximation and neglecting small diffusion) it is necessary to solve the following system of differential equations [37]

$$\left.\begin{aligned} \frac{\partial N}{\partial T} &= -N - \frac{\partial N}{\partial X} - \frac{\partial N_t}{\partial T}; \\ \frac{\partial N_t}{\partial T} &= \frac{\tau}{\tau_s} - \frac{\tau}{\tau_c} N_t \end{aligned}\right\} \tag{7.1}$$

with the boundary conditions

$$\left.\begin{aligned} N(x, 0) &= N_t(x, 0) = 0; \\ N(0, T) &= \Omega(T_1, T), \end{aligned}\right\} \tag{7.2}$$

where N is the concentration of electrons in the conduction zone, N_t is the concentration of electrons in the capture zone; $T = t/\tau$, $T_1 = t_1/\tau$, $X = x/vt$, t_1 is the duration of action of the initial impulse and $\Omega(T_1, T)$ is a unit step function.

If we introduce

$$f(x, z) = I_0(2\sqrt{azX}) \exp(-z\tau/\tau_c), \tag{7.3}$$

where $a = \tau^2/\tau_s\tau_c$, then a solution of (7.1) which satisfies (7.2) can be written as

$$N(X, T) = \exp\left[-X\left(1 + \frac{\tau}{\tau_s}\right)\right]$$

$$\times \begin{cases} 0, & T < X; \\ \frac{\tau}{\tau_c} D(X, T - X) + f(X, T - X), & X < T < T_1 + X; \\ \frac{\tau}{\tau_c} [D(X, T - X) - D(X, T - X - T_1)] & \\ \quad + f(X, T - X) - f(X, T - X - T_1), & T > X + T_1; \end{cases} \tag{7.4}$$

$$N_t(X, T) = \frac{\tau}{\tau_s} \exp\left[-X\left(1 + \frac{\tau}{\tau_s}\right)\right]$$

$$\times \begin{cases} 0, & T < X; \\ D(X, T - X), & X < T < T_1 + X; \\ D(X, T - X) - D(X, T - X - X_1), & T > X + T_1, \end{cases} \tag{7.5}$$

where for brevity we have introduced the notation

$$D(y, z) = \int_0^z I_0(2\sqrt{\xi y a}) \exp\left(-\frac{\tau}{\tau_c}\xi\right) d\xi. \tag{7.6}$$

The latter integral has been studied above, and can be expressed in terms of the incomplete Weber integral $\tilde{Q}(x, c)$:

$$D(y, z) = \frac{\tau_c}{\tau} \tilde{Q}_{0'}\left(\frac{y\tau}{\tau_s}, 2\sqrt{ayz}\right) \exp\left(-\frac{y\tau}{\tau_s}\right). \tag{7.7}$$

8. Damping of Radiation in High Temperature Plasma in the Presence of Shielding

In the determination of the intensity of radiation damping per unit volume in plasma with account taken of the shielding of the Coulomb field of the ions, a most important role is played by functions of the form [82]

$$F(\alpha, \gamma) = \int_0^\infty \frac{y}{(y + \gamma)^2} \exp\left[-\frac{\alpha}{2}\left(y + \frac{1}{y}\right)\right] dy, \tag{8.1}$$

where $\alpha = \omega/(2kT)$; $\gamma = h^2\varkappa^2/(2m\omega)$, T is the temperature of the plasma, ω the energy of radiation, $\varkappa^2 = 4\pi e^2 n/(kT)$ is the square of the reciprocal of the Debye radius, and m is the electron mass.

For $\gamma \to 0$ this function reduces to the MacDonald function[1],

$$F(\alpha, 0) = 2K_0(\alpha). \tag{8.2}$$

We show now that for $\gamma \neq 0$ F can be expressed by an incomplete Weber integral of a MacDonald function. To this end we substitute $y = 1/\xi$ in (8.1) and use the relation

$$\frac{1}{(1+\gamma\xi)^2} = \int_0^\infty t e^{-t(1+\gamma\xi)}\, dt. \tag{8.3}$$

Them after interchanging the order of integration we find

$$F(\alpha, \gamma) = \int_0^\infty e^{-t}\, t\, dt \int_0^\infty \exp\left[-\frac{\alpha}{2\xi}\left(\frac{\alpha}{2} + \gamma t\right)\xi\right]\frac{d\xi}{\xi}.$$

The inner integral reduces to the MacDonald function after a change of variable $\xi = u/(\alpha/\alpha + 2\gamma t)^{1/2}$, so that

$$F(\alpha, \gamma) = 2\int_0^\infty e^{-t} K_0\left(\sqrt{\alpha^2 + 2\alpha\gamma t}\right) t\, dt. \tag{8.4}$$

Moreover, introducing $\alpha^2 + 2\alpha\gamma t = \sigma^2$, we find

$$F(\alpha, \gamma) = (2x)^{-2}\, e^{\alpha/2\gamma} \int_\alpha^\infty (\sigma^2 - \alpha^2)\, K_0(\sigma)\, \sigma \exp\left(-\frac{\sigma^2}{4x}\right) d\sigma, \tag{8.5}$$

where $x = \alpha\gamma/2$. Now using the recursion

$$\sigma K_0(\sigma) = \sigma K_2(\sigma) - 2K_1(\sigma) \tag{8.6}$$

we can write (8.5) as

$$\begin{aligned} F(\alpha, \gamma) = \Bigg[& 2x \frac{e^x}{(2x)^3} \int_\alpha^\infty \sigma^3 K_2(\sigma) \exp\left(-\frac{\sigma^2}{4x}\right) d\sigma \\ & -2 \frac{e^x}{(2x)^2} \int_\alpha^\infty \sigma^2 K_1(\sigma) \exp\left(-\frac{\sigma^2}{4x}\right) d\sigma \\ & - \frac{\alpha^2}{2x} \frac{e^x}{2x} \int_\alpha^\infty \sigma K_0(\sigma) \exp\left(-\frac{\sigma^2}{4x}\right) d\sigma \Bigg] \exp\left(\frac{\alpha}{2\gamma} - x\right). \end{aligned}$$

By the definition of the incomplete Weber integral of the MacDonald functions, (c.f. § 3 of Chapter IV), we have

$$\begin{aligned} F(\alpha, \gamma) = \Big[& 2x \tilde{P}_2(x, a) - 2\tilde{P}_1(x, \alpha) \\ & - \frac{\alpha^2}{2x} \tilde{P}_0(x, \alpha) \Big] \exp\left(\frac{\alpha}{2\gamma} - x\right). \end{aligned} \tag{8.7}$$

---

[1] In reference [82] the multiplier "2" has been omitted.

Using the recursion relation (3.19) of Chapter IV, substituting $\nu = 0$, taking $n = 0$ and 1 and applying (8.6) again, we obtain finally the following expression for $F(\alpha, \gamma)$:

$$F(\alpha, \gamma) = \left(\alpha\gamma + 2 - \frac{\alpha}{\gamma}\right) \tilde{P}_0\left(\frac{\alpha\gamma}{2}, \alpha\right) \exp\left[\frac{\alpha}{2}\left(\frac{1}{\gamma} - \gamma\right)\right] \tag{8.8}$$
$$+ \frac{\alpha}{\gamma} K_0(\alpha) - \alpha K_1(\alpha).$$

Here $\alpha\gamma/2 = x$ and $\tilde{P}_0(x, \alpha)$ is the incomplete Weber integral

$$\tilde{P}_0(x, \alpha) = \frac{e^x}{2x} \int_\alpha^\infty t K_0(t) \exp\left(-\frac{t^2}{4x}\right) dt.$$

These formulae simplify the study of the functions considered. The parameter for the expansion of the incomplete Weber integral is $\alpha/(2x) = 1/\gamma$. Therefore for small values of $\gamma = h^2\varkappa^2/(2m\omega) \ll 1$ we can obtain, by means of (3.47) of Chapter IV the approximation

$$F(\alpha, \gamma) \approx 2K_0(\alpha) - 4\gamma K_1(\alpha) + 6\gamma^2 K_2(\alpha). \tag{8.9}$$

Conversely, for $\gamma \gg 1$ expansion (3.46) of the same Chapter gives

$$F(\alpha, \gamma) \approx \frac{2}{\gamma^2} K_2(\alpha). \tag{8.10}$$

For $\gamma = 1$ the integral $\tilde{P}_0(\alpha/2, \alpha)$ can be evaluated in closed form. In fact, from (4.24) of Chapter IV we have

$$\tilde{P}_0\left(\frac{\alpha}{2}, \alpha\right) = \frac{1}{\alpha} e^{\alpha/2} \int_0^\infty t K_0(t) \exp\left(-\frac{t^2}{2\alpha}\right) dt$$
$$= \frac{1}{2} e^{\alpha} \int_\alpha^\infty [K_0(u) + K_1(u)]\, e^{-u}\, du$$

and, taking account of (4.33) of the same chapter we find without difficulty

$$\tilde{P}_0\left(\frac{\alpha}{2}, \alpha\right) = \frac{1}{2} K_0(\alpha). \tag{8.11}$$

Thus $F(\alpha, \gamma)$ for $\gamma = 1$ has the form

$$F(\alpha, 1) = (1 + \alpha) K_0(\alpha) - \alpha K_1(\alpha). \tag{8.12}$$

The estimates of (8.8) allow one to find numerical values for $F(\alpha, \gamma)$ since as was shown in § 4 of Chapter V (see also Chapter XI), $\tilde{P}_0(x, \alpha)$ may be evaluated from available tables of special integrals.

9. Other Special Integrals

Here we take note of some *definite* integrals appearing in applications, which can still be identified with the class of incomplete cylindrical functions.

Integrals of the form

$$\left.\begin{aligned}&\int_0^{2\pi} e^{-\varrho(1-\cos\tau)}\,[1-\varrho(1-\cos\tau)]\cos(\mu-1)(\tau-\pi)\,d\tau;\\&\int_0^{2\pi} e^{-\varrho(1-\cos\tau)}\,[1-\varrho\cos\tau]\cos\mu(\tau-\pi)\,d\tau.\end{aligned}\right\} \tag{9.1}$$

arise often in the study of wave propagation in plasma [83]. These integrals can easily be expressed either by

$$\int_{-\pi}^{\pi} e^{-\varrho\cos\tau}\cos\nu\tau\,d\tau, \tag{9.2}$$

or by the derivative of this function with respect to ϱ. For integer values of ν, (9.2) is proportional to the modified Bessel function $I_\nu(-\varrho)$ (c.f. formula (4.16) of Chapter I), while for arbitrary ν, it may be expressed with the incomplete modified Bessel function $i_\nu(\omega, z)$, (c.f. § 1 of Chapter III).

$$\int_{-\pi}^{\pi} e^{-\varrho\cos\tau}\cos\nu\tau\,d\tau = \frac{1}{i}\int_{-i\pi}^{i\pi} e^{-\varrho\cosh u-\nu u}\,du = 2\pi i_\nu(i\pi, -\varrho). \tag{9.3}$$

Integrals of this type are usually analysed by expanding $\exp(-\varrho\cos\tau)$ in a Bessel function series (c.f., e.g. [82]). However, the theory of incomplete cylindrical functions presented here allows an analysis of these integrals, which, from our point of view, is more natural.

Integrals of a similar form were studied in [84]:

$$M_\nu(\varrho) = \int_0^\infty e^{-\varrho\cosh t}\sinh\nu t\,dt; \tag{9.4}$$

$$G_\nu(\varrho) = \int_0^\infty e^{-\varrho\cosh t+\nu t}\,dt, \tag{9.5}$$

between which the obvious relation

$$2M_\nu(\varrho) = G_\nu(\varrho) - G_{-\nu}(\varrho). \tag{9.6}$$

holds. Moreover, using the integral representation for the MacDonald function

$$K_\nu(\varrho) = \frac{1}{2}\int_{-\infty}^{\infty} e^{-\varrho\cosh t-\nu t}\,dt = \int_0^\infty e^{-\varrho\cosh t}\cosh\nu t\,dt, \tag{9.7}$$

it is not difficult to verify that

$$G_\nu(\varrho) = K_\nu(\varrho) + M_\nu(\varrho). \tag{9.8}$$

From (9.7) and this formula we find

$$G_\nu(\varrho) + G_{-\nu}(\varrho) = 2K_\nu(\varrho); \tag{9.9}$$

$$\frac{1}{\pi}\int_0^\pi e^{\varrho\cos\theta}\cos\nu\theta\,d\theta - \frac{\sin\nu\pi}{\pi}G_{-\nu}(\varrho) = I_\nu(\varrho). \tag{9.10}$$

From this it is clear that $G_\nu(\varrho)$ considered in detail above, is related to the class of incomplete cylindrical functions. The relation to $\varepsilon_\nu(w, z)$ and $\Phi_\nu(w, z)$ is given by:

$$\begin{aligned} G_\nu(\varrho) &= i\pi\left[\varepsilon_\nu\left(\frac{i\pi}{2}, i\varrho\right) - \varepsilon_\nu\left(-\infty + \frac{i\pi}{2}, i\varrho\right)\right]\exp\left(\frac{i\nu\pi}{2}\right) \\ &= i\pi\Phi_\nu\left(\frac{i\pi}{2}, i\varrho\right)\exp\left(\frac{i\nu\pi}{2}\right), \end{aligned} \tag{9.11}$$

which is readily obtained by substituting $t = i\pi/2 - u$ in (9.5).

Thus a complete analysis of $G_\nu(\varrho)$ and $M_\nu(\varrho)$ can be carried out with the general theory of incomplete cylindrical functions of the Bessel-Sonine-Schlaefli form.

Chapter XI

Compendium of Tables and Computation Formulae for Evaluation of Incomplete Cylindrical Functions

1. Incomplete Bessel Functions, Struve Functions, Anger Functions, and Weber Functions

In this chapter we summarize computation formulae for the incomplete cylindrical functions which we have considered and indicate tables for their direct numerical evaluation.

1. The incomplete Bessel and Struve functions of orders zero and one:

$$J_0(\alpha, \varrho) = \frac{2}{\pi} \int_0^\alpha \cos(\varrho \cos\theta)\, d\theta; \tag{1.1}$$

$$J_1(\alpha, \varrho) = \frac{2\varrho}{\pi} \int_0^\alpha \cos(\varrho \cos\theta) \sin^2\theta\, d\theta; \tag{1.2}$$

$$H_0(\alpha, \varrho) = \frac{2}{\pi} \int_0^\alpha \sin(\varrho \cos\theta)\, d\theta; \tag{1.3}$$

$$H_1(\alpha, \varrho) = \frac{2\varrho}{\pi} \int_0^\alpha \sin(\varrho \cos\theta) \sin^2\theta\, d\theta \tag{1.4}$$

may be evaluated from tables given in [8] for $\cos\alpha = -0.1\ (0.1)\ 1$ and $x = \varrho(1 - \cos\alpha) = 0\ (0.5)\ 12.5$.

Here and below the numbers in parenthesis indicate the step size of the table, and the bordering numbers show the range of the tables for the given variable.

Immediate values of these functions (with coefficient 1/2) is given in the appended tables for $\alpha = 0.2\ (0.2)\ 1.4,\ \pi/2$ and $\varrho = 0.2\ (0.2)\ 10$ [1]. From the same tables for $F_0^{\pm}(\alpha, \varrho)$ and $F_1^{\pm}(\alpha, \varrho)$, the incomplete Bessel and Struve functions of zero and first order for purely imaginary argu-

[1] More extensive six place tables with $\alpha = 0.01\ (0.01)\ \pi/2$ and $\varrho = 0.1\ (0.1)\ 10$ are published in [87].

ment, can be evaluated:

$$I_0(\alpha,\varrho) = \frac{2}{\pi}\int_0^\alpha \cosh(\varrho\cos\theta)\,d\theta = F_0^+(\alpha,\varrho) + F_0^-(\alpha,\varrho); \tag{1.5}$$

$$I_1(\alpha,\varrho) = \frac{2\varrho}{\pi}\int_0^\alpha \cosh(\varrho\cos\theta)\sin^2\theta\,d\theta = F_1^+(\alpha,\varrho) + F_1^-(\alpha,\varrho); \tag{1.6}$$

$$L_0(\alpha,\varrho) = \frac{2}{\pi}\int_0^\alpha \sinh(\varrho\cos\theta)\,d\theta = F_0^+(\alpha,\varrho) - F_0^-(\alpha,\varrho); \tag{1.7}$$

$$L_1(\alpha,\varrho) = \frac{2\varrho}{\pi}\int_0^\alpha \sinh(\varrho\cos\theta)\sin^2\theta\,d\theta = F_1^+(\alpha,\varrho) - F_1^-(\alpha,\varrho), \tag{1.8}$$

where

$$F_0^\pm(\alpha,\varrho) = \frac{1}{\pi}\int_0^\alpha \exp(\pm\varrho\cos\theta)\,d\theta; \tag{1.9}$$

$$F_1^\pm(\alpha,\varrho) = \frac{\varrho}{\pi}\int_0^\alpha \sin^2\theta\cdot\exp(\pm\varrho\cos\theta)\,d\theta. \tag{1.10}$$

2. Incomplete Anger functions and Weber functions of orders zero and one for real arguments are defined by (c.f. § 1 of Chapter III):

$$A_0(\alpha,\varrho) = \frac{1}{\pi}\int_0^\alpha \cos(\varrho\sin\theta)\,d\theta = \frac{1}{2}\left[J_0(\varrho) - J_0\left(\frac{\pi}{2}-\alpha,\varrho\right)\right]; \tag{1.11}$$

$$\begin{aligned} A_1(\alpha,\varrho) &= \frac{1}{\pi}\int_0^\alpha \cos(\varrho\sin\theta-\theta)\,d\theta \\ &= \frac{1}{2}\left[J_1\left(\frac{\pi}{2}-\alpha,\varrho\right) - J_1(\varrho)\right] + \frac{1}{\pi}\left(\frac{1}{\varrho}-\cos\alpha\right)\sin(\varrho\sin\alpha): \end{aligned} \tag{1.12}$$

$$B_0(\alpha,\varrho) = -\frac{1}{\pi}\int_0^\alpha \sin(\varrho\sin\theta)\,d\theta = \frac{1}{2}\left[H_0\left(\frac{\pi}{2}-\alpha,\varrho\right) - H_0(\varrho)\right]; \tag{1.13}$$

$$\begin{aligned} B_1(\alpha,\varrho) &= -\frac{1}{\pi}\int_0^\alpha \sin(\varrho\sin\theta-\theta)\,d\theta \\ &= \frac{1}{2}\left[H_1(\varrho) - H_1\left(\frac{\pi}{2}-\alpha,\varrho\right)\right] + \frac{1}{\pi}\left(1-\frac{1}{\varrho}\right) \\ &\quad + \frac{1}{\pi}\left(\frac{1}{\varrho}-\cos\alpha\right)\cos(\varrho\sin\alpha). \end{aligned} \tag{1.14}$$

These functions can be evaluated from the tables in [8] or from those appended here which also contain values of the Bessel functions $J_n(\varrho)/2 \equiv J_n(\pi/2,\varrho)/2$ and Struve functions $H_n(\varrho)/2 \equiv H_n(\pi/2,\varrho)/2$ for $n = 0$ and 1.

Incomplete Anger and Weber functions for purely imaginary z can also be evaluated from these tables by means of

$$A_0(\alpha, i\varrho) = \frac{1}{\pi}\int_0^\alpha \cosh(\varrho \sin\theta)\, d\theta$$
$$= \frac{1}{2}\left\{I_0(\varrho) - F_0^+\left(\frac{\pi}{2} - \alpha, \varrho\right) - F_0^-\left(\frac{\pi}{2} - \alpha, \varrho\right)\right\}; \tag{1.15}$$

$$A_1(\alpha, i\varrho) = \frac{1}{\pi}\int_0^\alpha \cosh(\varrho \sin\theta + i\theta)\, d\theta$$
$$= \frac{i}{2}\left[F_1^+\left(\frac{\pi}{2} - \alpha, \varrho\right) + F_1^-\left(\frac{\pi}{2} - \alpha, \varrho\right) - I_1(\varrho)\right] \tag{1.16}$$
$$+ \frac{1}{\pi}\left(\frac{1}{\varrho} - i\cos\alpha\right)\sinh(\varrho\sin\alpha);$$

$$B_0(\alpha, i\varrho) = -\frac{i}{\pi}\int_0^\alpha \sinh(\varrho\sin\theta)\, d\theta$$
$$= \frac{i}{2}\left[F_0^+\left(\frac{\pi}{2} - \alpha, \varrho\right) - F_0^-\left(\frac{\pi}{2} - \alpha, \varrho\right) - L_0(\varrho)\right]; \tag{1.17}$$

$$B_1(\alpha, i\varrho) = -\frac{i}{\pi}\int_0^\alpha \sinh(\varrho\sin\theta + i\theta)\, d\theta$$
$$= \frac{1}{2}\left[F_1^+\left(\frac{\pi}{2} - \alpha, \varrho\right) - F_1^-\left(\frac{\pi}{2} - \alpha, \varrho\right) - L_1(\varrho)\right] \tag{1.18}$$
$$+ \frac{1}{\pi}\left(1 + \frac{i}{\varrho}\right) - \frac{1}{\pi}\left(\frac{i}{\varrho} + \cos\alpha\right)\cosh(\varrho\sin\alpha).$$

Separation of real and imaginary parts of (1.16) and (1.18) allows immediate evaluation of the integrals

$$\int_0^\alpha \sinh(\varrho\sin\theta)\sin\theta\, d\theta = \frac{\pi}{2}\left[F_1^+\left(\frac{\pi}{2} - \alpha, \varrho\right)\right.$$
$$\left. - F_1^-\left(\frac{\pi}{2} - \alpha, \varrho\right) - I_1(\varrho)\right] - \cos\alpha\sinh(\varrho\sin\alpha); \tag{1.19}$$

$$\int_0^\alpha \cosh(\varrho\sin\theta)\sin\theta\, d\theta = \frac{\pi}{2}\left\{F_1^+\left(\frac{\pi}{2} - \alpha, \varrho\right)\right.$$
$$\left. - F_1^-\left(\frac{\pi}{2} - \alpha, \varrho\right) - L_1(\varrho)\right\} + 1 - \cos\alpha\cosh(\varrho\sin\alpha). \tag{1.20}$$

Here, as above,

$$L_n(\varrho) = F_n^+\left(\frac{\pi}{2}, \varrho\right) - F_n^-\left(\frac{\pi}{2}, \varrho\right);\ n = 0; 1$$

are determined from the same tables.

Tables for the incomplete Anger and Weber functions:

$$\left.\begin{aligned} \pi A_\nu(\alpha, \varrho) &= \int_0^\alpha \cos(\varrho \sin - \nu\theta)\, d\theta; \\ \pi B_\nu(\alpha, \varrho) &= -\int_0^\alpha \sin(\varrho \sin\theta - \nu\theta)\, d\theta \end{aligned}\right\} \tag{1.21}$$

for $\varrho = \nu = 0\ (0.05)\ 0.5$ and $\alpha/\pi = 0\ (0.05)\ 1$ are presented in [9].

2. Incomplete Hankel Functions and MacDonald Functions

1. The incomplete Hankel function of zero order for real arguments

$$\begin{aligned} \mathscr{H}_0(\beta, \varrho) &= -\frac{2i}{\pi}\int_0^\beta e^{i\varrho\cosh\theta}\, d\theta \\ &= \frac{2}{\pi}\int_0^\beta \sin(\varrho\cosh\theta)\, d\theta - \frac{2i}{\pi}\int_0^\beta \cos(\varrho\cosh\theta) d\theta \end{aligned} \tag{2.1}$$

may be evaluated from the tables in [4] and [28]. The real and imaginary parts of these functions are determined either by the formulae

$$\frac{2}{\pi}\int_0^\beta \sin(\varrho\cosh\theta)\, d\theta = \tanh\beta\left[N_0(\varrho)\, J_c\left(\lambda, \frac{\varrho}{\lambda}\right) - J_0(\varrho)\, Ns\left(\lambda, \frac{\varrho}{\lambda}\right)\right]; \tag{2.2}$$

$$\begin{aligned} \frac{2}{\pi}\int_0^\beta \cos(\varrho\cosh\theta)\, d\theta = \frac{2\beta}{\pi} J_0(\varrho) - \tanh\beta\Big[N_0(\varrho)\, Js\left(\lambda, \frac{\varrho}{\lambda}\right) \\ - J_0(\varrho)\, Nc\left(\lambda, \frac{\varrho}{\lambda}\right)\Big], \end{aligned} \tag{2.3}$$

(c.f. (1.7) of Chapter VII), where $\cosh(\beta) = 1/\lambda$, and the functions $Jc(\lambda, x)$, $Nc(\lambda, x)$, $Js(\lambda, x)$ and $Ns(\lambda, x)$ are tabulated [4] for $\lambda = 0.1\ (0.1)\ 1.0$; $x = 0\ (0.02)\ 2\ (0.1)\ 5$, or by the formulae

$$\frac{2}{\pi}\int_0^\beta \sin(\varrho\cosh\theta)\, d\theta = \frac{2}{\pi} S(\varrho, x); \tag{2.4}$$

$$\frac{2}{\pi}\int_0^\beta \cos(\varrho\cosh\theta)\, d\theta = \frac{2}{\pi}\{\beta - C(\varrho, x)\}, \tag{2.5}$$

(c.f. (1.19) of Chapter VII), where $x = \varrho \sinh\beta$, and $S(\varrho, x)$ and $C(\varrho, x)$ are generalized sine and cosine integrals, tabulated in [28] for $x, \varrho = 0\ (0.01)\ 1$, variable $2\ (0.05)\ 5\ (0.1)\ 10\ (0.2)\ 25$.

Evidently evaluation of the incomplete Hankel function $\mathscr{H}_0(\beta, \varrho)$ from the latter tables is considerably simpler; moreover, these tables cover larger intervals of β and ϱ.

2. The incomplete MacDonald function of zero order $K_0(\beta, \varrho)$ is readily evaluated from the tables in [28] by means of the formulae

$$K_0(\beta, \varrho) = \int_0^\beta e^{-\varrho\cosh\theta}\, d\theta = -E(\varrho, x) + \beta; \tag{2.6}$$

$$\begin{aligned} K_0(\beta, \varrho\sqrt{2}\, e^{\pm i\pi/4}) &= \int_0^\beta \exp\left(-\varrho\sqrt{2}\, e^{\pm i\pi/4} \cosh\theta\right) d\theta \\ &= \beta - Ec(\varrho, x) \mp iEs(\varrho, x), \end{aligned} \tag{2.7}$$

(c.f. (1.23) and (1.24) of Chapter VII). Here $x = \varrho \sinh\beta$ and $E(\varrho, x)$, $Ec(\varrho, x)$ and $Es(\varrho, x)$ are tabulated in the intervals ϱ, $x = 0$ (0.01) (variable) 2 (0.05) 5 (0.1) 10 (02.) 25.

The incomplete MacDonald function of first order may be evaluated from the tables of [47, 48] with the aid of the relations

$$K_1(\beta;\varrho) = \varrho\int_0^\beta e^{-\varrho\cosh\theta}\sinh^2\theta\, d\theta = K_1(\varrho) - e^{-x}\sinh\beta - \frac{e^{-x}}{\varrho} f(x, \sigma), \tag{2.8}$$

(see (2.15) of Chapter VII), where $x = \varrho\cosh\beta$; $\sinh\beta = \operatorname{ctg}\sigma$; $K_1(\varrho)$ is the MacDonald function, and $f(x, \sigma)$ is tabulated.

3. Incomplete Lipschitz-Hankel Integrals

1. The incomplete Lipschitz-Hankel integrals of the Bessel and Neumann functions of zero order are expressed by

$$Je_0(a, z) = \int_0^z e^{-\alpha t} J_0(t)\, dt;$$

$$Ne_0(a, z) = \int_0^z e^{-\alpha t} N_0(t)\, dt.$$

Tables available in the literature permit numerical evaluation of these integrals only for real $z = \varrho$ and purely imaginary $a = i\sigma$. For $\sigma \geq 1$ their real and imaginary parts can be evaluated by (c.f. expression (1.4) of Chapter VII):

$$\operatorname{Re}[Je_0(i\sigma, \varrho)] = \int_0^\varrho J_0(t)\cos\sigma t\, dt = \frac{1}{\sigma} Jc\left(\frac{1}{\sigma}, \sigma\varrho\right); \tag{3.1}$$

$$Im[Je_0(i\sigma, \varrho)] = -\int_0^\varrho J_0(t)\sin\sigma t\, dt = -\frac{1}{\sigma} Js\left(\frac{1}{\sigma}, \sigma\varrho\right); \tag{3.2}$$

$$\operatorname{Re}[Ne_0(i\sigma, \varrho)] = \int_0^\varrho N_0(t)\cos\sigma t\, dt = \frac{1}{\varrho} Nc\left(\frac{1}{\sigma}, \sigma\varrho\right); \tag{3.3}$$

$$\operatorname{Im}[Ne_0(i\sigma, \varrho)] = -\int_0^\varrho N_0(t)\sin\sigma t\, dt = -\frac{1}{\sigma} Ns\left(\frac{1}{\sigma}, \sigma\varrho\right), \tag{3.4}$$

where the functions on the right side are tabulated, (c.f. the comments following (2.2) and (2.3) of the preceeding paragraph.)

The first two integrals for these same values $a = i\sigma$, $\sigma > 1$ may also be evaluated by means of

$$\operatorname{Re}(Je_0) = \pm \frac{1}{\sqrt{\sigma^2 - 1}} [-J_0(\varrho) \sin \sigma\varrho + 2 \cos \sigma\varrho \, U_1(\lambda\varrho, \varrho) + 2 \sin \sigma\varrho U_2(\lambda\varrho, \varrho)]; \tag{3.5}$$

$$\operatorname{Im}(Je_0) = \pm \frac{1}{\sqrt{\sigma^2 - 1}} [-J_0(\varrho) \cos \sigma\varrho + 1 + 2 \cos \sigma\varrho U_2(\lambda\varrho, \varrho) - 2 \sin \sigma\varrho \cdot U_1(\lambda\varrho, \varrho)], \tag{3.6}$$

(c.f. (6.17) of Chapter IV) from tables of the Lommel function of two real variables [23]. To conform with our symbolism we have introduced λ and σ connected by $\lambda^2 - 2\lambda\sigma + 1 = 0$. The plus sign corresponds to $\lambda_1 = \sigma + \sqrt{\sigma^2 - 1}$, and the minus sign to $\lambda_2 = \sigma - \sqrt{\sigma^2 - 1}$. The tables cited give values of $U_1(\omega, \varrho)$ and $U_2(\omega, \varrho)$ for $z = \omega\,(0.01)\,10$ and $\omega = 0.5\,(0.021,\ 2\,(0.05)\,4\,(0.1)\,10$. To use these tables it is necessary for a fixed σ to determine λ from $\lambda = \sigma - \sqrt{\sigma^2 - 1}$ and to choose the minus sign in formulae (3.5) and (3.6).

None of these tables permits one to find values of the incomplete Lipschitz-Hankel integrals for $\sigma < 1$ ($a = i\sigma$). For these values of σ the tables appended to this book will serve, if one uses the formulae

$$\operatorname{Re}(Je_0) = \frac{\pi\varrho}{2\sqrt{1 - \sigma^2}} [H_0(\alpha, \varrho) J_1(\varrho) - H_1(\alpha, \varrho) J_0(\varrho)] + \varrho J_0(\varrho) \cos \varrho\sigma; \tag{3.7}$$

$$\operatorname{Im}(Je_0) = \frac{\pi\varrho}{2\sqrt{1 - \sigma^2}} [J_0(\alpha, \varrho) J_1(\varrho) - J_1(\alpha, \varrho) J_0(\varrho)] - \varrho J_0(\varrho) \sin \varrho\sigma; \tag{3.8}$$

$$\operatorname{Re}(Ne_0) = \frac{\pi\varrho}{2\sqrt{1 - \sigma^2}} [H_0(\alpha, \varrho) N_1(\varrho) - H_1(\alpha, \varrho) N_0(\varrho)] + \varrho N_0(\varrho) \cos \varrho\sigma; \tag{3.9}$$

$$\operatorname{Im}(Ne_0) = \frac{\pi\varrho}{2\sqrt{1 - \sigma^2}} [J_0(\alpha, \varrho) N_1(\varrho) - J_1(\alpha, \varrho) N_0(\varrho)] + \frac{2\alpha}{\pi\sqrt{1 - \sigma^2}} - \varrho N_0(\varrho) \sin \varrho\sigma, \tag{3.10}$$

c.f. (6.11) and (6.12) of Chapter II. Here as always $\alpha = \cos^{-1}\sigma$.

2. The incomplete Lipschitz-Hankel integrals involving $I_0(t)$ and $K_0(t)$:

$$Ie_0(a, \varrho) = \int_0^\varrho e^{-at} I_0(t)\, dt;$$

$$Ke_0(a, \varrho) = \int_0^\varrho e^{-at} K_0(t)\, dt$$

may be evaluated either from tables of Lommel functions of two purely imaginary variables, or from the tables in this book. These integrals have been evaluated for $-1 < a < 1$ by the formulae:

$$Ie_0(a, \varrho) = \frac{\pi\varrho}{\sqrt{1-a^2}} [F_0^-(\alpha, \varrho)\, I_1(\varrho) - F_1^-(\alpha, \varrho)\, I_0(\varrho)] + \varrho I_0(\varrho)\, e^{-a\varrho}; \tag{3.11}$$

$$Ke_0(a, \varrho) = \frac{\alpha}{\sqrt{1-a^2}} - \frac{\pi\varrho}{\sqrt{1-a^2}} [F_0^-(\alpha, \varrho)\, K_1(\varrho) + F_1^-(\alpha, \varrho)\, K_0(\varrho)] + \varrho K_0(\varrho)\, e^{-a\varrho}, \tag{3.12}$$

where $\alpha = \cos^{-1}\sigma$, and $-1 < a < 1$ (c.f. also (6.16) and (6.17) of Chapter II).

For $|a| > 1$ the first of these integrals can be calculated by means of

$$Ie_0(a, \varrho) = \frac{\pm 1}{\sqrt{a^2-1}} \{e^{-a\varrho} [I_0(\varrho) + 2Y_2(\lambda\varrho, \varrho) + 2Y_1(\lambda\varrho, \varrho)] - 1\} \tag{3.13}$$

(c.f. (6.14) of Chapter IV) from tables of Lommel functions of two purely imaginary variables. Again we have introduced a parameter λ, related to a by $\lambda^2 - 2\lambda a + 1 = 0$. The plus sign in (3.13) corresponds to $\lambda_1 = a + \sqrt{a^2-1}$, and the minus sign to $\lambda_2 = a - \sqrt{a^2-1}$. Available tables [22] give values of $Y_1(y, \varrho)$ and $Y_2(y, \varrho)$ for $y = 0\ (0.01)\ 1\ (0.1)\ 20$ and $\varrho = 0\ (0.01)\ 1\ (0.1)\ y$. To use the tables thus constructed, it is necessary to choose the plus sign in (3.13) and to take $\lambda = a + \sqrt{a^2-1}$.

The incomplete Lipschitz-Hankel integral $Ke_0(a, \varrho)$ cannot be immediately expressed with these functions for $a > 1$. However its values can be determined from those of $Ie_0(a, \varrho)$ and the incomplete MacDonald function $K_0(\beta, \varrho)$. In fact, from (1.28) and (4.22) of Chapter II we have

$$K_0(\beta, \varrho) = \beta I_0(\varrho) + \sqrt{a^2-1}\, [K_0(\varrho)\, Ie_0(a, \varrho) - I_0(\varrho)\, Ke_0(a, \varrho)],$$

from which we find

$$Ke_0(a, \varrho) = \frac{K_0(\varrho)}{I_0(\varrho)} Ie_0(a, \varrho) + \frac{1}{I_0(\varrho)\sqrt{a^2-1}} [\beta I_0(\varrho) - K_0(\beta, \varrho)], \tag{3.14}$$

where $\beta = \cosh^{-1}(\alpha)$. By (2.6) and (3.13) of this Chapter, numerical values of $Ie_0(\alpha, \varrho)$ and $K_0(\beta, \varrho)$ are available from other tables.

4. Incomplete Weber Integrals

1. The Incomplete Weber integrals of Bessel and Neumann functions

$$Q_0\left(\frac{cz}{2}, z\right) = (cz)^{-1}\, e^{cz/2} \int_0^z t J_0(t) \exp\left(-\frac{t^2}{2cz}\right) dt;$$

$$P_0\left(\frac{cz}{2}, z\right) = (cz)^{-1}\, e^{cz/2} \int_z^{\infty \exp i\lambda} t N_0(t) \exp\left(-\frac{t^2}{2cz}\right) dt$$

can be evaluated from our tables for $c = \exp i\,(\pi/2 - \alpha)$ by means of the formulae

$$\begin{aligned}\operatorname{Re} Q_0\left(\frac{c\varrho}{2}, \varrho\right) &= \frac{1}{2} + \frac{1}{2}(\varrho \sin\alpha - 1)\, J_0(\varrho) \cos(\varrho \cos\alpha) \\ &\quad + \frac{\pi\varrho}{4}\left[H_0(\alpha, \varrho)\, J_1(\varrho) - H_1(\alpha, \varrho)\, J_0(\varrho)\right]; \\ \operatorname{Im} Q_0\left(\frac{c\varrho}{2}, \varrho\right) &= \frac{1}{2}(\varrho \sin\alpha - 1)\, J_0(\varrho) \sin(\varrho \cos\alpha) \\ &\quad - \frac{\pi\varrho}{4}\left[J_0(\alpha, \varrho)\, J_1(\varrho) - J_1(\alpha, \varrho)\, J_0(\varrho)\right];\end{aligned} \tag{4.1}$$

$$\begin{aligned}\operatorname{Re} P_0\left(\frac{c\varrho}{2}, \varrho\right) &= \frac{1}{2}(1 - \varrho \sin\alpha)\, N_0(\varrho) \cos(\varrho \cos\alpha) \\ &\quad - \frac{\pi\varrho}{4}\left[H_0(\alpha, \varrho)\, N_1(\varrho) - H_1(\alpha, \varrho)\, N_0(\varrho)\right]; \\ \operatorname{Im} P_0\left(\frac{c\varrho}{2}, \varrho\right) &= \frac{1}{2} + \frac{1}{2}(1 - \varrho \sin\alpha)\, N_0(\varrho) \sin(\varrho \cos\alpha) \\ &\quad + \frac{\pi\varrho}{4}\left[J_0(\alpha, \varrho)\, N_1(\varrho) - J_1(\alpha, \varrho)\, N_0(\varrho)\right].\end{aligned} \tag{4.2}$$

Our tables permit the evaluation of the integrals under discussion only for those complex values of the variable c which lie on the unit circle. For purely imaginary values of this variable ($c = i\tau$) the tables of [4.23] may be used. To this end we note that because of the following relations resulting from (5.8) and (5.10) of Chapter IV,

$$Q_0\left(\frac{i\tau\varrho}{2}, \varrho\right) + Q_0\left(\frac{i\varrho}{2\tau}, \varrho\right) = 1 - J_0(\varrho) \cdot \exp \frac{i\varrho}{2}\left(\tau + \frac{1}{\tau}\right); \tag{4.3}$$

$$P_0\left(\frac{i\tau\varrho}{2}, \varrho\right) + P_0\left(\frac{i\varrho}{2\tau}, \varrho\right) = N_0(\varrho) \exp \frac{i\varrho}{2}\left(\tau + \frac{1}{\tau}\right). \tag{4.4}$$

it is sufficient to know values of these integrals in the interval $-1 < \tau < 1$.

We now record computation formulae for determining the real and imaginary parts of $Q_0(i\tau\varrho/2, \varrho)$ and $P_0(i\tau\varrho/2, \varrho)$ from the tables in [4]:

$$\begin{aligned}\operatorname{Re} Q_0\left(\frac{i\tau\varrho}{2}, \varrho\right) &= \frac{1}{2}\left[1 + \frac{1 - \tau^2}{1 + \tau^2} Js\left(\frac{1}{\sigma}, \sigma\varrho\right) - J_0(\varrho) \cos\sigma\varrho\right]; \\ \operatorname{Im} Q_0\left(\frac{i\tau\varrho}{2}, \varrho\right) &= -\frac{1}{2}\left[\frac{1 - \tau^2}{1 + \tau^2} Jc\left(\frac{1}{\sigma}, \sigma\varrho\right) + J_0(\varrho) \sin\sigma\varrho\right];\end{aligned} \tag{4.5}$$

$$\begin{aligned}\operatorname{Re} P_0\left(\frac{i\tau\varrho}{2}, \varrho\right) &= \frac{\ln\tau}{\pi} - \frac{1}{2}\left[\frac{1 - \tau^2}{1 + \tau^2} Ns\left(\frac{1}{\sigma}, \sigma\varrho\right) - N_0(\varrho) \cos\sigma\varrho\right]; \\ \operatorname{Im} P_0\left(\frac{i\tau\varrho}{2}, \varrho\right) &= \pm\frac{1}{2} + \frac{1}{2}\left[\frac{1 - \tau^2}{1 + \tau^2} Nc\left(\frac{1}{\sigma}, \sigma\varrho\right) + N_0(\varrho) \sin\sigma\varrho\right],\end{aligned} \tag{4.6}$$

(c.f. the comments following (2.2) and (2.3) of this Chapter.) Here the plus sign applies for $0 < \tau < 1$ and the minus sign for $\tau > 1$, while σ and τ are connected by

$$\sigma = \frac{\tau^2 + 1}{2\tau}. \tag{4.7}$$

The incomplete Weber integral $Q_0(i\varrho/2\tau), \varrho)$ can also be evaluated from tables of Lommel functions of real variables [23] by means of the following formulae, which result from (6.12) of Chapter IV:

$$\begin{aligned} \operatorname{Re} Q_0\left(\frac{i\varrho}{2\tau}, \varrho\right) &= -U_2(\tau\varrho, \varrho) \cos\sigma\varrho - U_1(\tau\varrho, \varrho) \sin\sigma\varrho; \\ \operatorname{Im} Q_0\left(\frac{i\varrho}{2\tau}, \varrho\right) &= U_1(\tau\varrho, \varrho) \cos\sigma\varrho - U_2(\tau\varrho, \varrho) \sin\sigma\varrho. \end{aligned} \tag{4.8}$$

Here as before σ is determined from (4.7). If $\tau < 1$, then values of $U_1(\tau\varrho, \varrho)$ and $U_2(\tau\varrho, \varrho)$ can be read from the cited tables. If $\tau > 1$, then it is necessary to use either the recomputation formulae in [23] or (4.3).

2. The incomplete Weber integrals of the functions $I_0(t)$ and $K_0(t)$:

$$\tilde{Q}_0\left(\frac{cz}{2}, z\right) = (cz)^{-1} e^{cz/2} \int_0^z t I_0(t) \exp\left(-\frac{t^2}{2cz}\right) dt;$$

$$\tilde{P}_0\left(\frac{cz}{2}, z\right) = (cz)^{-1} e^{cz/2} \int_z^\infty t K_0(t) \exp\left(-\frac{t^2}{2cz}\right) dt$$

may be evaluated from our tables for values $c = e^{i\alpha}$ lying on the unit circle by means of the formulae (c.f. also (5.16) and (5.17) of Chapter IV):

$$\begin{aligned} \operatorname{Re}\left[e^{-c\varrho} \tilde{Q}_0\left(\frac{c\varrho}{2}, \varrho\right)\right] &= \frac{1}{2}\left[1 - I_0(\varrho) e^{-\varrho\cos\alpha}\right]; \\ \operatorname{Im}\left[e^{-c\varrho} \tilde{Q}_0\left(\frac{c\varrho}{2}, \varrho\right)\right] &= -\frac{1}{2} \varrho \sin\alpha\, I_0(\varrho) e^{-\varrho\cos\alpha} \\ &\quad - \frac{\pi\varrho}{2}\left[F_0^-(\alpha, \varrho) I_1(\varrho) - F_1^-(\alpha, \varrho) I_0(\varrho)\right]; \end{aligned} \tag{4.9}$$

$$\begin{aligned} \operatorname{Re}\left[e^{-c\varrho} \tilde{P}_0\left(\frac{c\varrho}{2}, \varrho\right)\right] &= K_0(\varrho) e^{-\varrho\cos\alpha}; \\ \operatorname{Im}\left[e^{-c\varrho} \tilde{P}_0\left(\frac{c\varrho}{2}, \varrho\right)\right] &= \frac{1}{2} \varrho \sin\alpha\, K_0(\varrho) e^{-\varrho\cos\alpha} \\ &\quad - \frac{\pi\varrho}{2}\left[F_0^-(\alpha, \varrho) K_1(\varrho) + F_1^-(\alpha, \varrho) K_0(\varrho)\right]. \end{aligned} \tag{4.10}$$

Moreover, $\tilde{Q}_0(c\varrho/2, \varrho)$ may be evaluated for real values of $c = \lambda$ from tables of Lommel functions of two purely imaginary variables [22] by means of

$$\tilde{Q}_0\left(\frac{\lambda\varrho}{2}, \varrho\right) = \left[Y_1\left(\frac{\varrho}{\lambda}, \varrho\right) + Y_2\left(\frac{\varrho}{\lambda}, \varrho\right)\right] \exp\frac{\varrho}{2}\left(\lambda - \frac{1}{\lambda}\right). \tag{4.11}$$

(c.f. also expression (6.13) of Chapter IV).

If $\lambda < 1$, values of $Y_1(y, x)$ and $Y_2(y, x)$ may be taken from these tables. For $\lambda > 1$ it is necessary to use the appropriate recomputation formulae. The incomplete Weber integral $\tilde{P}_0(\lambda\varrho/2, \varrho)$, may be evaluated for the same parameter λ with the aid of (4.21) of Chapter V:

$$\tilde{P}_0\left(\frac{\lambda\varrho}{2}, \varrho\right) = \frac{e^{\lambda\varrho}}{2I_0(\varrho)}\left[K_0(\varrho) - K_0(\beta, \varrho)\right] - \frac{K_0(\varrho)}{I_0(\varrho)}\tilde{Q}_0\left(\frac{\lambda\varrho}{2}, \varrho\right), \qquad (4.12)$$

where $\beta = \ln(\lambda)$. The incomplete MacDonald function $K_0(\beta, \varrho)$ can be evaluated by (2.6), and $\tilde{Q}_0(\lambda\varrho/2, \varrho)$ by (4.11) both of this Chapter. The formulae and tables considered in this Chapter permit numerical evaluation over a wide parameter range of most of the incomplete cylindrical functions which appear in applications.

5. Tables of Incomplete Cylindrical Functions

Table 1 contains five-place values of the incomplete functions $1/2J_n(\alpha, \varrho)$ and $1/2H_n(\alpha, \varrho)$ for $n = 0$ and 1, $\alpha = 0.2\ (0.2)\ 1.4;\ \pi/2$, and $\varrho = 0.2\ (0.2)\ 10$. The general character of their dependence on α and ϱ is illustrated in Fig. 13—16.

In Table 2 numerical values of $F_n^{\pm}(\alpha, \varrho)$ are given to six decimal places for $n = 0$ and 1 in the intervals

$$\alpha = 0\ (0.2)\ 1.4;\ \pi/2,\quad \varrho = 0\ (0.2)\ 10.0.$$

In this table the numbers given in brackets indicate powers of ten; e.g.:

$$F_0^-(0, 4; 5, 2) = 0.812206\ (-3) = 0.000812206;$$

$$F_1^+(0.6; 3.4) = 0.154298\ (1) = 1.54298.$$

The variation of $F_0^{\pm}(\alpha, \varrho)$ and $F_1^{\pm}(\alpha, \varrho)$ is illustrated in Figs. 17—20.

Values of these functions for α and ϱ outside the given intervals can be determined with the aid of given representations (c.f. § 3 of Chapter II) or by means of series ane asymptotic developments (c.f. §§ 7, 10 of Chapter II)[1].

[1] Some other methods and new tables for computation of incomplete cylindrical functions are discussed in the following works: Agrest, M. M., Rikenglas, M. M.: Incomplete Lipshitz-Hankel Integrals; J. Comp. Math., Math. Phys. **6**, 1370 (1967). Agrest, M. M.: Evaluation of Incomplete Cylindrical Functions; J. Comp. Math., Math. Phys. **2**, 313 (1970).

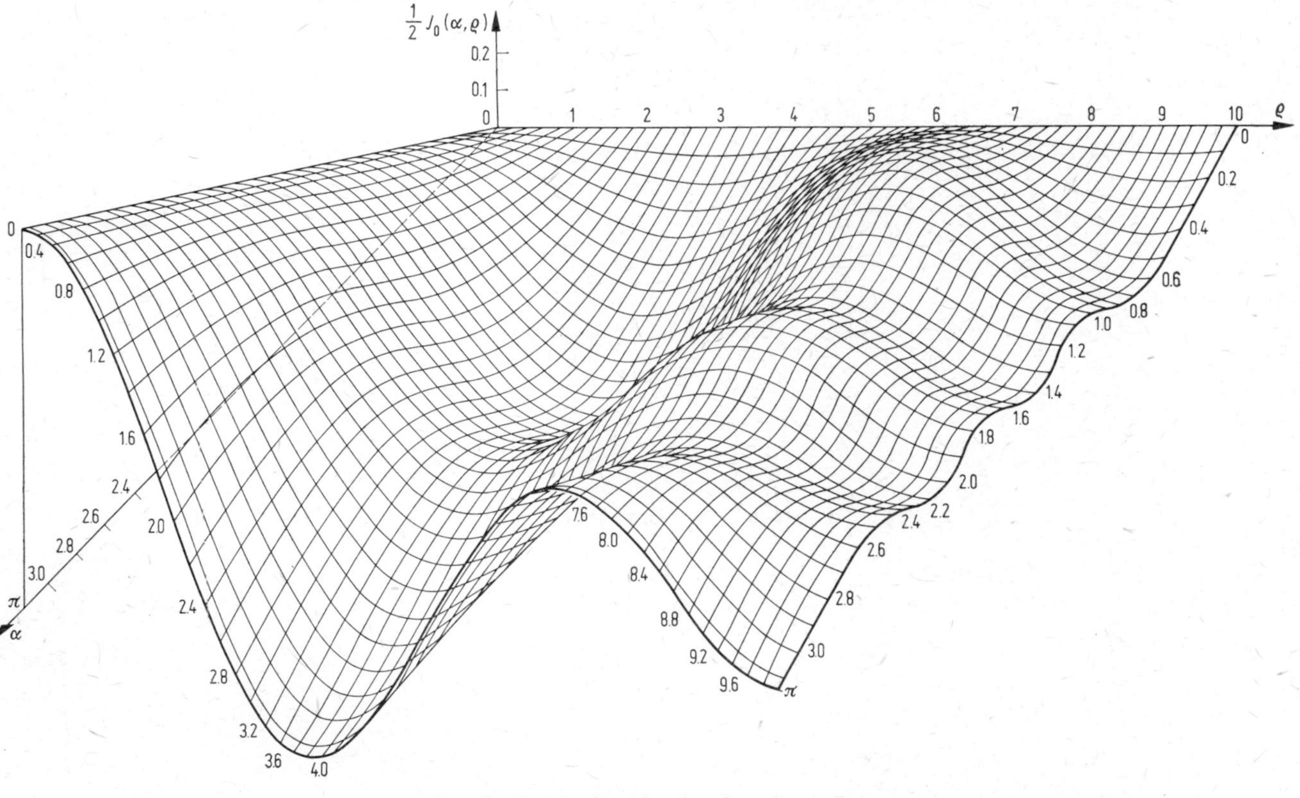

Fig. 13. $J_0(\alpha, \varrho)$ as a function of α and ϱ.

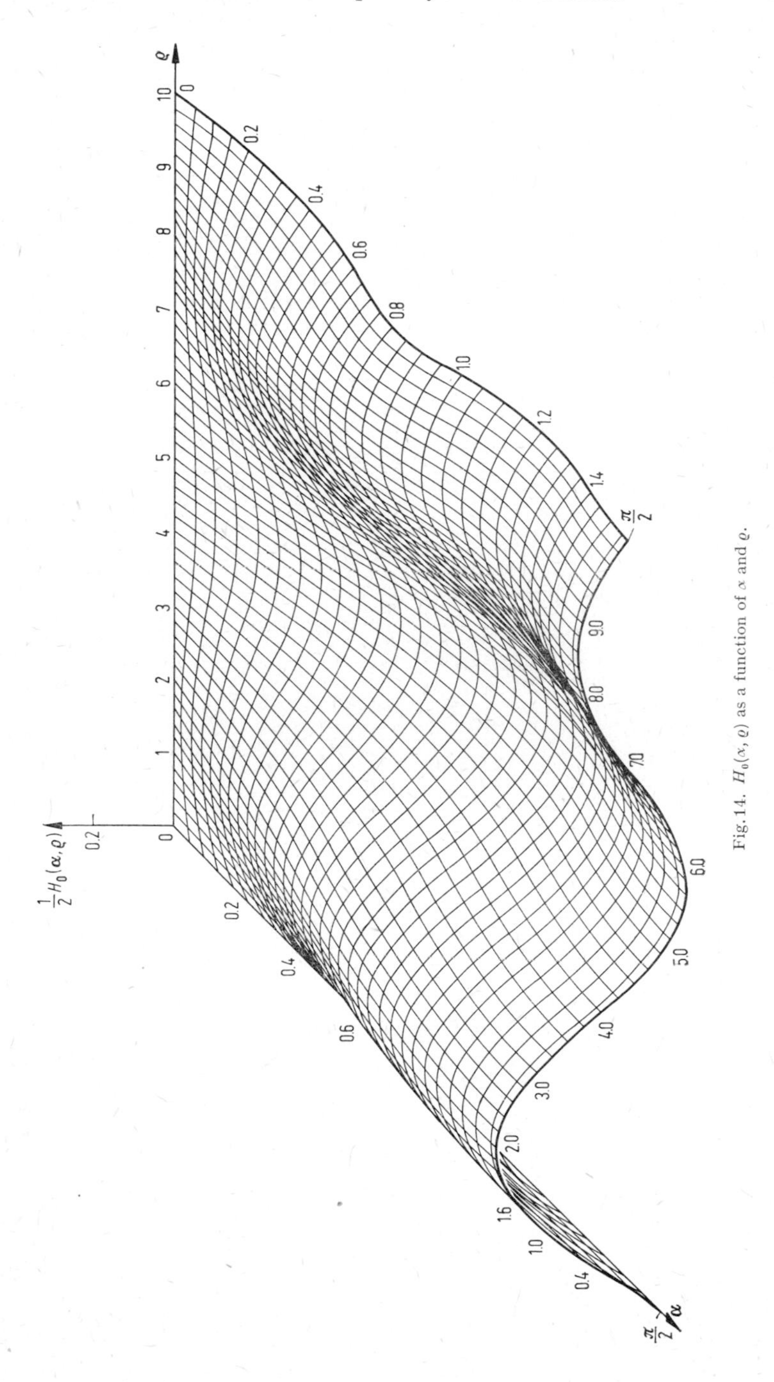

Fig. 14. $H_0(\alpha, \varrho)$ as a function of α and ϱ.

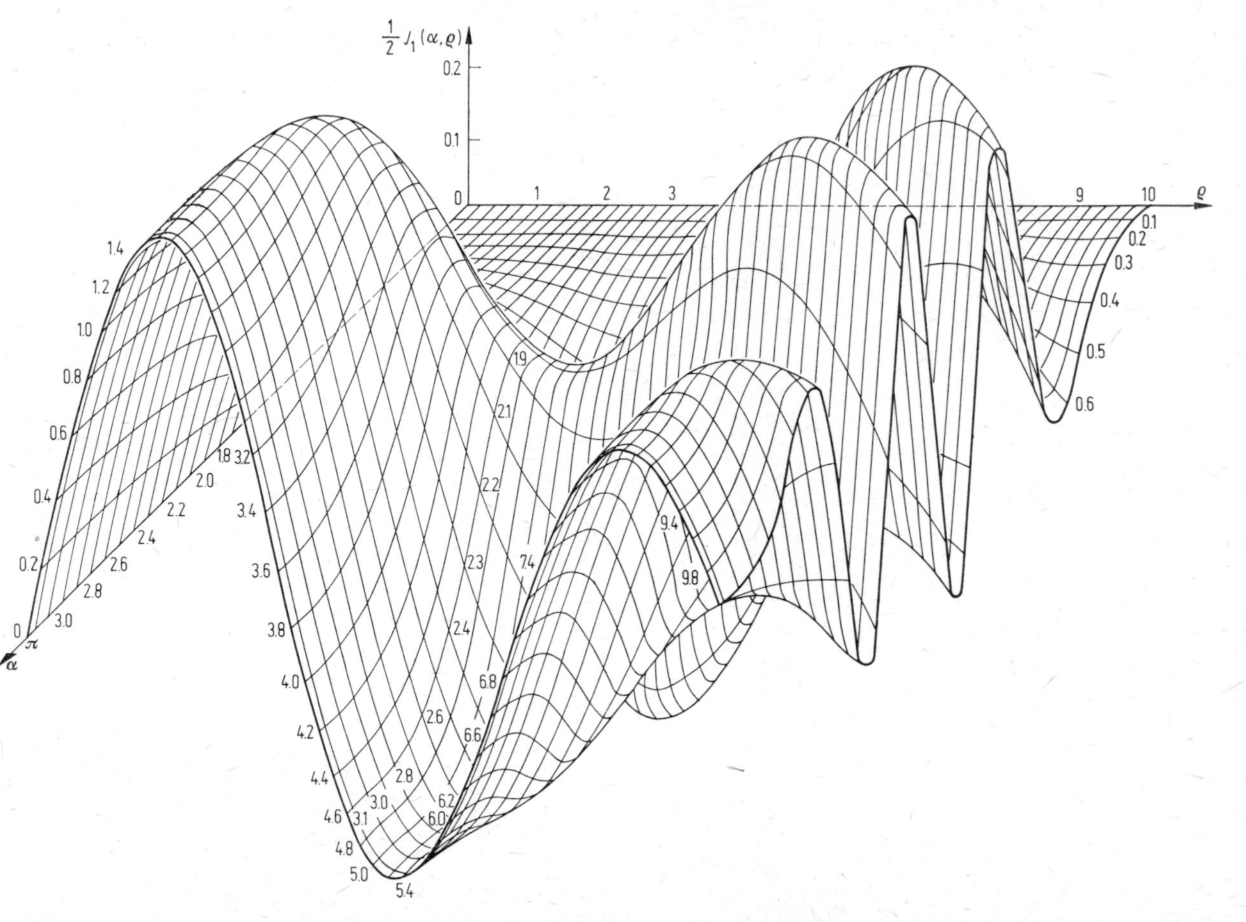

Fig. 15. $J_1(\alpha, \varrho)$ as a function of α and ϱ.

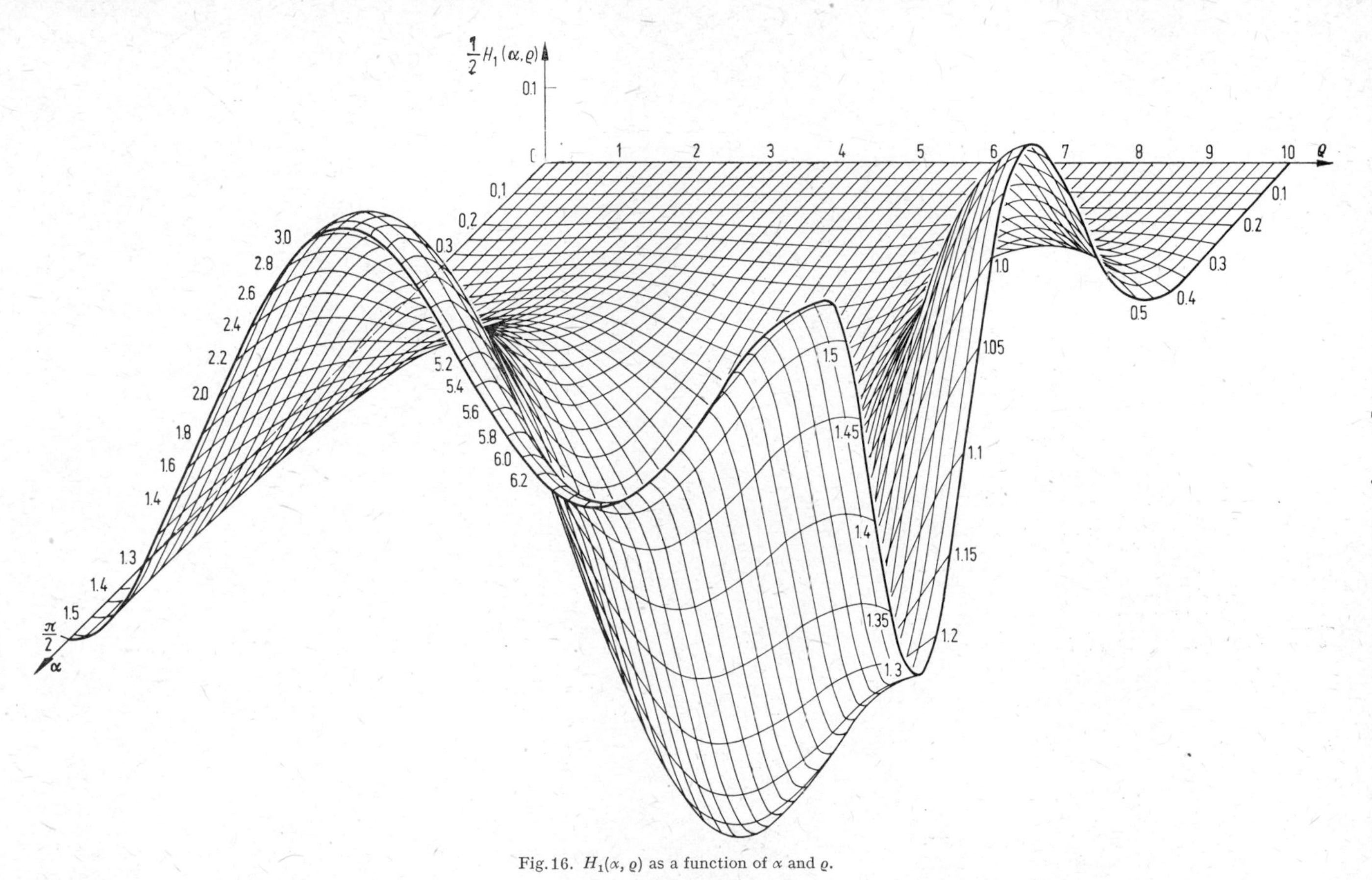

Fig. 16. $H_1(\alpha, \varrho)$ as a function of α and ϱ.

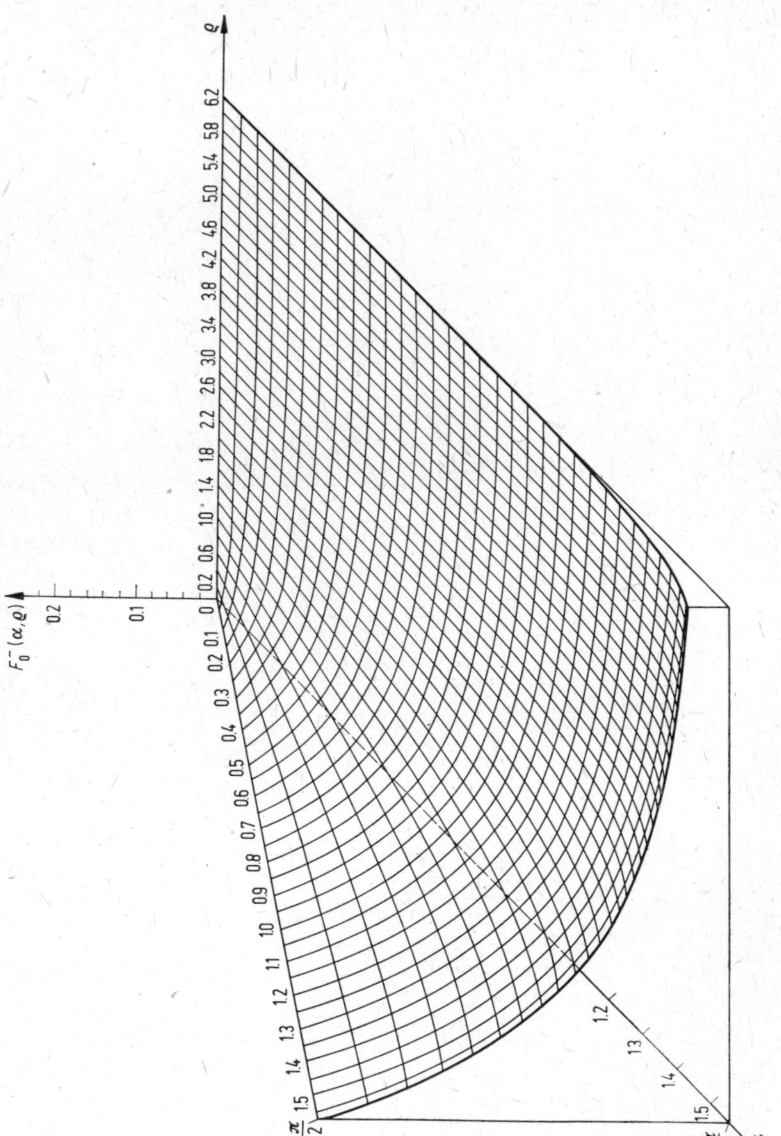

Fig.17. $F_0^-(\alpha, \varrho)$ as a function of α and ϱ.

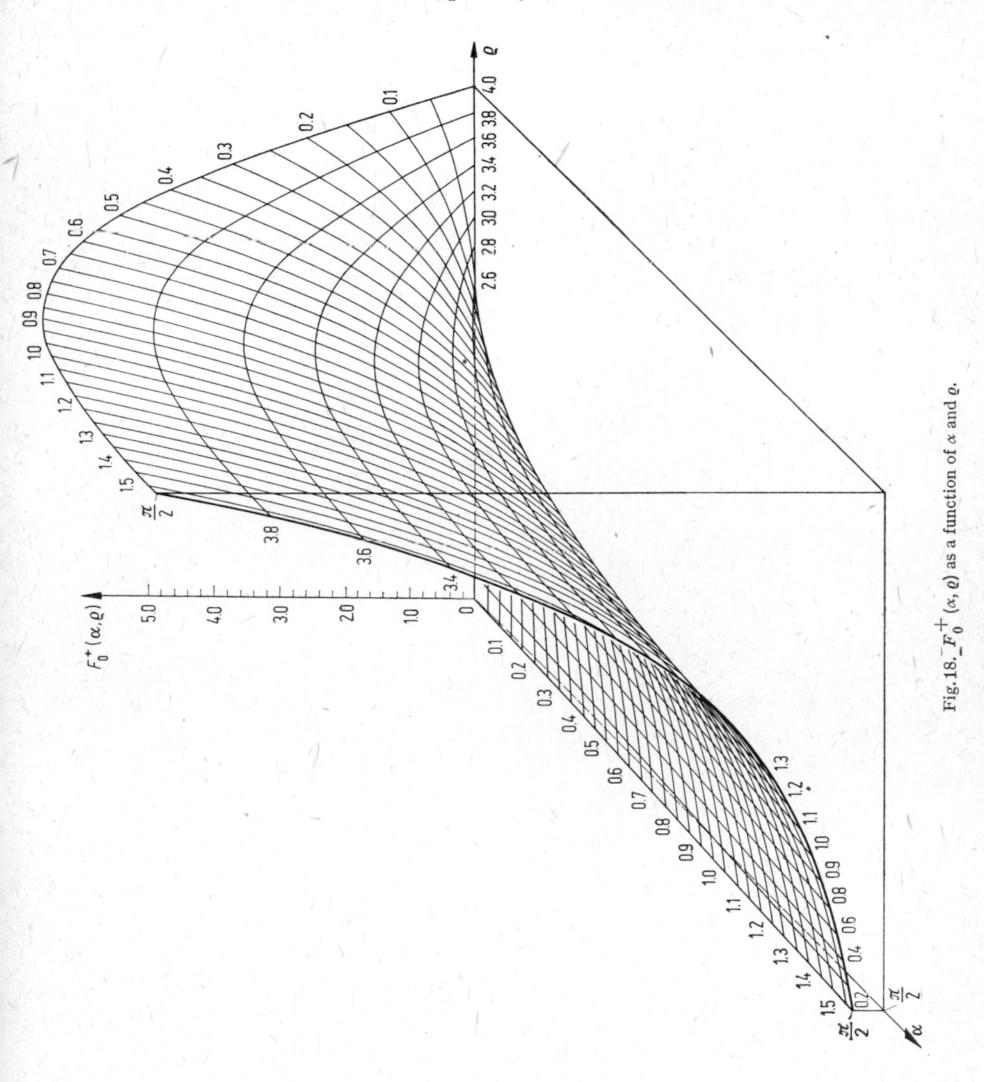

Fig. 18. $_{-}F_0^{+}(\alpha, \varrho)$ as a function of α and ϱ.

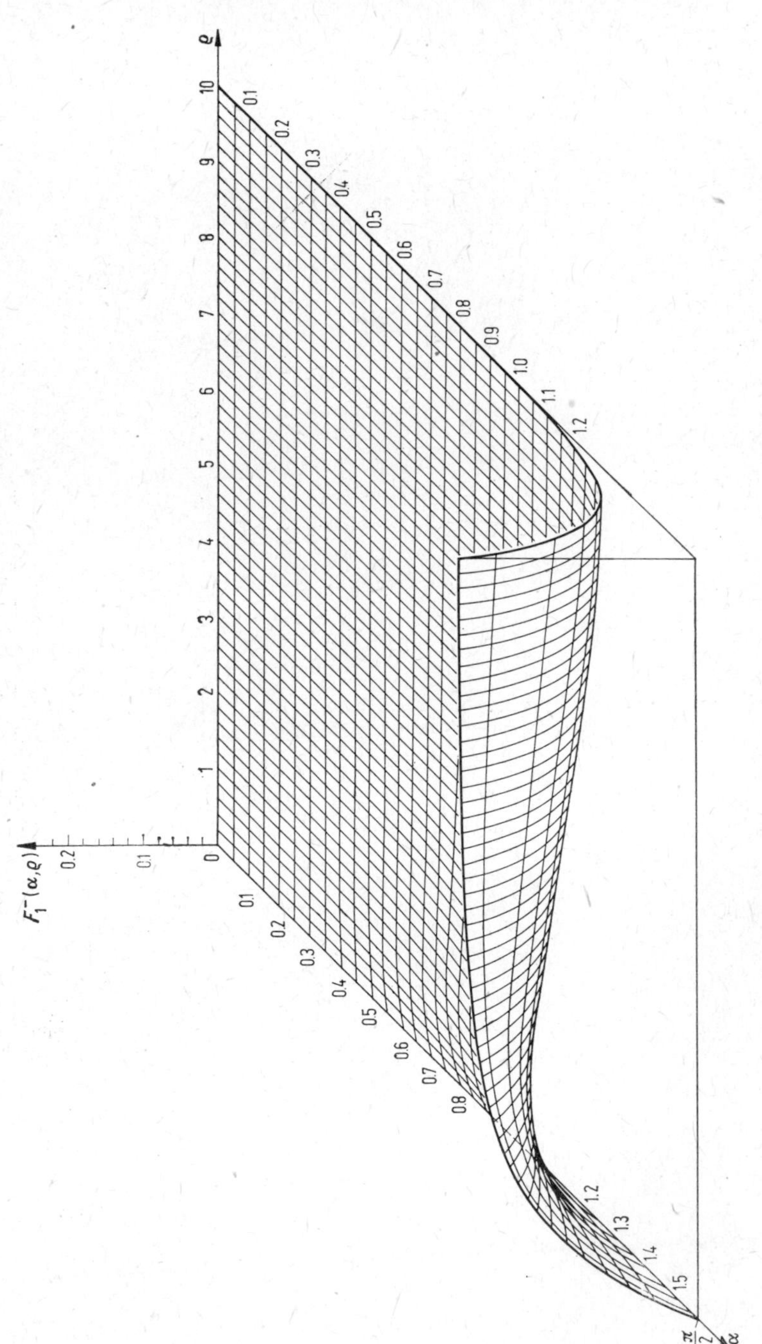

Fig. 19. $F_1^-(\alpha, \varrho)$ as a function of α and ϱ.

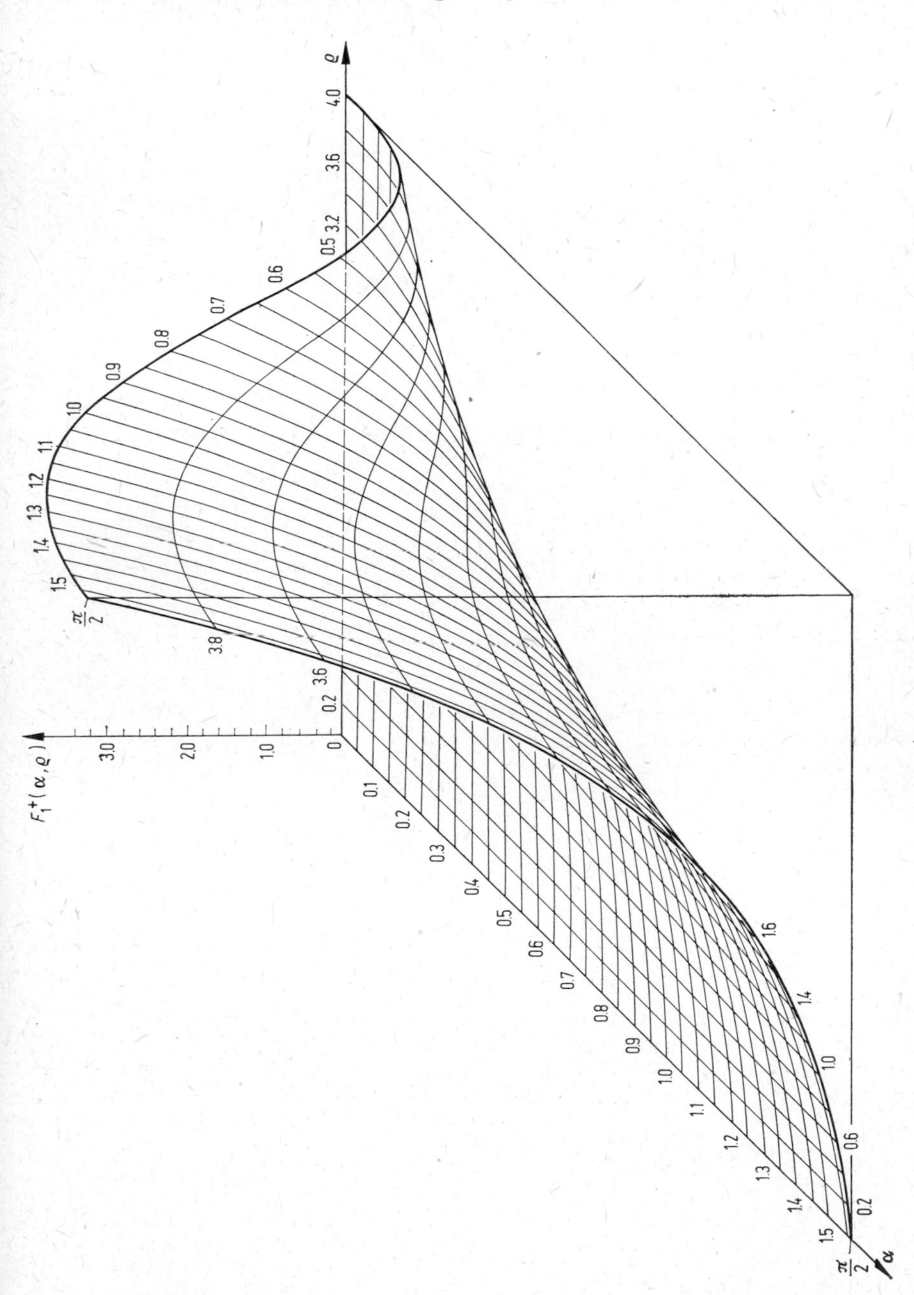

Fig. 20. $F_1^+(\alpha, \varrho)$ as a function of α and ϱ.

Table 1. *Incomplete Bessel and Struve Functions of Orders Zero and One*

ϱ	$\frac{1}{2} J_0(\alpha, \varrho)$	$\frac{1}{2} H_0(\alpha, \varrho)$	$\frac{1}{2} J_1(\alpha, \varrho)$	$\frac{1}{2} H_1(\alpha, \varrho)$
		$\alpha = 0.2$		
0.2	0.06241	0.01256	0.00017	0.00003
0.4	0.05870	0.02463	0.00031	0.00013
0.6	0.05269	0.03574	0.00042	0.00028
0.8	0.04460	0.04543	0.00047	0.00049
1.0	0.03475	0.05334	0.00046	0.00070
1.2	0.02354	0.05915	0.00038	0.00094
1.4	0.01140	0.06263	0.00022	0.00116
1.6	−0.00118	0.06365	−0.00001	0.00135
1.8	−0.01372	0.06216	−0.00031	0.00148
2.0	−0.02572	0.05823	−0.00066	0.00155
2.2	−0.03670	0.05201	−0.00105	0.00153
2.4	−0.04625	0.04874	−0.00145	0.00141
2.6	−0.05397	0.03375	−0.00184	0.00119
2.8	−0.05957	0.02244	−0.00219	0.00086
3.0	−0.06282	0.01024	−0.00249	0.00045
3.2	−0.06361	−0.00236	−0.00269	−0.00005
3.4	−0.06189	−0.01487	−0.00279	−0.00062
3.6	−0.05773	−0.02679	−0.00277	−0.00122
3.8	−0.05131	−0.03766	−0.00262	−0.00184
4.0	−0.04287	−0.04704	−0.00232	−0.00244
4.2	−0.03274	−0.05458	−0.00189	−0.00299
4.4	−0.02132	−0.05996	−0.00132	−0.00346
4.6	−0.00907	−0.06299	−0.00065	−0.00382
4.8	+0.00354	−0.06354	0.00012	−0.00404
5.0	0.01601	−0.06159	0.00095	−0.00410
5.2	0.02785	−0.05721	0.00181	−0.00399
5.4	0.03859	−0.05059	0.00265	−0.00369
5.6	0.04782	−0.04198	0.00345	−0.00321
5.8	0.05516	−0.03171	0.00416	−0.00256
6.0	0.06033	−0.02020	0.00474	−0.00175
6.2	0.06313	−0.00790	0.00515	−0.00082
6.4	0.06344	0.00473	0.00538	0.00022
6.6	0.06126	0.01715	0.00540	0.00131
6.8	0.05667	0.02890	0.00519	0.00241
7.0	0.04985	0.03951	0.00475	0.00349
7.2	0.04107	0.04857	0.00409	0.00447
7.4	0.03067	0.05672	0.00321	0.00533
7.6	0.01907	0.06067	0.00216	0.00602
7.8	0.00672	0.06324	0.00096	0.00649
8.0	−0.00589	0.06332	−0.00034	0.00672
8.2	−0.01827	0.06090	−0.00169	0.00669
8.4	−0.02993	0.06610	−0.00305	0.00638
8.6	−0.04041	0.04908	−0.00434	0.00579
8.8	−0.04930	0.04014	−0.00552	0.00494
9.0	−0.05625	0.02962	−0.00652	0.00384
9.2	−0.06099	0.01793	−0.00731	0.00254

Table 1 (cont'd)

ϱ	$\frac{1}{2} J_0(\alpha, \varrho)$	$\frac{1}{2} H_0(\alpha, \varrho)$	$\frac{1}{2} J_1(\alpha, \varrho)$	$\frac{1}{2} H_1(\alpha, \varrho)$
9.4	−0.06332	0.00554	−0.00733	0.00108
9.6	−0.06316	−0.00706	−0.00806	−0.00049
9.8	−0.06052	−0.01939	−0.00797	−0.00210
10.0	−0.05550	−0.03095	−0.00755	−0.00370
		$\alpha = 0,4$		
0.2	0.12492	0.02463	0.00129	0.00025
0.4	0.11779	0.04834	0.00244	0.00098
0.6	0.10620	0.07021	0.00332	0.00213
0.8	0.09060	0.08942	0.00381	0.00363
1.0	0.06966	0.10271	0.00381	0.00536
1.2	0.04844	0.11425	0.00327	0.00718
1.4	0.02626	0.12452	0.00216	0.00895
1.6	0.00167	0.12722	0.00049	0.01051
1.8	−0.02297	0.12512	−0.00170	0.01171
2.0	−0.04674	0.11828	−0.00432	0.01241
2.2	−0.06872	0.10698	−0.00725	0.01251
2.4	−0.08810	0.09164	−0.01034	0.01190
2.6	−0.10414	0.07284	−0.01344	0.01053
2.8	−0.11623	0.05130	−0.01636	0.00839
3.0	−0.12392	0.02783	−0.01891	0.00551
3.2	−0.12692	0.00333	−0.02091	0.00195
3.4	−0.12512	−0.02129	−0.02220	−0.00218
3.6	−0.11859	−0.04508	−0.02264	−0.00671
3.9	−0.10758	−0.06714	−0.02211	−0.01148
4.0	−0.09252	−0.08665	−0.02056	−0.01627
4.2	−0.07397	−0.10287	−0.01796	−0.02085
4.4	−0.05264	−0.11518	−0.01433	−0.02500
4.6	−0.02934	−0.12313	−0.00975	−0.02849
4.8	−0.00496	−0.12641	−0.00436	−0.03111
5.0	0.01958	−0.12492	0.00168	−0.03266
5.2	0.04336	−0.11870	0.00814	−0.03301
5.4	0.06547	−0.10800	0.01479	−0.03204
5.6	0.08508	−0.09323	0.02134	−0.02971
5.8	0.10145	−0.07496	0.02752	−0.02602
6.0	0.11396	−0.05388	0.03303	−0.02102
6.2	0.12215	−0.03078	0.03761	−0.01486
6.4	0.12571	−0.00656	0.04100	−0.00771
6.6	0.12451	0.01788	0.04300	0.00021
6.8	0.11861	0.04160	0.04344	0.00859
7.0	0.10823	0.06372	0.04221	0.01715
7.2	0.09378	0.08340	0.03926	0.02552
7.4	0.07581	0.09989	0.03463	0.03338
7.6	0.05500	0.11258	0.02842	0.04037
7.8	0.03214	0.12099	0.02078	0.04617
8.0	0.00811	0.12481	0.01195	0.05050

Table 1 (cont'd)

ϱ	$\frac{1}{2} J_0(\alpha, \varrho)$	$\frac{1}{2} H_0(\alpha, \varrho)$	$\frac{1}{2} J_1(\alpha, \varrho)$	$\frac{1}{2} H_1(\alpha, \varrho)$
8.2	−0.01618	0.12391	0.00222	0,05311
8.4	−0.03982	0.11832	−0.00807	0.05381
8.6	−0.06191	0.10828	−0.01853	0.05249
8.8	−0.08161	0.09416	−0.02877	0.04912
9.0	−0.09820	0.07650	−0.03837	0.04372
9.2	−0.11103	0.05599	−0.04694	0.03642
9.4	−0.11965	0.03341	−0.05410	0.02744
9.6	−0.12372	0.00961	−0.05951	0.01704
9.8	−0.12310	−0.01451	−0.06289	0.00557
10.0	−0.11784	−0.03802	−0.06403	−0.00657

$\alpha = 0.6$

ϱ	$\frac{1}{2} J_0(\alpha, \varrho)$	$\frac{1}{2} H_0(\alpha, \varrho)$	$\frac{1}{2} J_1(\alpha, \varrho)$	$\frac{1}{2} H_1(\alpha, \varrho)$
0.2	0.18760	0.03573	0.00420	0.00076
0.4	0.17757	0.07019	0.00799	0.00299
0.6	0.16126	0.10215	0.01099	0.00655
0.8	0.13925	0.13047	0.01285	0.01120
1.0	0.11232	0.15414	0.01331	0.01663
1.2	0.03144	0,17233	0.01216	0.02247
1.4	0.04772	0.18440	0.00928	0.02831
1.6	0.01237	0.18992	0.00467	0.03370
1.8	−0.02337	0.18870	−0.00159	0.03822
2.0	−0.05821	0.18080	−0.00930	0.04144
2.2	−0.09091	0.16652	−0.01817	0.04299
2.4	−0.12032	0.14638	−0.02783	0.04258
2.6	−0.14539	0.12111	−0.03785	0.03997
2.8	−0.16524	0.09162	−0.04772	0.03506
3.0	−0.17919	0.05898	−0.05694	0.02782
3.2	−0,18674	0.09244	−0.06496	0.01835
3.4	−0.18766	−0.01102	−0.07129	0.00688
3.6	−0.18193	−0.04587	−0.07547	−0.00627
3.8	−0.16678	−0.07896	−0.07708	−0.02067
4.0	−0.15167	−0.10911	−0.07584	−0.03581
4.2	−0.12826	−0.13528	−0.07154	−0.05109
4.4	−0.10042	−0.15653	−0.06410	−0.06590
4.6	−0.06916	−0.17214	−0.05356	−0.07958
4.8	−0.03560	−0.18158	−0.04013	−0.09148
5.0	−0.00095	−0.18453	−0.02411	−0.10101
5.2	0.03354	−0.18093	−0.00595	−0.10760
5.4	0.06665	−0.17094	0.01379	−0.11080
5.6	0.09720	−0.15495	0.03385	−0.11600
5.8	0.12413	−0.13357	0.05532	−0.10577
6.0	0.14648	−0.10758	0.07561	−0.09723
6.2	0.16350	−0.07796	0.09453	−0.08473
6.4	0.17460	−0.04576	0.11129	−0.06851
6.6	0.17943	−0.01217	0.12517	−0.04895
6.8	0.17787	0.02161	0.13549	−0.02659

Table 1 (cont'd)

ϱ	$\frac{1}{2} J_0(\alpha, \varrho)$	$\frac{1}{2} H_0(\alpha, \varrho)$	$\frac{1}{2} J_1(\alpha, \varrho)$	$\frac{1}{2} H_1(\alpha, \varrho)$
7.0	0.17000	0.05438	0.14170	−0.00209
7.2	0.15617	0.03496	0.14337	0.02376
7.4	0.13689	0.11229	0.14021	0.05010
7.6	0.11292	0.13541	0.13212	0.07603
7.8	0.08513	0.15353	0.11915	0.10058
8.0	0.05456	0.16605	0.10154	0.12285
8.2	0.02232	0.17255	0.07973	0.14194
8.4	−0.01043	0.17286	0.05430	0.15706
8.6	−0.04250	0.16703	0.02600	0.16751
8.8	−0.07276	0.15531	−0.00431	0.17275
9.0	−0.10014	0.13818	−0.03563	0.17241
9.2	−0.12368	0.11630	−0.06694	0.16630
9.4	−0.14258	0.09050	−0.09714	0.15441
9.6	−0.15621	0.06175	−0.12516	0.13697
9.8	−0.16413	0.03111	−0.14997	0.11438
10.0	−0.16611	−0.00031	−0.17061	0.08726

$\alpha = 0.8$

ϱ	$\frac{1}{2} J_0(\alpha, \varrho)$	$\frac{1}{2} H_0(\alpha, \varrho)$	$\frac{1}{2} J_1(\alpha, \varrho)$	$\frac{1}{2} H_1(\alpha, \varrho)$
0.2	0.25052	0.04542	0.00943	0.00156
0.4	0.23828	0.08933	0.01808	0.00615
0.6	0.21834	0.13030	0.02524	0.01352
0.8	0.19137	0.16697	0.03023	0.02326
1.0	0.15828	0.19814	0.03250	0.03481
1.2	0.12016	0.22279	0.03162	0.04752
1.4	0.07831	0.24014	0.02730	0.06061
1.6	0.03410	0.24963	0.01944	0.07329
1.8	−0.01098	0.25100	0.00811	0.08474
2.0	−0.05544	0.24423	−0.00646	0.09413
2.2	−0.09782	0.22960	−0.02384	0.10073
2.4	−0.13673	0.20763	−0.04345	0.10387
2.6	−0.17091	0.17911	−0.06459	0.10300
2.8	−0.19926	0.14503	−0.08642	0.09776
3.0	−0.22090	0.10655	−0.10804	0.08791
3.2	−0.23517	0.06499	−0.12847	0.07343
3.4	−0.24165	0.02175	−0.14678	0.05449
3.6	−0.24022	−0.02172	−0.16201	0.03145
3.8	−0.23100	−0.06397	−0.17330	0.00486
4.0	−0.21439	−0.10363	−0.17991	−0.02456
4.2	−0.19102	−0.13939	−0.18122	−0.05593
4.4	−0.16175	−0.17013	−0.17678	−0.08824
4.6	−0.12763	−0.19468	−0.16634	−0.12042
4.8	−0.08986	−0.21291	−0.14988	−0.15132
5.0	−0.04975	−0.22370	−0.12756	−0.17979
5.2	−0.00867	−0.22700	−0.09979	−0.20471
5.4	0.03199	−0.22282	−0.06718	−0.22505
5.6	0.07088	−0.21141	−0.03055	−0.23988

Table 1 (cont'd)

ϱ	$\frac{1}{2} J_0(\alpha, \varrho)$	$\frac{1}{2} H_0(\alpha, \varrho)$	$\frac{1}{2} J_1(\alpha, \varrho)$	$\frac{1}{2} H_1(\alpha, \varrho)$
5.8	0.10673	−0.19328	0.00912	−0.24841
6.0	0.13838	−0.16914	0.05070	−0.25007
6.2	0.16485	−0.13993	0.09295	−0.24448
6.4	0.18535	−0.10670	0.13456	−0.23147
6.6	0.19929	−0.07065	0.17420	−0.21116
6.8	0.20634	−0.09307	0.21054	−0.18387
7.0	0.20689	0.00476	0.24234	−0.15019
7.2	0.19959	0.04155	0.26847	−0.11091
7.4	0.18632	0.07606	0.28792	−0.06705
7.6	0.16717	0.10719	0.29991	−0.01980
7.8	0.14292	0.13393	0.30384	0.02954
8.0	0.11452	0.15549	0.29937	0.07954
8.2	0.08305	0.17124	0.28641	0.12872
8.4	0.04965	0.18079	0.26515	0.17562
8.6	0.01554	0.18396	0.23602	0.21879
8.8	−0.01811	0.18082	0.19971	0.25689
9.0	−0.03013	0.17162	0.15714	0.28870
9.2	−0.07946	0.15686	0.10945	0.31321
9.4	−0.10514	0.13719	0.05793	0.32956
9.6	−0.12639	0.11346	0.00403	0.33719
9.8	−0.14258	0.08659	−0.05075	0.33574
10.0	−0.15330	0.05764	−0.10485	0.32518
		$\alpha = 1.0$		
0.2	0.31369	0.05330	0.01717	0.00252
0.4	0.29998	0.10497	0.03322	0.00996
0.6	0.27762	0.15345	0.04706	0.02198
0.8	0.24731	0.19727	0.05771	0.03803
1.0	0.21001	0.23512	0.06430	0.05735
1.2	0.16690	0.26587	0.06615	0.07903
1.4	0.11933	0.28865	0.06274	0.10203
1.6	0.06879	0.30282	0.05380	0.12522
1.8	0.01684	0.30803	0.03926	0.14742
2.0	−0.03491	0.30422	0.01931	0.16745
2.2	−0.08489	0.29161	−0.00565	0.18418
2.4	−0.13158	0.27071	−0.03496	0.19657
2.6	−0.17361	0.24228	−0.06781	0.20370
2.8	−0.20975	0.20730	−0.10318	0.20483
3.0	−0.23900	0.16697	−0.13992	0.19942
3.2	−0.26057	0.12264	−0.17679	0.18715
3.4	−0.27395	0.07574	−0.21249	0.16793
3.6	−0.27890	0.02778	−0.24571	0.14193
3.8	−0.27545	−0.01971	−0.27518	0.10957
4.0	−0.26389	−0.06527	−0.29970	0.07149
4.2	−0.24480	−0.10751	−0.31821	0.02855
4.4	−0.21896	−0.14519	−0.32982	−0.01819

Table 1 (cont'd)

ϱ	$\frac{1}{2} J_0(\alpha, \varrho)$	$\frac{1}{2} H_0(\alpha, \varrho)$	$\frac{1}{2} J_1(\alpha, \varrho)$	$\frac{1}{2} H_1(\alpha, \varrho)$
4.6	−0.18738	−0.17725	−0.33382	−0.06750
4.8	−0.15121	−0.20283	−0.32976	−0.11808
5.0	−0.11175	−0.22130	−0.31739	−0.16849
5.2	−0.07036	−0.23229	−0.29675	−0.21730
5.4	−0.02844	−0.23567	−0.26812	−0.26309
5.6	0.01264	−0.23160	−0.23204	−0.30450
5.8	0.05157	−0.23045	−0.18928	−0.34028
6.0	0.08713	−0.20287	−0.14083	−0.36933
6.2	0.11829	−0.17964	−0.08786	−0.39074
6.4	0.14417	−0.15175	−0.03167	−0.40381
6.6	0.16411	−0.12032	0.02632	−0.40808
6.8	0.17766	−0.08656	0.08464	−0.40355
7.0	0.18461	−0.05171	0.14179	−0.38965
7.2	0.18502	−0.01701	0.19632	−0.36731
7.4	0.17913	0.01633	0.24686	−0.33688
7.6	0.16743	0.04721	0.29215	−0.29912
7.8	0.15058	0.07464	0.33109	−0.25503
8.0	0.12942	0.09780	0.36279	−0.20574
8.2	0.10490	0.11603	0.38653	−0.15252
8.4	0.07807	0.12889	0.40188	−0.09675
8.6	0.05004	0.13615	0.40861	−0.03982
8.8	0.02190	0.13780	0.40675	0.01686
9.0	−0.00527	0.13403	0.39656	0.07193
9.2	−0.03047	0.12524	0.37854	0.12411
9.4	−0.05283	0.11198	0.35338	0.17223
9.6	−0.07158	0.09500	0.32195	0.21531
9.8	−0.08614	0.07511	0.28525	0.25251
10.0	−0.09611	0.05326	0.24439	0.28321
		$\alpha = 1.2$		
0.2	0.37709	0.05905	0.02722	0.00343
0.4	0.36259	0.11644	0.05306	0.01357
0.6	0.33892	0.17053	0.07621	0.03005
0.8	0.30679	0.21981	0.09544	0.05223
1.0	0.26718	0.26293	0.10967	0.07925
1.2	0.22129	0.29873	0.11797	0.11008
1.4	0.17049	0.32626	0.11966	0.14351
1.6	0.11630	0.34487	0.11428	0.17824
1.8	0.06031	0.35416	0.10161	0.21288
2.0	0.00417	0.35405	0.08171	0.24603
2.2	−0.05051	0.34471	0.05489	0.27634
2.4	−0.10219	0.32663	0.02171	0.30250
2.6	−0.14942	0.30055	−0.01704	0.32335
2.8	−0.19096	0.26744	−0.06037	0.33789
3.0	−0.22574	0.22847	−0.10709	0.34530
3.2	−0.25295	0.18498	−0.15593	0.34501

Table 1 (cont'd)

ϱ	$\frac{1}{2} J_0(\alpha, \varrho)$	$\frac{1}{2} H_0(\alpha, \varrho)$	$\frac{1}{2} J_1(\alpha, \varrho)$	$\frac{1}{2} H_1(\alpha, \varrho)$
3.4	−0.27200	0.13840	−0.20550	0.33668
3.6	−0.28262	0.09025	−0.25437	0.32022
3.8	−0.28478	0.04204	−0.30114	0.29582
4.0	−0.27874	−0.00474	−0.34443	0.26389
4.2	−0.26502	−0.04870	−0.38300	0.22511
4.4	−0.24436	−0.08859	−0.41572	0.18035
4.6	−0.21772	−0.12331	−0.44166	0.13067
4.8	−0.18622	−0.15198	−0.46007	0.07728
5.0	−0.15111	−0.17394	−0.47046	0.02150
5.2	−0.11373	−0.18880	−0.47257	−0.03531
5.4	−0.07543	−0.19641	−0.46638	−0.09175
5.6	−0.03756	−0.19685	−0.45213	−0.14647
5.8	−0.00141	−0.19049	−0.43027	−0.19818
6.0	0.03184	−0.17789	−0.40151	−0.24571
6.2	0.06116	−0.15982	−0.36669	−0.28805
6.4	0.08568	−0.13723	−0.32685	−0.32436
6.6	0.10474	−0.11119	−0.28313	−0.35401
6.8	0.11793	−0.08285	−0.23675	−0.37660
7.0	0.12502	−0.05341	−0.18895	−0.39194
7.2	0.12605	−0.02407	−0.14098	−0.40008
7.4	0.12127	0.00402	−0.09404	−0.40130
7.6	0.11115	0.02978	−0.04921	−0.39607
7.8	0.09634	0.05229	−0.00747	−0.38504
8.0	0.07765	0.07074	0.03035	−0.36903
8.2	0.05601	0.08454	0.06363	−0.34896
8.4	0.03244	0.09327	0.09192	−0.32264
8.6	0.00801	0.09673	0.11498	−0.30068
8.8	−0.01620	0.09495	0.13278	−0.27458
9.0	−0.03917	0.08813	0.14550	−0.24852
9.2	−0.05994	0.07670	0.15349	−0.22724
9.4	−0.07765	0.06125	0.15726	−0.20017
9.6	−0.09160	0.04250	0.15748	−0.17939
9.8	−0.10125	0.02131	0.15491	−0.16162
10.0	−0.10623	−0.00139	0.15039	−0.14723
		$\alpha = 1.4$		
0.2	0.44066	0.06245	0.03898	0.00405
0.4	0.42587	0.12322	0.07649	0.01607
0.6	0.40173	0.18068	0.11110	0.03565
0.8	0.36895	0.23330	0.14148	0.06214
1.0	0.32850	0.27971	0.16644	0.09468
1.2	0.28158	0.31874	0.18498	0.13216
1.4	0.22957	0.34944	0.19630	0.17336
1.6	0.17401	0.37115	0.19986	0.21691
1.8	0.11647	0.38345	0.19534	0.26138
2.0	0.05862	0.38626	0.18274	0.30530

Table 1 (cont'd)

ϱ	$\frac{1}{2} J_0(\alpha, \varrho)$	$\frac{1}{2} H_0(\alpha, \varrho)$	$\frac{1}{2} J_1(\alpha, \varrho)$	$\frac{1}{2} H_1(\alpha, \varrho)$
2.2	0.00208	0.37974	0.16226	0.34724
2.4	0.05161	0.36436	0.13442	0.38583
2.6	−0.10101	0.34086	0.09992	0.41983
2.8	−0.14485	0.31021	0.05971	0.44813
3.0	−0.18206	0.27356	0.01489	0.46986
3.2	−0.21181	0.23224	−0.03328	0.48431
3.4	−0.23353	0.18769	−0.08346	0.49107
3.6	−0.24691	0.14141	−0.13427	0.48995
3.8	−0.25193	0.09492	−0.18431	0.48102
4.0	−0.24883	0.04969	−0.23223	0.46460
4.2	−0.23813	0.00710	−0.27680	0.44125
4.4	−0.22055	−0.03158	−0.31690	0.41174
4.6	−0.19704	−0.06528	−0.35156	0.37702
4.8	−0.16872	−0.09310	−0.38006	0.33820
5.0	−0.13683	−0.11440	−0.40186	0.29648
5.2	−0.10268	−0.12878	−0.41666	0.25313
5.4	−0.06763	−0.13607	−0.42443	0.20944
5.6	−0.03301	−0.13640	−0.42533	0.16666
5.8	−0.00010	−0.13009	−0.41980	0.12599
6.0	0.02994	−0.11773	−0.40843	0.08849
6.2	0.05606	−0.10009	−0.39205	0.05511
6.4	0.07743	−0.07810	−0.37159	0.02658
6.6	0.09340	−0.05282	−0.34813	0.00347
6.8	0.10355	−0.02541	−0.32278	−0.01389
7.0	0.10767	0.00293	−0.29671	−0.02540
7.2	0.10581	0.03101	−0.27106	−0.03114
7.4	0.09822	0.05769	−0.24693	−0.03146
7.6	0.08539	0.08190	−0.22530	−0.02686
7.8	0.06798	0.10270	−0.20705	−0.01804
8.0	0.04679	0.11932	−0.19288	−0.00584
8.2	0.02277	0.13116	−0.18330	0.00880
8.4	−0.00305	0.13781	−0.17866	0.02485
8.6	−0.02960	0.13908	−0.17906	0.04128
8.8	−0.05580	0.13500	−0.18443	0.05702
9.0	−0.08061	0.12579	−0.19447	0.07110
9.2	−0.10307	0.11189	−0.20872	0.08258
9.4	−0.12233	0.09387	−0.22654	0.09069
9.6	−0.13767	0.07250	−0.24716	0.09477
9.8	−0.14855	0.04864	−0.26970	0.09433
10.0	−0.15461	0.02321	−0.29322	0.08907
		$\alpha = \pi/2$		
0.2	0.49501	0.06338	0.04975	0.00423
0.4	0.48020	0.12507	0.09801	0.01680
0.6	0.45600	0.18346	0.14335	0.03729
0.8	0.42314	0.23700	0.18442	0.06506

Subject Index